Springer Aerospace Technology

The series explores the technology and the science related to the aircraft and spacecraft including concept, design, assembly, control and maintenance. The topics cover aircraft, missiles, space vehicles, aircraft engines and propulsion units. The volumes of the series present the fundamentals, the applications and the advances in all the fields related to aerospace engineering, including:

- structural analysis,
- aerodynamics,
- aeroelasticity,
- aeroacoustics,
- flight mechanics and dynamics
- orbital maneuvers,
- avionics,
- systems design,
- materials technology,
- launch technology,
- payload and satellite technology,
- space industry, medicine and biology.

The series' scope includes monographs, professional books, advanced textbooks, as well as selected contributions from specialized conferences and workshops.

The volumes of the series are single-blind peer-reviewed.

To submit a proposal or request further information, please contact: Mr. Pierpaolo Riva at pierpaolo.riva@springer.com (Europe and Americas) Mr. Mengchu Huang at mengchu.huang@springer.com (China)

The series is indexed in Scopus and Compendex

K. F. Long

Stellar Engines

Advanced Spacecraft Propulsion through Extreme Aerospace Engineering Design

Springer

K. F. Long
Astrophysicist
London, UK

ISSN 1869-1730 ISSN 1869-1749 (electronic)
Springer Aerospace Technology
ISBN 978-3-032-18061-2 ISBN 978-3-032-18062-9 (eBook)
https://doi.org/10.1007/978-3-032-18062-9

This Springer imprint is published by the registered company Springer Nature Switzerland AG
The registered company address is: Gewerbestrasse 11, 6330 Cham, Switzerland

I dedicate this book to the dreamers. To the aerospace engineers that want to go far beyond the existing paradigms of technological performance. To those that pursue a vision for humanity far into the future where we are living at peace with each other and with other life forms out there in this vast and wonderful cosmos. The dreamers are often not taken seriously, since their thoughts are of the worlds that are to come. Yet, this is also how the future is made. Although some of the concepts explored in this book may seem like dreams, it is entirely possible that one day some of them will be built and fly. This is my wildest hope, and my heart's desire, for dreams to become reality, and my children live to see that day come to pass.

"And God made two great lights; the greater light to rule the day, and the lesser light to rule the night: he made the stars also."

Genesis 1:16 [1]

Foreword

We live in an age where we are mastering the art of getting to space. The colloquial expression "*it's not exactly rocket science*" is often used to highlight the challenges associated with designing, building, and flying launch systems capable of delivering mass from Earth's surface into orbit.

Within this domain of getting to space, the current reigning champion is, of course, SpaceX, with its Falcon-9 and Falcon Heavy families of launch vehicles. In 2024, these two systems accounted for over half of all global launches. At the time of writing (2025), SpaceX is well on its way to fielding a super-heavy, fully reusable launch vehicle known as *Starship*, capable of lofting more than 100 metric tons to Low Earth Orbit (LEO) in a reusable configuration, or up to 250 metric tons in an expendable mode. That represents a true step-change in launch capability.

This system was born from Elon Musk's vision of making humanity a multi-planetary species by establishing a sustained human presence on Mars. Whether that vision is ultimately realized or not is immaterial to this rocket scientist. *Starship* serves as a touchstone for how far we have come in our quest to master the art of leaving Earth's surface—a thirty-story building producing roughly 17 million pounds of thrust at liftoff, powerful enough to lift itself against gravity and climb all the way to orbit. No. Small. Feat.

As impressive as that is, we must bring some sobriety to the table to appreciate how hard the next challenge will be. The next great hurdle is mastering the art of moving *through* space. Once we reach LEO, where do we go next, and most importantly, how fast do we want to get there?

As most readers know, humanity has long studied how to send crewed missions to Mars. It has been analyzed hundreds of times by space agencies and likely thousands more in academia. Because of its proximity to Earth, its atmosphere, its geologic record of surface water, and its tantalizing hints of past life—not to mention its cultural gravity—Mars has remained the centerpiece of human spaceflight planning. While a return to the Moon may occur within a few years, Mars holds the deeper romantic pull, perhaps because of something innate—an exploration gene, if you will—that recognizes the importance of this next chapter for humanity.

Now let us get down to brass tacks and look at the energy bill. The delta-v required to place a payload into LEO is roughly 9.4 km/s. The round-trip delta-v for a Mars mission (about 200-day transits each way, with aerocapture and ascent) adds another 13–14 km/s, for a total mission delta-v of 23–24 km/s, just over twice the energy cost of reaching orbit. A quick encounter with the rocket equation, using today's chemical propulsion inputs, makes clear why rockets are mostly propellant with only a small fraction left for payload.

Getting to space requires a propulsion system capable of generating enough thrust to lift the entire launch vehicle—structure, engines, propellant, and cargo. If thrust-to-weight falls below one, you do not fly; you just make noise and heat. That is why the systems we use in gravity wells are chemical. Combustion provides immense instantaneous power without requiring an external energy source; the chemical binding energy in the propellants does the work.

To appreciate the magnitude: the effective jet power of a Falcon 9 at liftoff exceeds 10 GW. For the roughly three minutes of its first-stage burn, the rocket's power output is comparable to about 2% of the entire U.S. electrical grid (about 500 GW in 2025). Yes, kids, this is why rocket science is hard.

Now hold on to your hats, this is where the problem really bites. Suppose, instead of Mars, we wanted to send humans to Saturn in 200 days, roughly the same transfer time as a typical Mars mission. Using a quick-look "desert-island physics" approach we often used at NASA for rapid assessments—turning gravity off, neglecting orbital mechanics, and focusing purely on energy—we find that to accelerate from Earth (1 AU) to Saturn (9.58 au) and capture there within 200 days requires a delta-v of roughly 150 km/s. That is over six times the round-trip Mars mission energy, and that is just one-way. It is a simplified estimate but probably within 15% of a high-fidelity number, making it an excellent first-order sizing tool.

Applying the rocket equation, $m_0/m_f = e^{(\Delta v/v_e)}$, and using Falcon 9's exhaust velocity ($v_e = 2.77$ km/s, or 282 s Isp), yields a mass fraction of about 2.3×10^{23}. In plain terms, for every kilogram of payload, you would need about 38,500 Earth masses of propellant. You can pick yourself up off the floor now.

So, is there no hope? Are we doomed to dream about Mars while the rest of the Solar System stays out of reach? Hardly. The rocket equation is nonlinear, not only in the mission Δv, but also in the exhaust-velocity term. If we can boost exhaust velocity, the equation begins to bend in our favor.

In propulsion, we often speak not in exhaust velocity but in *specific impulse* (I_{sp}), related by $I_{sp} = v_e/g$. The Merlin engines, for example, deliver 282 s. Let us aim for a Saturn cruiser with a total mass fraction similar to Falcon 9's 25:1. Plugging that into our 150 km/s target delta-v, we find we need a propulsion system with an I_{sp} of roughly 6,500 s, an exhaust velocity of about 64 km/s. The best chemical rocket ever flown, the RL10B-2, delivers about 466 s. So combustion alone will not cut it; the specific heat of chemical reactions simply does not provide enough energy.

We must turn to electric propulsion, accelerating ionized propellant through electric or magnetic fields to form a plasma jet. This approach achieves far higher exhaust velocities and therefore much greater specific impulse. A high-performance ion thruster such as NEXIS, which accelerates xenon through electrostatic grids, gets us into the right range. But there is a catch: power.

Chemical propulsion carries its energy internally through chemical bonds. Electric propulsion requires an external power source, solar arrays, nuclear reactors, perhaps someday fusion. And the thrust-to-power ratio matters: NEXIS produces about 4.4 mN/kW. Even at 2 MW, roughly the power of a terrestrial locomotive, the thrust is only about 9 N. That is not going to lift off a launch pad anytime soon.

Still, once you are in space, low-thrust electric propulsion shines, since in the vacuum of space you no longer require a thrust-to-weight ratio greater than one. Nearly all commercial satellites now rely on it because its efficiency is a mission enabler. Think of it this way: chemical propulsion dumps all its energy in minutes, a burst of ultra-high power. Electric propulsion spreads that energy out over days, weeks, or months. Sustained thrust from a persistent power source opens the Solar System to crewed travel.

NASA and others have studied Nuclear Electric Propulsion (NEP) architectures combining a compact fission reactor with ion or Hall thrusters. Such systems, based on known physics and engineering, could make the entire Solar System accessible to human exploration.

But what about interstellar? Aye, there is the rub, lad (as Scotty would say). Just as Saturn compared to Mars demands a new paradigm, interstellar travel demands one again. Our nearest stellar neighbor, Proxima Centauri, lies 4.26 light-years away. Voyager-1, launched in 1977, would take roughly 70,000 years to cover that distance, a stark reminder of chemical propulsion's limits.

To make that crossing in meaningful timeframes, cruise velocities must reach fractions of light speed. For a 5% c flyby mission, the delta-v jumps to about 15,000 km/s, roughly a hundred times greater than the Saturn case. To achieve a 100:1 mass ratio, the propulsion system would need a specific impulse of roughly 330,000 s, around 1% c in exhaust velocity. NEP cannot get us there. For that, we must step to the frontiers of engineering and explore fusion propulsion.

With this background and a quantified sense of the challenge ahead, I now refer the reader to the book you hold in your hands, or view on your screen.

Dr. Kelvin Long has done a masterful job assembling the most comprehensive and disciplined survey to date of humanity's propulsion options beyond chemical and electric systems. He guides the reader from the foundations of rocket and fusion physics through architectures for interstellar craft, sails driven by light or laser, fusion micro-pellet engines, and concepts that merge plasma physics with spacetime dynamics. What distinguishes Kelvin's work is that it never drifts into wishful thinking; every speculative thread is tied to first-order equations, energetics, and design logic. He reminds us that interstellar flight is not a metaphor; it is a solvable engineering problem once we accept that power, not distance, is the governing variable.

In this way, Long extends the engineering lineage established by NASA's early fusion studies, including the Discovery-II MCF and his own Discovery-III ICF concepts that carried those frameworks forward to the outer planets. From there, the discussion expands to civilization-scale constructs—rotating world ships, asteroid-hulled arks, and even black-hole-powered vessels—each treated not just as fantasy but as a physical system constrained by the same conservation laws and boundary conditions that govern launch vehicles today. The result is a design compendium for the far frontier, one that teaches how to frame interstellar travel as a tractable physics and engineering problem. Long characterizes this as "extreme aerospace engineering"—an apt description, since every relevant parameter, from power density to temperature and velocity, operates orders of magnitude beyond conventional systems. His chapters close with worked problems, reinforcing that this is not a coffee-table survey but a rigorous textbook intended to train the next generation of propulsion physicists and systems engineers.

Stellar Engines is both textbook and manifesto: a quantitative roadmap for the era when humanity chooses to master not just getting to space, but moving through it, and ultimately, between the stars.

Houston, TX, USA
November, 2025

Dr. Harold "Sonny" White

Preface

Nuclear fusion can be traced back to 1905 with the publication of Albert Einstein and his Special Theory of Relativity in which the relationship between matter and energy was first described as a transmutable process. Then in 1920 the British astrophysicist Arthur Eddington first proposed that the Sun was powered by nuclear fusion reactions. In this process, hydrogen nuclei fuse to form helium, releasing enormous amounts of energy. That energy then gradually propagates as photons of light through the solar atmosphere over tens of thousands of years to become the bright and luminescent object we see in our skies.

The precise mechanism for how the Sun produces its energy was worked out by the physicist Hans Bethe in 1939. His pioneering work on stellar nucleosynthesis, for which he later received the Nobel Prize, described the proton-proton chain and the carbon-nitrogen-oxygen cycle (CNO) through which lighter elements are used to form heavier ones. The amount of energy that is released in the process is significantly large.

This is also the basis of the first thermonuclear weapon—known as the Hydrogen Bomb. But whilst some were considering the applications of this technology for mass destruction, others realised there are other uses, such as the generation of energy to power our vast and expanding cities. In principle, since the main fuel was hydrogen based (water) this would open an era of unlimited energy for humankind and solve all our energy needs. Indeed, in 1953 the US President Dwight Eisenhower gave an *Atoms for Peace* speech to the United Nations General Assembly, urging for peaceful cooperation with nuclear energy among nations. Since this time scientists have continued to work on the problem of fusion energy to power our cities.

Historically, there have been two main routes to achieving fusion within a reactor. The first involves the use of large magnets to magnetically confine a plasma for a few seconds at low plasma density. The main configuration of the magnetic fields has tended to be a Tokamak, and this is the basis of outstanding physics programmes like the Joint European Torus (JET) at Oxfordshire in the United Kingdom. It is also what has led to the design of the International Thermonuclear Reactor (ITER) which is under construction in France. This will be a full-scale experimental power reactor designed to demonstrate commercial feasibility, useful as a prototype for a future reactor that can deliver electrical energy to a national grid.

The other main approach to achieving fusion is inertial confinement fusion (ICF) which typically involves the use of an energy driver (usually a laser of some form) to compress a fuel capsule for a brief instant to extremely high density and temperature. This is the physics basis of the National Ignition Facility (NIF) in Livermore, California, which finally achieved ignition in December 2022. They not only produced the same amount of energy that went into the reaction (breakeven) but also more (energy gain). Although it wasn't a large amount of energy, it was enough to demonstrate the feasibility of this method, and for the first time, we can claim to have created a mini star in a physics laboratory.

There are many other approaches to achieving fusion that have been considered by theoreticians and experimentalists, and indeed many of those approaches are still a part of active programmes, such as the Z-Pinch approach and electrostatic confinement. Given we have so many approaches to fusion on the table, it seems highly likely that in the coming decades the production of thermonuclear energy to power our cities will become a reality. Especially since fusion reactions do not produce radioactive materials from atomic decay. This is unlike its counterpart fission where materials must be stored away for tens of thousands of years due to the harmful effects to life of radioactive decay in atoms.

While achieving fusion power on Earth is a vital milestone, this book explores a more audacious application—using fusion to propel spacecraft across the vast distances of space. Since a large amount of energy is released in the reactions, this has the potential to lead to a very high exhaust velocity approaching $\sim$10,000 km/s, which will then lead to very high spacecraft velocities. This has two huge implications for missions in space.

Firstly, with such an engine achieving a high energy capability, this means that very large payloads can be carried around the Solar System. This would be essential for a Mars colony, for example where regular supply ships to and from Earth may be required to sustain it in the early days of its formation. In addition, with the current age of space telescopes and the success of the Hubble Space Telescope (HST) and the James Webb Space Telescope (JWST) to name a few, these same large mass payloads could in principle be carried far outside of the Earth's orbit to capture high resolution images of nearby stellar objects. This might be a deep space mission to the gravitational focal point at $\sim$550–1,000 au for example.

Second, since the velocity of a spacecraft may be very high, this means we can get about quicker, such as crewed missions to the gas giants in months. A high velocity may also be helpful in the mitigation of an incoming asteroid which could be on an impact trajectory to Earth, and so getting a spacecraft out to it early to deploy some defence strategy would be critical. Yet we can go further, since missions that go outside of our solar system also become possible, beyond the distance of the highly successful NASA/JPL Voyager 1 and 2 probes launched in 1977—outside of the solar heliosphere and into the interstellar medium.

Indeed, with fusion propulsion, we can meet the stars on their own terms. For it would be possible to send spacecraft to the exoplanets around other stars out to perhaps $\sim$10–20 light years travelling at speeds of around $\sim$10–15% of the speed of light which would imply trip times of order $\sim$50–200 years. Once our robotic ambassadors have made the trip they can fully characterise those nearby stars and any planets around them. They could also deploy orbiters, atmospheric penetrators and even ground landers. The information would be relayed back to Earth through a deep space communication network, even if it took years for the information to get to us travelling at the speed of light.

These spacecraft would also be equipped with an artificial intelligence capability and so would be fully autonomous. They would be our ambassadors to the stars and in effect will be us, since they will contain all of the knowledge and experience of the human journey. Since they are not biological in nature, they would be better equipped to handle the demands of long-distance space flight. The results from their discoveries transmitted back to Earth for our enjoyment and enlightenment, and helping us to map interstellar space as a means for risk mitigation.

Next, after our robots, comes the possibility of sending human beings to those same stars. At this point, all of the dreams of science fiction would become a reality, and the true age of interstellar flight would begin. Since the destinations would have been fully characterised by the earlier probes, we would know exactly what was there when we got there, so that we could immediately begin the task of building our first colonies.

In terms of defining interstellar flight, it is worth quoting one of the leaders of the field who pioneered much original research in the 1970s and 1980s:

"The problems with stellar flight can be stated as follows: The reaction mass and energy required to make an interstellar flight are too great to be contained in a rocket. Alternatively, we can say the distances to be traversed are too large for reasonable flight times. If we master these, we find the speed of light is too low to achieve reasonable Earth-elapsed flight times.....No theoretical objection prohibits the projection of a payload across interstellar space. The problems are exclusively human, due to the short life span of the species and the amount of human effort required to mount such a programme of exploration" Alan Bond, 1971 [2].

This is a vision of humans out there in the Cosmos, living up to our full potential, and becoming a starfaring species. It sounds fantastic, but it is also entirely possible. This is thanks to the efforts of many scientists, engineers and physicists who over many decades (and it could be argued even over centuries) have pushed the envelope on what may be possible in our distant future. That within the minds of these futuristic pioneers are ideas, and when combined with rigorous calculations emerge as potential concepts for how we might some say make the journey.

This book considers fusion as a means for propelling a spacecraft on these sorts of missions, through interplanetary and interstellar space. Yet we will go beyond just fusion and look at some other advanced propulsion methods and all placed within different types of mission architectures. This includes discussions on world ships, asteroid ships, interstellar ramjets and even a black hole engine. Although this text is dominated by the issues of advanced space propulsion, since it is a critical driver of any mission, we will also place the discussions within the wider context of spacecraft vehicle design. The main propulsion subjects discussed pertain to reaction type engines associated with the nuclear pulse method, but in our discussions of vehicle design we will also touch on a few other areas to include solar sails and laser sails.

Over many years the author has published papers in this field gradually building up the knowledge for how interstellar flight could be achieved theoretically. So, these discussions represent the culmination of this work, and it is the hope that it serves an educational purpose. The greatest goal of this book is that you, the reader, will start to perform calculations of your own and figure out how to design fantastic machines that may be a part of our plausible future. This book then is intended to be a guide for aerospace design practice in regard to spacecraft in deep space that are beyond the current state-of-the-art capability.

As well as covering fundamental mathematical and physical theory relating to advanced propulsion, we will also develop spacecraft concept designs as a potential blueprint for the sorts of missions that can be conducted in space. We will also go beyond the existing theory into advanced and even exotic applications, showing how some of the advanced ideas solve some of the technical problems, but perhaps also open new opportunities for scientific research and technological innovation.

It will not be easy, and there is still a lot of work to be done both theoretically and experimentally before we can really consider the application of fusion physics or other methods to spacecraft propulsion. Yet steady progress has been made and some of that progress will be described. For the first time, we will lay out the theoretical basis for advanced space propulsion, how the engines can be designed and the various physics and engineering parameters calculated through the fundamental equations.

It is a part of the aim of this book to discuss some visionary ideas so that you the reader can get a glimpse of the potential. Enjoy the ride and welcome to the physics of advanced spacecraft propulsion. It is my sincere hope that this book might assist you in becoming a starship designer so you too can bring your own ideas to fruition for a future that is yet to be engineered.

London, UK
December, 2025

Kelvin F. Long

"opta ardua pennis astra sequi."
(Desire to pursue the hard to reach stars on wings.)

Aenied by Publius Virgil, 19 BCE [3].

"non est ad astra mollis e terris via."
(There is no easy way from the Earth to the stars.)

Hercules Furens by Seneca the Younger, 40–65 CE [4].

Acknowledgements During the writing of this book, I had many helpful discussions about advanced space propulsion. I would like to give thanks to Professor Greg Matloff, Professor Phil Lubin, Professor Roman Kezerashvili, Professor Hideki Nakashima, Dr Al Jackson, Robert Swinney, Robert Freeland and Michel Lamontagne (ML). Michel also produced some of the awesome coloured graphics for Project Icarus which are shown in this book. A huge thank you to Dr Harold "Sonny" White for writing the foreword. As good friends we both share the dream of a future where star travel is possible and we are kindred spirits of the stars. Sonny is also busy pioneering his own advanced propulsion research for which I am certain will make great impacts on our future. I also give thanks to the staff of the National Ignition Facility at LLNL for their work on inertial confinement fusion. This author has had the pleasure of visiting NIF several times and was thrilled to hear they had finally achieved thermonuclear ignition in late 2022. A huge thanks to NASA for hosting me at their various facilities over the years and the many good discussions I enjoyed with their staff. When I gave presentations at their laboratories on interstellar flight they did not ridicule the ideas but seemed genuinely interested and I found these encounters inspiring. Thanks to the British Interplanetary Society for their support over the years, particularly with their in-house technical journal and multiple symposia. I thank my family for their patients when I am working. My two daughters Amelia and Jemima and my wife Gemma. The undertaking of large writing tasks such as this is not possible without the background love, stability and order you provide to my creativity and constant calculations. A huge thanks also for the love and support of my Mother Susan Kelly, and Parents-in-law Maureen and Michael Perrin. Finally, a big thank you to my publisher Springer, who also published my first propulsion book back in 2012 and so we have come full circle.

Competing Interests The author has no competing interests to declare that are relevant to the content of this manuscript.

Contents

About the Author

Dr Kelvin F. Long holds a bachelor's degree in Aerospace Engineering, a master's degree in Astrophysics, both from Queen Mary and Westfield College, University of London. He also holds a Doctor of Philosophy in Physics from the University of Warwick. He is a Fellow of the British Interplanetary Society, the former Chief Editor of its academic technical journal, The Journal of the British Interplanetary Society. He is also a Chartered Member of the Institute of Physics. He previously worked as a physicist with the UK Ministry of Defence and in this capacity worked with the United States Los Alamos National Laboratory and the Lawrence Livermore National Laboratory. He has also presented at every National Aeronautics and Space Administration centre in the United States as well as to the European Space Agency and has served as a consultant to the USAF and the broader aerospace industry. He has been a teacher of Engineering Design at an East London technical college for two years, a continuing education teacher in space science at the University of Reading for three years and a Visiting Lecturer at the International Space University, Strasbourg France for five years where he also instigated and co-supervised thesis projects. He has published many papers in peer-reviewed journals about interstellar studies and in particular relating to advanced space propulsion. He has been involved in many exciting interstellar projects including the BIS Project Icarus (which he initiated), the Breakthrough Initiatives Project Starshot (serving on the advisory committee for seven years), the Johns Hopkins Applied Physics Laboratory Interstellar Probe Decadal study and an International Academy of Astronautics study into an interstellar precursor probe that went beyond the Voyagers. He has also been involved with the co-founding of many non-profit and for-profit companies in the UK and the USA dedicated to the vision of realising interstellar flight. He is the current lead for the Interstellar Research Centre and the UK aerospace consultancy company Stellar Engines.

Acronyms

ACF	Antimatter Catalysed Fusion
ADD	Arkani-Hamed Dimopoulos Dvali model
AMU	Atomic Mass Unit
BAIR	Black Hole Augmented Interstellar Ramjet
BIS	British Interplanetary Society
CRFP	Carbon Fiber Reinforced Polymer
CTBT	Comprehensive (Nuclear) Test Ban Treaty
CTR	Controlled Thermonuclear Reaction
DFD	Direct Fusion Drive
DOE	Department of Energy
DPSSL	Diode Pumped Solid State Laser
ESA	European Space Agency
FOM	Figure of Merit
GDM	Gas Dynamic Mirror
HiPER	High Power Laser for Energy Research
HST	Hubble Space Telescope
IAC	International Astronautical Congress
ICF	Inertial Confinement Fusion
IFAR	In Flight Aspect Ratio
IFR	Inertial Fusion Rocket
ISM	Interstellar Medium
ITER	International Experimental Thermonuclear Reactor
JET	Joint European Torus
JPL	Jet Propulsion Laboratory
JHUAPL	John Hopkins University Applied Physics Laboratory
JWST	James Webb Space Telescope
KST	Kepler Space Telescope
LASER	Light Amplification by Stimulated Emission of Radiation
LEO	Low Earth Orbit
LHC	Large Hadron Collider
LLNL	Lawrence Livermore National Laboratory

LMF	Laboratory Microfusion Facility
LMJ	Laser Megajoule
MCF	Magnetic Confinement Fusion
NAS	National Academy of Sciences
NASA	National Aeronautics & Space Administration
NEO	Near Earth Object
NIF	National Ignition Facility
NPT	Non-Proliferation Treaty
OST	Outer Space Treaty
PCLM	Power Cycle Line Map
PIC	Particle In Cell
RAIR	Ram Augmented Interstellar Ramjet
SETI	Search for Extraterrestrial Intelligence
TAU	Thousand Astronomical Unit
TNT	Trinitrotoluene
TRL	Technology Readiness Level
VAST	Voyager for Advanced Space Travel
VISTA	Vehicle for Interplanetary Transport Applications

Constants and Conversions

AU	Astronomical Unit, 1.496×10^{11} m
A_v	Avogadro Number, 6.022×10^{23} mol^{-1}
c	Vacuum Speed of Light 2.9979×10^{8} ms^{-1}
g_o	Gravitational acceleration, 9.81 m/s^2
G	Gravitational Constant 6.674×10^{-11} m^3/kgs^2
h	Planck's Constant 6.6261×10^{-34} Js
$\hbar$	Reduced Planck's Constant 1.0546×10^{-34} Js
e	Electron Charge 1.602×10^{-19} J
ϵ_o	Permittivity of Free Space, 8.885×10^{-12} F/m
μ_o	Permeability of Free Space, 1.245×10^{-6} NA^2
k_B	Boltzmann Constant 1.3806×10^{-23} J/K
ly	Distance in Light Years, $1\,ly = 9.4605 \times 10^{15}$ m
m_i	Ion Mass 1.6726×10^{-27} kg
m_e	Electron Mass 9.109×10^{-31} kg
N_A	Avogadro Number, 6.022×10^{23} mol^{-1}
Yr	Time in Years, $1\,Yr = 3.1536 \times 10^{7}$ s
σ	Steffan-Boltzmann Constant, 5.67×10^{-8} W/m^{2}K^4
π	3.14159265
1 keV	11604525 K
1 ton	1,000 kg
1 kg TNT Equivalent	4.184×10^{6} J

References

1. Genesis (1991) The Holy Bible. King James Version, Genesis 1:16. Harper-Collins, New York
2. Bond A (1971) Problems of interstellar propulsion. Spaceflight 13:245–251
3. Virgil P (2003) Aenied (19 BCE). Penguin Classics, London
4. Seneca LA (2023) Hercules furens: a tragedy (40–65 CE). Lagare Street Press

Chapter 1
Introduction

The development of a spacecraft with the engineering and scientific sophistication and capability of Voyager extends engineers, scientists and managers to the frontiers of creativity and technology.

— A. Thomas Young, 1978 [1]

1.1 Advanced Space Propulsion

The vision of starflight has been around for as long as humans could look up at the stars at night. In 1610 Johannes Kepler was dreaming big when he said: *"When ships to sail the void between the stars have been built, there will step forth men to sail these ships"* [2]. This vision has persisted up to the modern age, when even the Russian inventor of the rocket Konstantin Tsiolkovsky showed great optimism for what might be possible: *"When the Sun has exhausted its energy, it would be logical to leave it and look for another, newly kindled star still in its prime"* [3].

The NASA/JPL space probes Voyager 1 and Voyager 2 were launched on a "Grand Tour" of the planets in 1977. Travelling at a mere 17 km/s or $\sim$3.6 AU/Yr these probes have now gone beyond the solar heliosphere and have entered the interstellar medium as a part of their extended interstellar mission. Although it is not intended these spacecraft will travel to the nearby stars, at this speed it would take them on the order of tens of thousands of years to reach that distance—the nearest stars are at around 270,000 AU or 4.3 LY from the Sun. The astronomer Carl Sagan worked with the Voyager programme and showed great optimism about the potential discoveries in deep space when he said: *"The sky calls to us. If we do not destroy ourselves, we will one day venture to the stars"* [4].

Although considerations are already under way for missions that will go beyond the Voyager probes, perhaps to 1000 AU, there are currently no plans to send space missions even further than this. The limiting factor on why this is not feasible today is the immaturity of our space propulsion technology. However, the discovery of thousands of exoplanets in recent decades is slowly changing the paradigm in terms

K. F. Long, *Stellar Engines*, Springer Aerospace Technology,
https://doi.org/10.1007/978-3-032-18062-9_1

of the potential scientific exploration mission opportunities; this needs to occur in parallel with the continued improvement in our technological capabilities.

The Universe is filled with stars. Those stars function under the delicate balance of hydrostatic equilibrium where the gravitational attraction of the mass opposes the radiation pressure generated by the nuclear reactions at the core. It seems logical, that if we are to truly extend the range of our technological robotic probes to other stars, that we must also utilise the very energy source that powers them—fusion reactions. Since there are many stars, it would seem serendipitous that we may utilise the same energy generation mechanism that powers them to effectively travel among them.

The ability to send spacecraft out to deep space and beyond is strongly dependent on the type of propulsion system that is utilised [5]. Indeed, much research has been conducted on this topic exploring the various physics and engineering issues of interstellar flight [6]. Although not yet mature in a laboratory environment, we understand enough about the physics of fusion reactions to know that in principle this can be used to propel a spacecraft to extremely high velocity on the order of $\sim 10\%c$, where c is the speed of light. This places several dozen stars (and their exoplanets) potentially within the reach of future missions.

Unlike fission, fusion offers the hope of large energy supply but for minimum radioactive consequences to our environment and delicate ecological systems. In essence any technology that utilises fusion reactions must figure out how to contain a hot plasma for a minimum duration and at a minimum density to achieve thermonuclear ignition and then energy multiplication gain with respect to the initial input energy. Today, many schemes have been invented for how fusion energy may also be harnessed on the Earth for the purposes of energy supply to our great and vast civilisation. Assuming this can be successfully achieved in the scientific laboratories, and then applied to our cities, next comes the additional applications of this technology.

One of these is to advanced propulsion of spacecraft that can potentially take our robotic probes to the stars in mission durations that are equivalent to half a lifetime of a human being; which means that it would be possible for a scientist to begin their career at the outset of the mission and to see the results come in as their career comes to a fruitful close. This is astonishing and has parallels with the Voyager probes where some of the scientists involved in their undertaking are still working on the missions.

Fusion reactions have a potential specific energy release of around 340 million MJ/kg which is substantial. In terms of nuclear fuels, there are other alternatives to fusion, including the use of fission energy which will have an associated specific energy of around 82×10^6 MJ/kg so is much lower in performance than fusion. Chemical reactions are even worse with an estimated specific energy 13 MJ/kg so are not even in the list of considerations.

Fusion reactions are also relatively clean compared to their counterpart fission reactions, where radioactive materials must be disposed of and stored for millennia. In addition, unlike fission reactions which require the resource of materials with a high atomic element at low abundance, fusion reactions require the resource of materials with a low atomic element at high abundance; the universe is dominated by hydrogen gas. Although unlike the Sun which confines the plasma using its enormous gravitational field over billions of years, an artificial star created in a laboratory must re-create the correct conditions upon which "stardom" becomes feasible, which suggests isotopes of hydrogen such as Deuterium and Tritium since these have larger reaction cross sections and at lower temperature. Other reactant combinations may also be used although any involving Helium-3 may require lunar or gas giant mining due to the low abundance on Earth [7, 8].

There are many other alternative options for sending spacecraft into deep space that we can eventually develop. This includes propulsion methods like *laser beaming* and *microwave beaming* systems, such as with the historical Starwisp concept [9, 10] and the current Project Starshot [11, 12]. Whilst these may permit faster transit times to the interstellar targets on the order of a few decades, they are also likely to be utilised only for the application of low payload masses. This is because the beam power requirement scales linearly as a function of the mass. Although fusion-based systems also come at high energy expense, it is possible to envisage spacecraft for high payload masses that can reach the nearby stars in a reasonable time.

There is also the more exotic possibility of matter-antimatter with an associated 90×10^9 MJ/kg but this is currently highly speculative physics. There is also antimatter catalysed fusion (ACF) [13, 14] reactions which would achieve higher fractions of light speed, require substantially less fuel whilst also produce far greater energy in terms of 'bang per buck'. Where exactly ACF would lay in between fusion and antimatter reaction energy release is currently too speculative to give an accurate estimate.

There is also the historical nuclear (bomb) pulse propulsion such as was pursued in Project Orion [15] and written about by Freeman Dyson [16, 17]. This technology could have been built decades ago and in principle the use of atomic bombs will propel a spacecraft to around $\sim 5\%c$ and the use of hydrogen bombs (thermonuclear weapons) will propel a spacecraft to around $\sim 10\%c$, which means that missions to the nearest stars are achievable in time durations on the order of $\sim$100–150 years. However, this technology is controversial to say the least and so it may be better to pursue other options. That said, the American astronomer Carl Sagan stated in the *Cosmos* television series that he couldn't think of a better application of this technology.

Effectively, if we shrink a Project Orion propulsion system down and move away from an external detonation and pusher plate to a reaction chamber with a much smaller detonation, then in the limit we end up with laser fusion propulsion, where the ignition of the fuel is initiated by using a set of lasers. So, it should be clear that ICF propulsion technology is the clear answer to a no Orion policy, achieving similar principles of pulsed detonation without the political and environmental drawbacks. Indeed, it may even be far more efficient in the conversion of energy to useful work in propulsion. The two methods are illustrated in Fig. 1.1.

Except for Orion, many of these alternative propulsion options are at much lower technology readiness level [18, 19] than fusion-based systems and the latter have for

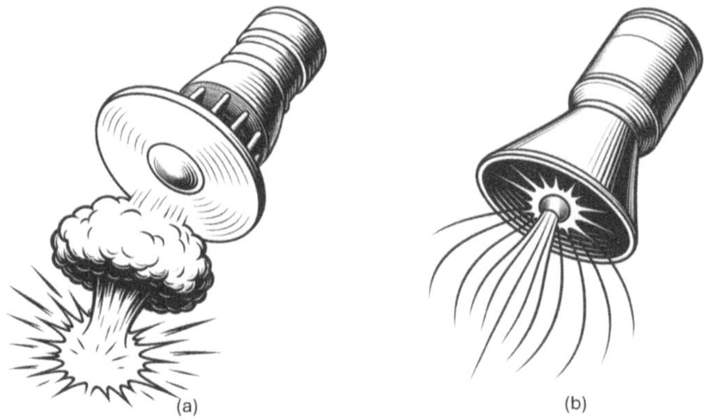

Fig. 1.1 Comparison of the ignition methods of (**a**) external nuclear pulse propulsion and (**b**) laser driven ICF propulsion[KFL]

many decades been on an experimental footing giving us substantial insights into the physics operation at work. The various methods of advanced spacecraft propulsion have the goal of getting payloads to a destination in as quick a time as possible for minimum mass and using as least energy. In application to interstellar flight, propulsion via fusion offers near term realisation and high-performance benefits and we should continue to invest in this technology. This also means that any theoretical spacecraft models we construct have a higher level of credibility and validation if they can be directly linked to realistic knowledge about fusion physics operation and performance.

If such engines can be constructed then they have the potential for application to space exploration missions focussed on high energy capability, using high specific energy to deliver much higher dv from nuclear fusion reactions. This is unlike our current low energy capabilities in space which yield low specific energy and lead to conversion of energy to low performance designs with dv ~10 km/s. Although with electric and nuclear electric systems it is conceivable that we will achieve dv ~ 100s km/s within the first part of the current century. Such a high velocity would allow space probe missions to the edge of our known solar heliosphere at ~100 AU within ~5 years and missions to the beginning of the Oort cloud at ~1000 AU within ~50 years. Whilst these are critical 'stepping stones'out to significant distances and will permit us to attain much science data to help us understand the boundaries of our planetary system, they are clearly not interstellar missions.

Unless we are prepared to accept extremely long journey times where people are born and die on-board the ship as a part of a large-scale colony (a world ship [20, 21]) or where people are put asleep for the duration of the journey (such as on a hibernation ship). We clearly need to put more effort into our technological power and propulsion development. Instead of people, our focus could remain only on robotic probes, but even then, it comes down to the survivability of the materials and the maintenance of the outdated technology.

Whilst the propulsion technology is being matured towards the higher energy applications of full interstellar missions it can also be applied to lower energy missions. This includes interplanetary missions to the other planets of our solar system and especially to the gas giants, perhaps to $\sim$100 AU. Missions that go beyond our Solar System at these sorts of ranges have been proposed by others [22–26] and some have been proposed only in the last decade [27–29]. Some of the proposals include interstellar precursor missions using fusion technology [30–32].

On paper, the energy released from the mass difference between the reactants and the products of a nuclear fusion reaction, may suggest a value for the mass difference of $\alpha \sim 0.004$ which through $E = mc^2$ would imply exhaust velocities on the order of v_{ex} $\sim$26,000 km/s. However, there are significant losses associated with this method of energy production. These are losses associated with the energy generation mechanism from a laser. Losses associated with absorption, reflection and scattering of light around a capsule. Losses due to asymmetric implosion symmetries. Losses associated with how well any reaction can be confined sufficiently long and at high density and temperature prior to disassembly. All these act to degrade the realistic performance of any machine that proposes to use this reaction to generate power and propulsion thrust. This means that any exhaust velocity may be significantly less than this value, on the order of $\sim$10,000 km/s.

We must also keep in mind here that we are discussing the mission utilisation phase. Before this can take place, there must also come the research and development phase and then the construction and production phase. For an interstellar mission each of these is likely to be elongated and these long-time scales must be considered as a part of any strategic and technological roadmap in terms of a first launch date [33–35].

Even if we had workable fusion reactors on Earth-today that successfully achieved ignition and energy gain, the implementation of this technology to space propulsion will take longer. The design phase may take up to a decade. The manufacture and production may also take another decade, although there is the possibility of using in-space additive manufacturing techniques and artificial intelligence robots to speed up that process. But realistically, from the decision point of planning any interstellar mission of this type, we may be looking at several decades before the mission utilisation phase even begins. Given that successful ignition has only recently been achieved in an Earth-based laboratory [36–38] we are still some way off from this goal. Given the current technology readiness of our engineering systems, this suggests that the first launch may likely be in the first half of the twenty-second century.

When one thinks of fusion research in commercial industry this is usually associated with magnetic confinement fusion (MCF), which utilises a low plasma density ($n \sim 10^{14}\,\text{cm}^{-3}$) of fusion particles at high temperature (T $\sim$10 keV) for a relatively long (τ $\sim$s) confinement time. The goal of any fusion method is to achieve the Lawson criteria where; $n\tau T \geq 10^{21} m^{-3}$ s keV. This is also known as the fusion triple product [39].

However, a problem with this type of magnetic fusion technology in its application to space travel is the high mass associated with its various components such as the magnets. In this book we have instead focussed on another method of igniting fusion fuels, known as inertial confinement fusion (ICF). This uses a high plasma density ($n \sim 10^{23}\,\text{cm}^{-3}$) of particles, in which a small sub-gram capsule of low mass isotopes (typically Hydrogen, Deuterium, Tritium and Helium) is compressed to extremely high density (ρ/ρ_o ~1000s) and temperature (T ~10s keV), whilst being inertially confined in that state for a sufficiently short period (τ ~ns) so that fusion ignition can be initiated within a hot spot region.

The physics processes that take place are highly complex, from the interaction of the laser beam with the capsule and the resulting laser plasma interactions, to the hydrodynamic compression of the fuel with high symmetry and minimum instabilities, to the successful ignition of the fusion fuel, to the thermonuclear burn wave and disassembly of the capsule as it explodes outwards in a high temperature plasma state [40–42]. Whilst the technology is still being investigated in Earth based laboratories for the application of national grid commercial power generation, here we are considering the extended application of this technology to the advanced propulsion of a spacecraft. In addition to correctly understanding the ignition physics of the capsule and its predicted output performance, one must also understand the interaction of the output effects onto the spacecraft vehicle structure—mainly through x-ray Bremsstrahlung radiation and high energy neutrons which will radiate the materials causing significant damage and heat.

As mentioned, the Project Orion system was an external nuclear pulse engine. Another option is to have an internally nuclear pulse engine where capsules (or units) are detonated internal to a reaction chamber, and its products are then used for thrust generation by passing through a magnetic nozzle. Alternatively, the reaction chamber could be half-internal, but the rest exposed to the vacuum of space, and this would be a hybrid internal/external nuclear pulse engine. The reaction still occurs within a reaction chamber, but the products are immediately expelled into the vacuum. Both systems may work based on laser induced irradiation of the capsule and this is performed by a symmetrically positioned array of energy drivers.

Historically, those interested in researching this field are propulsion scientists who make assumptions about the energy output from a given capsule design, such as the exhaust velocity to be expected; but they know little about the complexities of fusion physics. Conversely, those who are experts in fusion physics know little about space propulsion and in most cases have little interest in this application. In addition, many of the fusion experts are based at the national weapons laboratories and use numerical codes which are not accessible to the public because they are validated by underground nuclear testing; so much of the interesting fusion research that may be relevant and helpful remains confidential and cannot be used by the space propulsion researcher. This prevents tangible progress in assessing a class of deep spacecraft designs that may be a realistic part of our near-term future. A part of the purpose of this book has been an attempt to bridge that gap and so present advanced space propulsion assessments as more credible.

In this book, although we are mainly discussing the potential development of advanced propulsion technology such as fusion based systems, we will also go beyond into more exotic types of engines that also utilise a reaction engine. It is useful to step away from the technical details of this work and to examine the broader picture in terms of the implications for developing this technology for the use of humankind. This has ramifications for both our exploration strategies in space but also our defence strategies of Earth. To keep things simple, we will focus just on fusion propulsion in our discussion for the first few chapters as we begin our design calculations and expand our discussions a little from there.

Throughout this book the reader will see the words *capsule*, *pellet* and *target* and it is helpful to define them at the outset. The term capsule refers only to a thermonuclear fuel(s) component. When the fuel is taken together along with any other materials such as a high-density ablator, collectively this is referred to as a pellet. For a direct drive scheme this would also be equivalent to a target. However, in the case of indirect drive it is the addition of the pellet and the external radiation cavity within which it sits that constitutes the target. Similarly, for any pellet designs that include hydrogen propellant layers around it (i.e. such as for VISTA) this would be referred to as a target. The target is the point of focus of the collimated laser energy beams or generated radiation field. The author may use these phrases interchangeably and at this level of analysis it is not so important. Yet it is important to at least appreciate the distinction in definition.

1.2 Fundamental Theory

The motion of an interstellar flyby spacecraft will be determined by two distinct phases of operation. The first is the boost phase where the engine is operating to burn some fuel for the purpose of thrust generation. The spacecraft will then gradually accelerate up to a high velocity, although this may be performed in several burn stages in either a serial or parallel thrust mode. This is then followed by the cruise phase once the target velocity has been attained. In the case of a rendezvous mission there may also be a third phase which is the deceleration of the spacecraft (also referred to in this text as deburn) to a lower velocity for the purposes of increasing encounter time at the target. Alternatively, the spacecraft may be fully decelerated to facilitate full orbital insertion around a stellar or exoplanetary body. It is also possible to design a mission with no cruise phase at all, where the spacecraft simply accelerates and then decelerates.

The transport of a spacecraft into deep space and beyond the boundary of the Solar System will take substantial energy. This can be sourced from on-board propellant mass, and the specific energy of that propellant will depend upon the choice of fuel adopted for the mission.

In a conventional rocket engine, the exhaust velocity is directly related to the temperature inside the reaction chamber, where the temperature of the expanding gas is a measure of how fast the molecules are moving. For this we adopt the concept of specific enthalpy h which describes how much work is in a system due to thermal and internal pressure and this is related to the kinetic energy $\sim v^2/2$ so that $h+v^2/2$ = constant for a steady, adiabatic flow. This leads to a statement of the first law of thermodynamics for an engine

$$h_{ch} + \frac{v_{ch}^2}{2} = h_{ex} + \frac{v_{ex}^2}{2}. \tag{1.1}$$

Where h_{ch} is the enthalpy in the combustion chamber, h_{ex} is the enthalpy at the nozzle exit, v_{ch} is the flow velocity in the combustion chamber and v_{ex} is the exhaust velocity. Then we can consider the combustion chamber velocity $v_{ch} \approx 0$. This is because in the combustion chamber the flow is high pressure and high temperature, but also subsonic which means it is moving very slow and close to stagnant. This is to facilitate complete mixing and combustion. In essence the kinetic energy of the combustion products is negligible compared to the enthalpy. This is essentially a thermodynamic reservoir which defines the initial temperature and pressure for the expansion, but it also contributes almost no energy to the exhaust stream velocity. For example, we might have a situation where v_{ch} = 1–10 m/s, and v_{ex}= 1000 m/s, in which case $v_{ch}^2/v_{ex}^2 \approx 10^{-4} - 10^{-6}$. Hence it is a good approximation to assume the combustion chamber velocity is negligible. Therefore

$$h_{ch} = h_{ex} + \frac{v_{ex}^2}{2}. \tag{1.2}$$

Then since $h = c_p T$, where c_p is the specific heat at constant pressure, we find

$$c_p T_{ch} = c_p T_{ex} + \frac{v_{ex}^2}{2}. \tag{1.3}$$

Then re-arranging for the exhaust velocity

$$v_{ex} = \sqrt{2c_p(T_{ch} - T_{ex})}. \tag{1.4}$$

Next, we want to express the exhaust temperature as a function of the exhaust pressure and for this we use the equations that describe an isentropic expansion for an ideal gas as a function of the ratio of specific heats $\gamma = c_p/c_v$ as follows

$$\frac{T_{ex}}{T_{ch}} = \left(\frac{P_{ex}}{P_{ch}}\right)^{(\gamma-1)/\gamma}. \tag{1.5}$$

Therefore

$$T_{ex} = T_{ch}\left(\frac{P_{ex}}{P_{ch}}\right)^{(\gamma-1)/\gamma}. \tag{1.6}$$

If we substitute this into the equation for the exhaust velocity we find

$$v_{ex} = \sqrt{2c_p T_{ch}\left(1 - \left(\frac{P_{ex}}{P_{ch}}\right)^{(\gamma-1)/\gamma}\right)}. \tag{1.7}$$

Then for an ideal gas we can also express the specific heat capacity as a function of the specific gas constant R and the specific heat ratio γ as follows

$$c_p = \frac{\gamma R}{\gamma - 1}. \tag{1.8}$$

Then substituting this into the equation we find

$$v_{ex} = \sqrt{\frac{2\gamma}{\gamma - 1} R T_{ch}\left(1 - \left(\frac{P_{ex}}{P_{ch}}\right)^{(\gamma-1)/\gamma}\right)}. \tag{1.9}$$

Where P_{ex} is the exhaust pressure at the nozzle exit and P_{ch} is the chamber pressure. We can also substitute for the specific gas constant for the exhaust gas $R = R_u/M$ where M is the molar mass of the exhaust gas (kg/mol) so that

$$v_{ex} = \sqrt{\frac{2\gamma}{\gamma - 1} \frac{R_u T_{ch}}{M}\left(1 - \left(\frac{P_{ex}}{P_{ch}}\right)^{(\gamma-1)/\gamma}\right)}. \tag{1.10}$$

We can see that for a lighter exhaust gas the exhaust velocity will be higher. However, in the vacuum of space this becomes the fully expanded ideal limit, since there is no atmospheric pressure to push against and so there is no pressure thrust, and our main interest is in the expansion of the hot gases and the associated velocity of the exhaust products. This then reduces to

$$v_{ex} = \sqrt{\frac{2\gamma R_u T_{ch}}{(\gamma - 1)M}}. \tag{1.11}$$

Where the higher the combustion chamber temperature, and the lower the molecular weight, the higher the exhaust velocity. We might refer to this as a thermal rocket, where any combusted products are heated to an equilibrium temperature with other materials and then expand through the nozzle.

This equation applies for a rocket engine that is undergoing thermal isentropic (adiabatic and reversible) expansion of a gas according to ideal gas behaviour, such as the flow through a De Laval nozzle. This might be applied in the use of hot gases like H_2O or CO_2 for example. However, in a fusion engine the exhaust velocity is not determined by the thermodynamic expansion of a gas, but instead by the direct kinetic energy of the fusion reaction products. In a chemical rocket engine, it is driven by thermal gas expansion but in a fusion engine it is the kinetic energy of the charged particles via direct particle kinematics. The kinetic energy E of particles with mass m is given as a function of the velocity in the exhaust v_{ex} as follows

$$E = \frac{mv_{ex}^2}{2} \tag{1.12}$$

We can then define the non-relativistic exhaust velocity for a monoenergetic exhaust as

$$v_{ex} = \left(\frac{2E}{m}\right)^{1/2}. \tag{1.13}$$

However, this will be substantially reduced due to several inefficiencies in the operation of the system. This includes due to the firing of the laser, the coupling of the laser to the capsule, the capsule implosion symmetry, the escape of energy from the capsule due to Bremsstrahlung radiation and electron conduction, and then also due to the design of the magnetic nozzle. As a result, the maximum theoretical exhaust velocity will be reduced and it is useful to demonstrate this.

Let us suppose we adopted a D^3He fuel. This is because the fuel combinations of DD and DT lead to the production of substantial neutrons which then suggests a need for large shielding mass. Whilst the D^3He reaction will also lead to neutron production, via DD self-burn secondary reactions, it is largely aneutronic. The charged particle productions from the reaction, protons and ^4He, can also be directed magnetically for thrust, although the reaction is more difficult to ignite than DD or DT and requires a higher temperature. For a D^3He fuel (with mass defect α = 0.0039), with a specific energy on the order of 3.5×10^{11} J/g, this computes to a maximum theoretical exhaust velocity given by the Albert Einstein equation for energy

$$E = \alpha mc^2. \tag{1.14}$$

Then substituting this into the exhaust velocity follows

$$v_{ex} \approx (2\alpha)^{1/2}c = 26,500\,\text{km/s}. \tag{1.15}$$

However, considering the burn-up fraction (the amount of fuel burned through fusion reactions and converted into usable charged particles for thrust) like the value adopted for the Project Daedalus study (discussed in Chap. 9) of ϕ = 13.3–17.5%, this significantly reduces the potential exhaust velocity to by the introduction of a further inefficiency term

$$v_{ex} \approx (2\alpha\phi)^{1/2}c = 9650\text{–}11,076\,\text{km/s}. \tag{1.16}$$

Yet, then we must also factor in the divergence of the exhaust jet which is related to the energy losses from the system. For the Daedalus engine this was near perfect

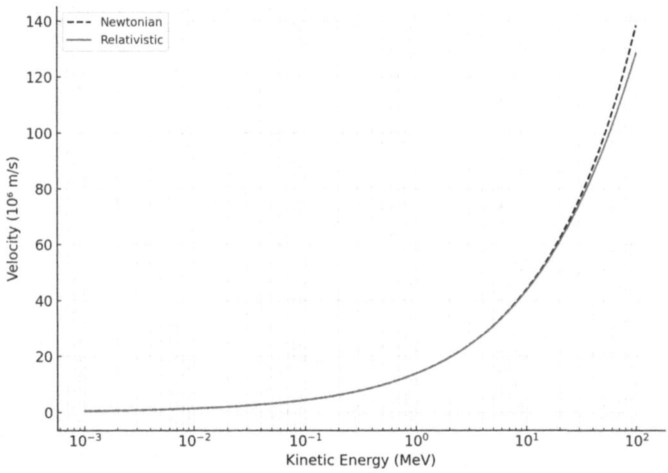

Fig. 1.2 Comparison of Newtonian vs relativistic proton velocity

at around $\mu_j = 91\%$ and so taking this into account we can see how the exhaust velocity is reduced further still which brings us approximately to the two exhaust velocities adopted for the Daedalus study

$$v_{ex} \approx (2\alpha\phi\mu_j)^{1/2}c = 9210\text{–}10,600\,\text{km/s}. \tag{1.17}$$

The burn-up fraction ϕ and jet efficiency η_j are discussed in more detail in later chapters when we estimate values for them from capsule physics and magnetic nozzle optics. We could also have included a charged particle energy fraction term f_c in this equation but we neglect it for this analysis.

Therefore, in deriving a new design, if the burn-up fraction or jet efficiency are higher, towards a more ideal rocket, we can expect a higher exhaust velocity. Similarly, if the burn-up fraction or jet efficiency are lower, then this indicates there are more energy losses from the system and so we should expect a lower exhaust velocity. If the exhaust velocity is lower for the same velocity increment, we must aim for a design that has a higher mass ratio to give an equivalent performance, which means a leaner propellant design.

Next, we must consider corrections due to the Special Theory of Relativity. Figure 1.2 shows a comparison between Newtonian physics (non-relativistic) and relativistic flight. In relativity, the total energy of a particle is given by

$$E^2 = (pc)^2 + (mc^2)^2. \tag{1.18}$$

Where m is the rest mass and p is the momentum. If we solve this for the momentum we get

$$p = \frac{(E^2 - m^2c^4)^{1/2}}{c}. \tag{1.19}$$

In relativistic mechanics the particle momentum and total energy can be expressed

$$p = \gamma m v, \tag{1.20}$$

$$E = \gamma m c^2. \tag{1.21}$$

Therefore

$$\frac{P}{E} = \frac{\gamma m v}{\gamma m c^2} = \frac{v}{c^2} \tag{1.22}$$

So that rearranging we end up with a result for the velocity, which we can express as an equivalence to the exhaust velocity (for a photon $v_{ex} \equiv v \equiv c$) since in the relativistic limit the exhaust is dominated by the photon radiation where $E >> mc^2$. Therefore, for particles with mass

$$v = \frac{pc^2}{E} \tag{1.23}$$

Dividing both sides by c

$$\frac{v}{c} = \frac{pc}{E}, \tag{1.24}$$

$$\beta = \frac{pc}{E}. \tag{1.25}$$

Where

$$\beta = \sqrt{1 - \frac{1}{\gamma^2}}, \tag{1.26}$$

Note that in the non-relativistic limit we would end up with the classical velocity $v_{ex} = p/m = v$, since the energy is given by $E = mc^2 + mv^2/2$. Then substituting the equation for the momentum into the velocity we find

$$v = \frac{pc^2}{E} = \frac{(E^2 - m^2c^4)^{1/2}}{E}c. \tag{1.27}$$

Then simplifying

$$v = \sqrt{1 - \left(\frac{mc^2}{E}\right)^2}c. \tag{1.28}$$

We can also relate this equation to the mass defect α from $\alpha = E/mc^2$ and where for relativistic flight $E = (\gamma - 1)mc^2$ so that $\gamma = 1 + \alpha$ and therefore

$$E = (1 + \alpha)mc^2. \tag{1.29}$$

Therefore

$$\frac{mc^2}{E} = \frac{1}{1 + \alpha}. \tag{1.30}$$

Then substituting and expressing as our exhaust velocity

$$v_{ex} = \left(1 - \left(\frac{1}{1 + \alpha}\right)^2\right)^{1/2}c. \tag{1.31}$$

The mass defect fraction α for a given particle is given by the particle energy as a fraction of the rest mass energy

$$\alpha = \frac{E}{mc^2}. \tag{1.32}$$

We can calculate this for a D^3He reaction where the main output is a proton of energy E = 14.7 MeV. The rest mass of a proton is $m = 1.0073u \times 931.5 =$ 938.272 MeV/c^2, so that α = 14.7/938.272 = 0.0157. Then

$$v = \left(1 - \left(\frac{1}{1.0157}\right)^2\right)^{1/2}c = 0.175c = 52{,}500\,\text{km/s}. \tag{1.33}$$

Similarly for a DT reaction we have a ^4He (alpha particle) which we can direct through the magnetic nozzle, where the energy is E = 3.5 MeV and the rest mass is m = 4.0015u×931.5 MeV/c^2/u = 3727.4 MeV/c^2. Therefore mc^2/E = 3727.4/3.5 = 1065. Then this computes to a value of $v_{ex} = 0.043c$ = 12,900 km/s. For the two DD reactions the exhaust velocity computes to a value of 24,000 km/s (proton) and 7230 km/s (^3He).

For a ^3He particle the rest mass energy will be 4.016u × 931.494 MeV/u = 2810.94 MeV, and so α = 0.82 MeV/2810.94 MeV = 0.00292. For a ^4He particle the rest mass energy will be 4..0026u × 931.494 MeV/u = 3728.4 MeV, and so α = 3.5 MeV/3728.4 MeV = 0.000939. For a D^3He reaction we get the products ^4He(3.6 MeV) and a proton (14.7 MeV) and so the total released energy Q = 3.6 + 14.7 = 18.3 MeV. The total initial rest mass energy is then 2.014102u + 3.016029u

= 5.030131u, and we can convert that to MeV via $E = 5.030131 \times 931.494 =$ 4684.7 MeV. The total kinetic energy release fraction is then given as $\alpha = Q/E$ = 18.3 MeV/4684.7 MeV =0.00391.Therefore in the context of a propulsion engine utilising a D^3He fuel with α = 0.0039 results in

$$v_{ex} = \left(1 - \left(\frac{1}{1.0039}\right)^2\right)^{1/2} c = 0.088c = 26{,}400\,\text{km/s}. \tag{1.34}$$

This would represent a theoretical upper bound. However, if we include a loss parameters and assume a burn-up fraction of ϕ = 13.3% and jet efficiency μ_j = 91% (see Chap. 9 for Daedalus) then the equation becomes modified as

$$v_{ex} = \left[\phi\mu_j\left(1 - \left(\frac{1}{1+\alpha}\right)^2\right)\right]^{1/2} c. \tag{1.35}$$

This results in a much-reduced exhaust velocity

$$v_{ex} = \left[0.133 \times 0.91 \times \left(1 - \left(\frac{1}{1.0039}\right)^2\right)\right]^{1/2} c = 0.0306c = 9184\,\text{km/s}. \tag{1.36}$$

Comparing this to the value computed earlier of v_{ex} = 9210 km/s for the non-relativistic case, we see that this is a difference of only ~26 km/s and that is due to the relatively low energy of the particles. As shown in Fig. 1.2 it is only when the energy becomes substantially large. On the figure a divergence begins to become visible at an energy of ~10s MeV. We can convert this to a velocity if we treat E_{kin} as the kinetic energy of the particles as distinct from the total energy $E = \gamma mc^2$ where

$$E_{kin} = E - mc^2 = (\gamma - 1)mc^2. \tag{1.37}$$

So that

$$\gamma = 1 + \frac{E_{kin}}{mc^2}, \tag{1.38}$$

$$\beta = \frac{v}{c} = \left(1 - \frac{1}{\gamma^2}\right)^{1/2}. \tag{1.39}$$

For a value of 10 MeV, γ = 1.0107, β = 0.145 with a proton of energy mc^2 = 937 MeV we find

$$v = \beta c = 0.145c = 43{,}500\,\text{km/s}. \tag{1.40}$$

For comparison if instead we assumed a ^4He particle at 10 MeV with mc^2 = 3727 MeV, then γ = 1.0027, β = 0.073 and v_{ex} = 0.073c = 21,900 km/s. In summary, this shows us that for low speeds a non-relativistic approximation is sufficient, but when one begins to approach high fractions of the speed of light, special relativistic corrections must be considered.

1.3 Rocket Equation

The basis of a rocket is the ejection of mass rearward, opposite to the flight direction, to facilitate momentum transport, and therefore the generation of a force which we call thrust. For any spacecraft the initial mass (wet mass) m_o will be made up of the structure mass m_s, payload mass m_{pay}, power system mass m_{pow}, radiator mass for thermal management m_{rad}, propellant tank mass m_{tank}, engine mass m_{eng} and the mass of any additional systems m_{add}.

It is possible to examine how this influences the possible performance of the spacecraft using the ideal rocket equation

$$\Delta v = v_{ex} ln\left(\frac{m_o}{m_f}\right). \tag{1.41}$$

Where m_f is the final mass and taken as a fraction this is known as the mass ratio R and the exhaust velocity can also be defined in terms of a specific impulse I_{sp} (measured in seconds) with respect to acceleration due to gravity on Earth g_o, so that $v_{ex} = g_o I_{sp}$.

For a rendezvous mission the mass ratio would go as R^2 consistent with a much larger propellant requirement for the acceleration and deceleration phases of the mission, provided these two phases are symmetric. We see how the velocity increment Δv is dependent upon the logarithm of the mass ratio but also the exhaust velocity discussed earlier. For a two-staged system the velocity increment is given by

$$\Delta v_1 = v_{ex1} ln\left(\frac{m_{0,1}}{m_{f,1}}\right), \tag{1.42}$$

$$\Delta v_2 = v_{ex2} ln\left(\frac{m_{0,2}}{m_{f,2}}\right), \tag{1.43}$$

$$\therefore v = \Delta v_1 + \Delta v_2. \tag{1.44}$$

For n number of stages

$$v = \sum_{i=1}^{n} \Delta v_i. \tag{1.45}$$

Alternatively with an initial mass $m_{0,i}$ and final mass $m_{f,i}$ at stage i after expending all the propellant an stages to leave only the payload and any remaining structure

$$\Delta v = \sum_{i=1}^{n} v_{ex,i} ln\left(\frac{m_{0,i}}{m_{f,i}}\right). \tag{1.46}$$

However, we note two restrictions on this relationship. The first is the constant speed of light, for which it would not be possible to exceed. Second, if one adds in more structural stages (and even series engines) then each stage will get smaller and deliver less energy than the last and so one reaches a limit of diminishing performance returns for a fixed payload mass. One can begin to also shrink the payload to provide for more structural mass or propellant, but this also will reach a limit, when the mass of the payload is no longer sufficient to meet the mission requirement and carry the appropriate instruments. We might express this as a limit equation where $\Delta v \equiv v_c$, the cruise velocity, such that

$$\lim_{n \to \infty} v_c \to v_{max}. \tag{1.47}$$

For any staged system the total mass ratio would be equal to the product of each stage

$$R_{tot} = R_1 \times R_2 \times R_3 \ldots \times R_n. \tag{1.48}$$

Alternatively

$$R_{tot} = \prod\nolimits_{i=1}^{n} R_i. \tag{1.49}$$

Where $R_i = m_{0,i}/m_{f,i}$. Then for a decelerating stage, assuming equivalent exhaust velocities and symmetric Δv for the burn and deburn stages, the mass ratio would be found from

$$R_{deburn} = \sqrt{R_{tot}}. \tag{1.50}$$

In terms of the mass ratio in the rocket equation, for the purpose of analysis it is useful to simplify this as much as possible and just assume a propellant mass and a structure mass, and where all other systems are contained within the structure mass to include the payload. On this basis the mass ratio for a single staged system is then approximated by

$$R = \frac{m_p + m_s}{m_s}. \tag{1.51}$$

The inverse of this effectively defines a structural coefficient

$$\epsilon_s \equiv \frac{m_s}{m_p + m_s}. \tag{1.52}$$

Assuming all of the propellant is used on the mission, it is then possible to examine the mass ratio under a set of scenarios, as approximations to define a likely mass ratio range for our performance.

For an interplanetary mission scenario out to $\sim$100s AU, where the mass is dominated by the spacecraft and its power plant, the quantity of propellant is minimised and assuming $m_p << m_s$ then

$$R = \frac{m_p + m_s}{m_s} \approx \frac{m_s}{m_s} \rightarrow 1. \tag{1.53}$$

For an interstellar precursor mission scenario out to $\sim$1000s–10,000s AU, where the mass of propellant mass is equivalent in scale to the structure mass, such that $m_p \sim m_s$, then

$$R = \frac{m_p + m_s}{m_s} \approx \frac{2m_s}{m_s} \rightarrow 2. \tag{1.54}$$

For an interstellar mission scenario out to $\sim$100,000s AU, where the propellant mass is much greater than the structure mass, such that $m_p >> m_s$ (and in this scenario the payload mass fraction would be on the order of $\sim$1% or less), then

$$R = \frac{m_p + m_s}{m_s} = \frac{m_p}{m_s} + 1. \tag{1.55}$$

Although, the structure mass will be some fraction ξ of the propellant mass so that $m_s = \xi m_p$, in which case

$$R = \frac{m_p}{\xi m_p} + 1 = \frac{1}{\xi} + 1. \tag{1.56}$$

Then given that the structure mass for an interstellar mission is likely to be around $\sim$5% of the propellant mass, and is not likely to exceed around $\sim$10% of the propellant mass, the approximate likely range for the mass ratio will be on the order of

$$R = \frac{1}{\xi} + 1 \approx 11 - 21. \tag{1.57}$$

Staging the design would then further reduce the mass ratio to a value somewhat less than $R \lesssim 10$ per stage.

A typical example of such a probe that would utilise ICF technology would be the Project Daedalus study mentioned earlier. This was a probe that travelled at 0.122c to 5.9 ly in around 50 years, carrying a 450 tons payload mass and propelled by 50,000 tons D^3He fuel in two vehicle structurul stages. The propellant mass was divided into 46,000 tons for the first stage and 4000 tons for the second stage. This design would have mass ratios for the two stages of $R_1 = 7.9$ and $R_2 = 5.1$ respectively, which corresponds to a total mass ratio of $R_{tot} = R_1 \times R_2 = 40.3$. It would have a payload mass fraction to wet mass of 0.85% and a payload mass fraction to dry mass of 14.05%. Had the probe been a single stage but for the same total propellant mass and structure mass, the total mass ratio would be on the order of $R \sim 20$, which is near the upper limit of our range.

In general, it would be fair to assume that the mass ratio for a fusion driven staged spacecraft over an interstellar trajectory would be in the approximate range $R \sim 5$–10. Then taking the likely estimated exhaust velocity above of ~8000–10,000 km/s which leads to a likely velocity range of ~13,000–23,000 km/s. If the possibility of a double stage is considered so that twice these velocity additions are possible then one computes a revised range of ~26,000–46,000 km/s. This means that practical propulsion via fusion is likely limited to around $\sim 0.15c$. It may be possible to extend this a little further with additional staging but as with all rockets there is a diminishing return on performance with increased stages. It is doubtful that fusion could be pushed as far as $0.2c$.

Each time a fusion capsule is detonated within a reaction target chamber, the number of these detonations occurring per second corresponds to the engine pulse frequency. This may be as little as 1 Hz or as many as 1000 Hz. At the lower limit this implies a very low mass flow rate and so a low acceleration of the spacecraft leading to an elongated mission profile. At the higher limit this puts considerable stress on the engineering structure and may lead to material fatigue from the heat loads for example. So, this goes from the extremes of practical but not very useful, to very useful but not very practical. Pulse frequencies for next generation reactors on the planet which have the aim of generating electricity for a national grid may typically be in the range ~1–50 Hz. Spacecraft on an interstellar trajectory may typically be in the range ~ 100–200 Hz, or even higher.

For a pulse detonation system, the number of units or capsules N_{cap} will be important since this is related to the total propellant mass m_p and the capsule design mass m_{cap} as follows

$$N_{cap} = \frac{m_p}{m_{cap}}. \tag{1.58}$$

In general, the larger the mass of the spacecraft the harder it will be to push. This is simply a question of the propulsion which is a function of the mass flow rate $\dot{m}$ and the exhaust velocity v_{ex}. The mass flow rate is also a function of the capsule mass m_{cap} and the engine pulse frequency f_{hz} and these are described in more detail elsewhere [43, 44]. It is also useful to write it in this form since we are dealing with nuclear pulse propulsion.

$$T = m_{cap} f_{hz} v_{ex}. \tag{1.59}$$

We note that this can also be written in terms of the burn time t_b to achieve the acceleration phase as a function of the total number of capsules N_{cap} such that

$$f_{hz} = \frac{N_{cap}}{t_b}. \tag{1.60}$$

and so

$$T = \frac{m_{cap} N_{cap} v_{ex}}{t_b}. \tag{1.61}$$

From this equation it can be observed that to derive maximum thrust, one needs to ideally increase the exhaust velocity, pulse detonation frequency and the capsule mass as much as possible. However, the exhaust velocity is energy limited since it depends on the type of primary fusion reactions being utilised. There is also a maximum pulse frequency since there comes a point where it is so high that it is impractical to maintain and the system is then tending towards a continuous propulsion system rather than pulse detonation. Further, there is also a limit to the mass since there comes a point where the capsules become so large that it is not practical to ignite them using any reasonable method and instead the capsule ignition engine is tending more towards the operation of the units associated with nuclear pulse propulsion such as in Project Orion.

The mass flow rate can be expressed as a function of the capsule mass and the pulse frequency and in the case of a parallel staged thrust system also the number of engines n_{eng} as follows

$$\dot{m}_{series} = m_{cap} f_{hz}. \tag{1.62}$$

$$\dot{m}_{parallel} = m_{cap} f_{hz} n_{eng}. \tag{1.63}$$

For a propulsion system that includes expellant (propellant augmentation) such as via the use of propellant injection into the exhaust stream, the mass flow rate is then modified to include an augmented mass term m_{exp} as follows

$$\dot{m} = \left(m_{cap} + m_{exp}\right) f_{hz} n_{eng}. \tag{1.64}$$

Therefore the thrust equation, which has the same units as Newton's Second Law $F = ma$ is

$$T = \dot{m} v_{ex}. \tag{1.65}$$

The exhaust kinetic power is that carried by the exhaust in the rocket frame and we term this the jet power P_j (the power in the exhaust jet) and if we divide this into the total propulsion (engine) mass m_{eng} we also get the specific power

$$P_j = \frac{1}{2}\dot{m}v_{ex}^2, \tag{1.66}$$

$$P_{sp} = \frac{P_j}{m_{eng}}. \tag{1.67}$$

Yet the maximum thrust must also be traded with the requirements for a minimum burn time on the assumption that the pulse detonation frequency was high. Sustaining a high frequency for a prolonged period would place considerable stress on the engine and may lead to material fatigue in the surrounding structure due to the exposure to large heat loads and the continued dose of x-ray and neutron radiations.

The thrust power P_{th} requirement is separate from the jet power and defines an average power over the burn duration of the acceleration phase of the mission at burnout m_f. For a constant power system, it is given by

$$P_{th} = Tv_c = \frac{m_f v_c^2}{2t_b}. \tag{1.68}$$

In considering the ideal rocket equation, it is useful to examine this derivation in a little more detail since an interesting implication emerges from our understanding of its form. Suffice it to say the rocket equation that comes out is given as the additive velocity increment as a function of the exhaust velocity, initial vehicle mass m_o and final vehicle mass m_f as follows

$$\Delta v = v_{ex}ln\left(\frac{m_o}{m_f}\right) + v_n. \tag{1.69}$$

We note the inclusion of an additional velocity term v_n. This term will be made up of many components, and they are the external forces on the vehicle. For a conventional rocket moving through the atmosphere this will be aerodynamic drag and the gravitational force, both of which retard the ascent and must be taken into consideration as a correction to the orbital velocity equation

$$v_{orb} = \left(\frac{GM}{R}\right)^{1/2}. \tag{1.70}$$

In this scenario, the external force is in general seen as a negative thing which must be mitigated. In fact, it is argued that a different perspective is required when moving through deep space and that the external force term can be seen as a positive tool for augmenting the vehicle propulsion, at least for a vehicle on a solar system escape trajectory.

There are times when external forces can also generate a positive effect on the vehicle, such as with a gravity assist manoeuvre. In such a manoeuvre a spacecraft is seen to be on a collision course with a large gravitating object, and the planet can transfer kinetic energy to the spacecraft. Typically, for a direct Jovian like approach (example only) the escaping velocity will be $\sim$2–3 times the encounter velocity. So, this is an example of an external force which can add to a vehicle's velocity, and this usually requires complex trajectory modelling calculations.

As mentioned earlier, the way designers have tended to approach the interstellar flight challenge is to focus on single propulsion solutions, such as fusion, or laser beam, or microwave beam or solar sails or one of the many exotic solutions. Each eventually reaches the performance limitation and we are led to consider one of the alternatives. What would happen if in fact we considered many of these solutions in parallel?

An examination of the problem soon demonstrates that the Konstantin Tsiolkovsky equation is in fact not the general function for interstellar flight or even the space-time problem. It is in fact what you might call in mathematical terms a particular solution, bounded by initial conditions. This is the same way that a general solution to a differential equation when bounded by input values may yield the particular solution for the problem being examined. This is also referred to as an initial value problem, where a solution of a differential equation with its constants undermined is called a general solution, but the solution of an initial value problem complete with the values of the constants is called a particular solution.

The inclusion or exclusion of the internal and external forces acting on a vehicle will act as the bounding conditions and thus determine the solutions possible. It is argued here that this approach has the effect of producing a form of tunnel vision on our thinking and that we need to step back from the Tsiolkovsky equation and search for a more general form. Now in theory this could be the entire laws of physics which is obviously not a practical approach to consider. But we can at least make some headway by considering various propulsion solutions in parallel.

The claimed benefit to such an approach is to elucidate new directions in the design solution space. All current approaches are a trade-off with vehicle configuration requirements and performance requirements and designers will find a limit of diminishing returns for a given design type. But by combining multiple solutions, so-called hybridized designs, this has the potential to permit new levels of performance trade-off, and perhaps new windows of optimality in the design space not yet explored.

First, we must revisit the derivation of the rocket equation to fully appreciate the point being made in this section. We can derive the ideal rocket equation from a rocket thrust description. We begin by the definition of Newton's second law of motion

$$F = m\frac{dv}{dt}. \tag{1.71}$$

For a conventional launch rocket, the force on the rocket is the net difference between the thrust of the rocket engine, aerodynamic drag and the mass of the vehicle. We are assuming here that there are no external forces

$$T = m\frac{dv}{dt}. \tag{1.72}$$

The Thrust is related to the exhaust velocity v_{ex} and the mass flow rate $\dot{m}$ by

$$T = v_{ex}\dot{m}. \tag{1.73}$$

If the propellant mass is m_p and the structure mass is m_s (which we assume includes the payload mass). The mass flow rate is then given as

$$\dot{m} = -\frac{dm_p}{dt} = -\frac{dm}{dt}. \tag{1.74}$$

Equating the above

$$T = -v_{ex}\frac{dm}{dt}. \tag{1.75}$$

Then substituting into the thrust equation

$$-v_{ex}\frac{dm}{dt} = m\frac{dv}{dt}, \tag{1.76}$$

$$-\frac{dm}{m} = \frac{dv}{v_{ex}}. \tag{1.77}$$

Then integrating

$$-\int_{m_o}^{m_f} \frac{1}{m}dm = \frac{1}{v_{ex}}\int_{v_o}^{v_b} dv. \tag{1.78}$$

After the substitution of the limits, we get the ideal rocket equation

$$\Delta v = v_{ex} ln\left(\frac{m_o}{m_f}\right). \tag{1.79}$$

Let us now re-evaluate the derivation again but including external forces f_1 and f_2. In which case the Thrust equation is now

$$T + f_1 + f_2 = m\frac{dv}{dt}. \tag{1.80}$$

And substituting in for the Thrust expressed as a function of the exhaust velocity and mass flow rate we get

$$\dot{m} = -\frac{dm_p}{dt} = -\frac{dm}{dt}. \tag{1.81}$$

Equating the above equations

$$T = -v_{ex}\frac{dm}{dt}. \tag{1.82}$$

Then substituting into the $T + f_1 + f_2$ equation

$$-v_{ex}\frac{dm}{dt} + f_1 + f_2 = m\frac{dv}{dt}. \tag{1.83}$$

Where $f_1 = mdv_{b,1}/dt$ and $f_2 = mdv_{b,2}/dt$

$$-v_{ex}\frac{dm}{dt} + m\frac{dv_{b,1}}{dt} + m\frac{dv_{b,2}}{dt} = m\frac{dv}{dt}. \tag{1.84}$$

Then integrating we get

$$-v_{ex}\int_{m_o}^{m_f}\frac{1}{m}dm + \int_{v_{1,o}}^{v_{b,1}} dv_{b,1} + \int_{v_{2,o}}^{v_{b,2}} dv_{b,2} = \int_{v_o}^{v_b} dv. \tag{1.85}$$

Leading to

$$\Delta v = v_{ex}ln\left(\frac{m_o}{m_f}\right) + \delta v_{b,1} + \delta v_{b,2}. \tag{1.86}$$

Alternatively, we can express an external force F_{ext} as follows

$$\Delta v = v_{ex}ln\left(\frac{m_o}{m_f}\right) + \int_{t_o}^{t_f}\frac{F_{ext(t)}}{m(t)}dt. \tag{1.87}$$

or as an approximation

$$\Delta v = v_{ex}ln\left(\frac{m_o}{m_f}\right) + \frac{1}{m_{ave}}\int_{t_o}^{t_f} F_{ext}(t)dt. \tag{1.88}$$

Where the average mass $m_{ave} = (m_o + m_f)/2$, $\int F_{ext}(t)dt$ is the impulse from the external forces over the duration of the burn. Since there may be multiple external forces, we can re-express this as an extended rocket equation where each impulse is a contribution to the total impulse

$$\Delta v = v_{ex} ln\left(\frac{m_o}{m_f}\right) + \frac{1}{m_{ave}} \int_{t_o}^{t_f} \left(\sum_i F_{ext,i}(t)\right) dt. \tag{1.89}$$

It can be seen then that the ideal rocket equation is merely a particular solution, with the assumptions of motion in a vacuum, no gravity, no drag, no external forces, constant exhaust velocity and one-dimensional motion. It is a simplified version of a more general problem, although these assumptions under idealized conditions are highly useful in simplifying calculations.

In contrast, the extended rocket equation accounts for the possibility of multiple external forces such as gravity, drag, thrust augmentation. It is also time dependent. We can therefore refer to this as a general rocket equation with a full description of the momentum evolution under force and mass loss. The impulse may be generated on-board or from some external interaction.

We might consider this generalized rocket equation as the foundation framework for all space propulsion systems. Since it is possible within this equation to include chemical rockets, ion engines, solar sails, laser sails, fusion engines and many others. Certainly, for the case of interstellar flight, it should be seen as a more foundational coherent framework.

The author originally proposed a version of this argument at a conference in Texas in 2013 and published an article in a journal [45], but the insight was not widely noticed. Yet this is a unifying principle that allows for the subset of other propulsion options within a broad momentum framework extended to include the other external forces. On this basis, it is possible to incorporate reactive and non-reactive propulsion methods. It also opens the possibility of designing hybrid propulsion systems by combining internal fuel driven systems with external thrust augmentation systems. This will be demonstrated later in Chap. 7.

Finally, once a vehicle is moving in flight it is important to be able to describe its basic motion using fundamental equations and any *Rocket Scientist* should be familiar with these. From the standard theory of kinematics, we can write out the equations for constant acceleration, and these are critical for any designer to know in performing rapid back-of-envelope calculations. These describe the relationship of final velocity v, initial velocity v_o, acceleration a, distance x and time t as follows

$$v = v_o + at, \tag{1.90}$$

$$v^2 = v_o^2 + 2ax, \tag{1.91}$$

$$x = v_o t + \frac{1}{2} a t^2. \tag{1.92}$$

It is important to gain familiarity with these equations, and it is amazing how useful they are in calculating first order problems relating to space flight. Although since they are equations of constant acceleration, they do not allow for a change in mission thrust. It is also important to know their calculus form which can be used in problems of non-constant acceleration which derive from the derivatives of

$$v = \frac{dx}{dt}, \tag{1.93}$$

$$a = \frac{dv}{dt}, \tag{1.94}$$

$$vdv = adx. \tag{1.95}$$

So that they can also be written in integral form

$$\int_{t_o}^{t} vdt = \int_{x_o}^{x} dx, \tag{1.96}$$

$$\int_{t_o}^{t} adt = \int_{v_o}^{v} dv. \tag{1.97}$$

$$\int_{v_o}^{v} vdv = \int_{x_o}^{x} adx. \tag{1.98}$$

To become comfortable with using these equations you should solve them to derive the equations of constant acceleration and work out different problems in relation to a spacecraft mission.

Earlier we showed the distance obtained under the assumption of constant acceleration. But for a rocket, we must consider the loss of mass during acceleration due to the use of propellant fuel, the distance attained under acceleration is given by a logarithmic relation which is a function of the burn time t_b, exhaust velocity v_{ex}. This is obtained by integration of one of the equations of motion

$$x = \int_{0}^{t_b} v(t)dt = v_{ex} \int_{0}^{t_b} ln\frac{m_o}{m(t)}dt. \tag{1.99}$$

Then substitution of $m = m(t) = m_o - \dot{m}t$ which implies $dt = -dm/\dot{m}$

$$x = -\frac{v_{ex}}{\dot{m}} \int_{m_o}^{m_f} ln\frac{m_o}{m}dm = \frac{v_{ex}}{\dot{m}} \int_{m_f}^{m_o} (ln(m_o) - ln(m))dm. \tag{1.100}$$

This leads to

$$x = \frac{v_{ex}}{\dot{m}}\Big[(m_o - m_f) - m_f ln(R)\Big]. \tag{1.101}$$

Then writing in the context of a burn-stage the distance x_b in time t_b and using $\dot{m} = (m_o - m_f)/t_b = m_f(R-1)/t_b$ we get

$$x_b = v_{ex} t_b \left(1 - \frac{ln(R)}{R-1}\right). \tag{1.102}$$

Where R is the mass ratio for the stage which is the ratio of the initial mass m_o to the final mass m_f between staged burns. The total distance attained is then found by adding up the sum of the staged boost distance and the cruise distance and the total time duration under boost is defined

$$x_{tot} = \sum_{i=1}^{n} x_{b,i} + x_c, \tag{1.103}$$

$$t_{b,i} = \sum_{i=1}^{n} \frac{m_{p,i}}{\dot{m}_i}. \tag{1.104}$$

The time duration under cruise is given by $t_c = x_c/v_c$ and the total time duration of the mission is then

$$t_{tot} = \sum_{i=1}^{n} t_{b,i} + t_c. \tag{1.105}$$

This then summarises our introduction to some of the basic rocket equations which can also be used for interstellar problems. Indeed, in some of the later chapters we will use these equations in the design of various vehicles for different missions, so the reader is advised to become familiar with them now and to become fluent in their use. However, for an interstellar flight there is not just the rocket but the gravitational interaction of the stars and we discuss this in the next chapter and show where the appropriate approximations are valid.

Problems of Chap. 1

1.1 The Voyager probes were launched in 1977 and currently have a hyperbolic excess velocity of around 17 km/s (i) How long would it take such probes to reach the nearest stars at 4.3 ly distance, assuming they were headed in that direction? (ii) Estimate the speed required of such probes if the mission was to be completed in 100 years. Express your answers in km/s and AU/Year and neglect time for acceleration.

1.2 A space probe is sent on a mission to 10,000 AU and it must reach there within 30 years of launch by accelerating and then decelerating (i) Calculate the average acceleration required? (ii) What will be the peak velocity of the spacecraft at midpoint in the mission and what is this as a percentage of light speed?

1.3 The 10,000 AU space mission is redesigned with an engine that does not have constant acceleration but instead increases linearly with time by an amount $a(t) = kt$ for the first half of the mission (up to $t_{1/2}$ = 15 years = 4.733×10^8 s), and then decreases by the same amount towards the destination. In this k has the units of m/s^3 (i) For the same total mission time of 30 years, find the value of k (ii) Find the midpoint speed v_{peak} and the peak acceleration a_{peak}.

References

1. Young AT (1978) Quoted in voyager. NASA Mission Status Bull 14
2. Kepler J (1610) Dissertatio cum Nuncio Sidereo. Johnson Reprint Corp., New York. Translated by Edward Rosen in Kepler's conversation with Galileo's sidereal messenger, 1965, p 39
3. Tsiolkovsky K (1911) The will of the universe: the unknown intelligence. Kaluga
4. Sagan C (1980) Cosmos. Random House, New York
5. Shepherd LR (1952) Interstellar flight. J Br Interplanetary Soc 11(4):149–167
6. Long KF (2012) Deep space propulsion, a roadmap to interstellar fligh. Springer, Berlin
7. Parkinson RC (1978) Project Daedalus: Propellant Acquisition Techniques. J Br Interplanetary Soc Special Supplement:S83–S89
8. Palaszewski B (2012) Atmospheric mining in the outer solar system: resource capturing, storage and utilization, AIAA 2012-3742. Presented 48th AIAA/ASME/SAE/ASEE joint propulsion conference & exhibit, Atlanta, Georgia, 30 July–01 August 2012
9. Forward RL (1984) Roundtrip interstellar travel using laser-pushed lightsails. J Spacecraft 21(2):187–195
10. Forward RL (1985) Starwisp: an ultra-light interstellar probe. J Spacecraft 22(3):345–350
11. Lubin PA (2016) A roadmap to interstellar flight. J Br Interplanetary Soc 69(2/3):40–72
12. Parkin KLG (2018) The breakthrough starshot system model. Acta Astronaut 152:370–384
13. Gaidos G, Lewis RA, Meyer K, Schmidt T, Smith GA (1999) AIMSTAR: antimatter initiated microfusion for precursor interstellar missions. Acta Astronaut 44(2–4):183–186
14. Kammash T (2002) Antiproton driven magnetically insulated inertial confinement fusion (MICF) propulsion system. NIAC 98-02 Final Report
15. Everett CJ, Ulam SM (1955) On a method of propulsion of projectiles by means of external nuclear explosions. LAMS. Los Alamos Scientific Laboratory of the University of California, Los Alamos
16. Dyson FJ (1968) Interstellar transport. Phys Today 21(10):41–45
17. Dyson G (2002) Project orion - the atomic spaceship 1957–1965. The Penguin Press, London
18. Mankins JC (1995) Technology readiness levels. A White Paper, NASA
19. Schmidt GR, Patterson MJ (2010) In-space propulsion technologies for the flexible path exploration strategy. Presented 61st IAC Prague, IAC-10.C4.6.2
20. Martin AR (1984) World ships - concept, cause, cost, construction and colonisation. J Br Interplanetary Soc 37(6):243–253
21. Bond A, Martin AR (1984) World ships - an assessment of the engineering feasibility. J Br Interplanetary Soc 37(6):254–266
22. Jaffe LD, Ivie C, Lewis C, Lipes R, Norton HN, Stearns JW, Stimpson LD, Weissman P (1980) An interstellar precursor mission. J Br Interplanetary Soc 33(1):3–26
23. Nock KT (1987) TAU - A mission to a thousand astronomical units. AIAA-87-1049, presented at the AIAA/DGLR/JSASS international electric propulsion conference, Colorado Springs, 11–13 May 1987
24. Fearn DG (2008) Technologies to enable near-term interstellar precursor missions; Is 400 au accessible? J Br Interplanetary Soc 61(8):279–283

25. Matloff GL, Johnson L, Maccone C (2007) Helios and prometheus: a solar/nuclear outer-solar system mission. J Br Interplanetary Soc 60(12):439–442
26. Swinney RW, Long KF, Hein A, Galea P, Mann A, Crowl A, Obousy R (2012) Project icarus: exploring the interstellar roadmap using the icarus pathfinder and starfinder probe concepts. J Br Interplanetary Soc 65(7/8):244–254
27. McNutt Jr RL, Wimmer-Schweingruber RF, Gruntman M, Krimigis SM, Roelof EC, Brandt PC, Vernon SR, Paul MV, Stough RW, Kinnison JD (2022) Interstellar probe - destination: universe! Acta Astronaut 196:13–28
28. McNutt Jr RL, Brandt PC et al ()2019 Interstellar probe study 2019 report. Interstellar Probe Study Team, NASA Task Order NNN06AA01C
29. Napolillo DH, Paul MV, McNutt Jr RL (2018) Interstellar probe, humanity's first step to interstellar space is possible. Johns Hopkins Applied Physics Laboratory, Poster Presentation, COSPAR 2018, Los Angeles
30. Cohen SA, Swanson C, Mcgreivy N, Raja A, Evans E, Jandovitz P, Khodak M, Pajer G, Rognlien TD, Thomas S, Paluszek M (2019) Direct fusion drive for interstellar exploration. J Br Interplanetary Soc 72(2):37–50
31. Aime P, Gajeri M, Kezerashvili RY (2021) Exploration of trans-neptunian objects using the direct fusion drive. Acta Astronaut 178(7):257–264
32. Gajeri M, Aime P, Kezerashvili RY (2021) A titan mission using the direct fusion drive. Acta Astronaut 180:429–438
33. Long KF (2011) Project icarus: the first unmanned interstellar mission, robotic expansion & technological growth. J Br Interplanetary Soc 64(4):107–115
34. Long KF (2020) A rapid study on the development of an interstellar roadmap and planning ahead for technology maturation. J Br Interplanetary Soc 73(1):6–14
35. Long KF (ed) (2015) Beyond the boundary, exploring the science and culture of interstellar spaceflight. Initiative for Interstellar Studies, Oak Ridge
36. Zylstra AB, Hurricane OA, Callahan DA et al (2022) Burning plasma achieved in inertial fusion. Nature 601(7894):542–548
37. Zylstra AB, Kritcher AL, Hurricane OA et al (2022) Experimental achievement and signatures of ignition at the national ignition facility. Phys Rev E 106:025202
38. Christopherson AR, Hurricane OA, Weber C et al (2023) Alpha-heating analysis of burning plasma and ignition experiments on the national ignition facility. Phys Plasmas 30(6):062705
39. Lawson JD (1957) Some criteria for a power producing thermonuclear reactor. Technical Report for the Atomic Energy Research Establishment, Harwell, England, UK, AERE GP/R 1807, December 1955. Published Proc Phys Soc B 70(1)
40. Lindl JD (1972) Laser compression of matter to super-high densities. Nature 239
41. Kruer WL (2003) The physics of laser plasma interactions, frontiers in physics. Westview Press Books, Boulder
42. Pfalzner SP (2006) An introduction to inertial confinement fusion. Taylor & Francis, Milton Park
43. Long KF (2022) On the application of pulse propulsion frequency in inertial fusion space mission design. J Br Interplanetary Soc 75(4):127–141
44. Long KF, Obousy RK, Hein A (2011) Project icarus: optimisation of nuclear fusion propulsion for interstellar missions. Acta Astronaut 68(11–12):1820–1829
45. Long KF (2015) The 'Invention' of the starship and revisiting tsiolkovsky. Axiom J Initiative Interstellar Stud 1(1)

Chapter 2
Interstellar Trajectory

Interstellar travel may still be in its infancy, but adulthood is fast approaching, and our descendants will someday see childhood's end.

—Eugene Mallove and Gregory Matloff, 1989, p15 [1]

2.1 Equation of Motion

As early as the 1940s the rocket scientist Samuel Herrick first coined the term "Astrodynamics", as a way to describe the fusion of dynamical astronomy, space propulsion, celestial navigation and mission planning under a single subject [2]. In this book we do not plan to discuss astrodynamics in detail, since it is a complicated subject that goes far outside the scope of this text. That said, there is a part of the theory of astrodynamics that is highly relevant to the physics of advanced space propulsion and for which needs discussion. The work presented in this chapter is an original derivation by the author which can also be expanded upon by others to include a more sophisticated model.

Individual researchers will often advocate for one propulsion system over another. Sometimes this is a valid argument since a specific propulsion method may be more appropriate for a specific mission. However, this is also often an error in logic, since each propulsion system may be better for a different mission application, and ultimately it will depend on the sort of payload mass that it can carry. To do work in physics a minimum amount of energy must be expended, and this includes to the application of the propulsion of spacecraft. Thus, rather than asking what propulsion method should be used, a better question relates to the minimum energy (or velocity) that one needs to accomplish the mission.

Interstellar space begins where the solar magnetic field interacts with the galactic magnetic field within the interstellar medium. However, operationally it is more convenient to think of missions to the edge of the solar heliosphere as interstellar precursor missions, and the Voyager probes were the first of such spacecraft.

One could define an interstellar mission as when the spacecraft exceeds $\sim 1\ ly$, but this seems rather arbitrary. Instead, a good way to think of having successfully

K. F. Long, *Stellar Engines*, Springer Aerospace Technology,
https://doi.org/10.1007/978-3-032-18062-9_2

'went interstellar', is when the gravitational influence of the target destination star becomes greater than the gravitational influence of the origin star, our sun. This is manifest in the gravitational potential of each star given by

$$V = -\int_{\infty}^{r} \frac{GMm}{r^2} = -\frac{GMm}{r}. \tag{2.1}$$

Where G is the gravitational constant, M is the mass of the origin or destination star, m is the spacecraft mass and r is the distance from the stars. The point on the journey in which $V_{target} > V_{origin}$ we consider that as the crossing point of an interstellar voyage and is stellar mass dependent. This equation is important when considering the trajectory of a spacecraft across interstellar space and the derivation of the minimum path solution and this will be the focus of this chapter.

In the work that follows this is an original derivation by the author which has not been previously published. For assistance in understanding the mathematics the reader is referred to the books by Stroud [3, 4]. For a more comprehensive treatment of the Lagrange formalism see Spiegel [5].

In the discussion that follows we will change the notation as being $x \equiv \theta$ for distance or displacement, where θ is a radial line of centre coordinate (not an angle), $v = dx/dt \equiv d\theta/dt = \dot{\theta}$ for velocity and $a = dv/dt \equiv d\dot{\theta}/dt = \ddot{\theta}$ as acceleration. The derivative notation was introduced by Gottfried Leibniz (1646–1716) and the dot notation was introduced by Isaac Newton (1643–1727) who both developed calculus independently of each other. We adopt this notation since it is more convenient when considering the minimum path solution.

It is useful to see the motion of a spacecraft during its flight as the change in kinetic energy K and potential energy V. This is an alternative to using the standard equations of motion from classical mechanics discussed earlier, and instead describes the motion by use of an equation for a Lagrangian L as follows

$$L = K - V. \tag{2.2}$$

This is solved using the Euler-Lagrange formulation as a set of derivatives respect to time t, displacement position θ and velocity $\dot{\theta}$ as follows

$$\frac{d}{dt}\left(\frac{\partial L}{\partial \dot{\theta}}\right) = \frac{\partial L}{\partial \theta}. \tag{2.3}$$

Note that in this equation we have two types of derivatives to measure the functional change in a variable. An ordinary derivative is expressed like d/dt, but a partial derivative like $\partial L/\partial \theta$ measures how a multi-function variable changes with respect to one variable whilst holding all the others fixed.

As shown in Fig. 2.1, for a spacecraft of mass m moving from the gravitational influence of the Sun with mass $M_1 = 1.9885 \times 10^{30}$ kg at distance r_1, towards the gravitational influence of another star with mass $M_2 = 2.1452 \times 10^{30}$ kg at distance r_2, the total potential energy is given by a two-body gravitational field description

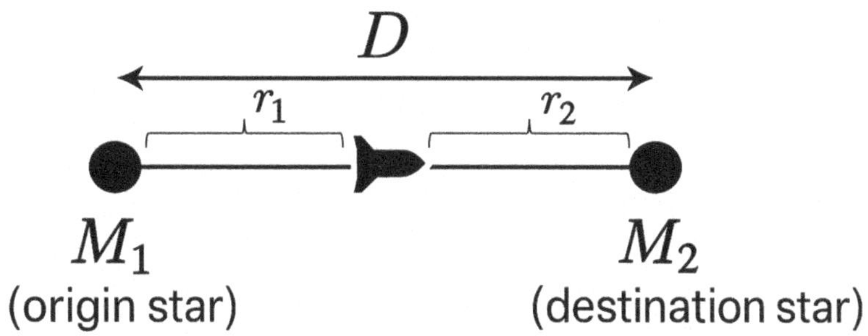

Fig. 2.1 Illustration of interstellar mission distance for Lagrangian model[KFL]

acting on the smaller mass and so constitutes a three-body problem described by

$$V = -\frac{GM_1 m}{r_1} - \frac{GM_2 m}{r_2}. \tag{2.4}$$

Where we have neglected any gravitational influence of the planets, other stars, assumed the motion is constrained to the radial dimension so have neglected tangential motion and orbital inclination out of the galactic plane. These assumptions are satisfactory provided that the velocity of the spacecraft is a lot greater than the velocity of those astronomical bodies. This represents a straight-line, line-of-centres co-ordinate model in which the planetary and tangential terms are neglected, for orbit transfers instead of conic sections for the trajectory. We also neglect gravity losses and optimum steering. This helps to simplify the model, so that the spacecraft is subject to a force given by

$$F = -\nabla V = -\nabla\left(-\frac{GM_1 m}{r_1} - \frac{GM_2 m}{r_2}\right). \tag{2.5}$$

Along the radial line of flight, the spacecraft is pulled gravitationally by the two stars but in opposite directions to each other, with a kinetic energy of

$$K = \frac{1}{2} m\dot{\theta}^2. \tag{2.6}$$

So that the Lagrangian can be expressed as

$$L = \frac{1}{2} m\dot{\theta}^2 + \frac{GM_1 m}{r_1} + \frac{GM_2 m}{r_2}. \tag{2.7}$$

We note that r is the independent variables but they are also a function of θ, the distance along the straight line that connects to the stars. It may therefore be more informative to express the Lagrangian as

$$L = \frac{1}{2}m\dot{\theta}^2 + \frac{GM_1 m}{r_1(\theta)} + \frac{GM_2 m}{r_2(\theta)}. \tag{2.8}$$

Then taking derivatives

$$\frac{\partial L}{\partial \dot{\theta}} = m\dot{\theta}, \tag{2.9}$$

$$\frac{d}{dt}\Big(\frac{\partial L}{\partial \dot{\theta}}\Big) = m\ddot{\theta}. \tag{2.10}$$

We then solve for the right-hand side of the Lagrange Equation

$$\frac{\partial L}{\partial \theta} = \frac{\partial}{\partial \theta} = \Big(\frac{1}{2}m\dot{\theta}^2 + \frac{GM_1 m}{r_1(\theta)} + \frac{GM_2 m}{r_2(\theta)}\Big). \tag{2.11}$$

Which leads to

$$\frac{\partial L}{\partial \theta} = \frac{\partial}{\partial \theta}\Big(\frac{GM_1 m}{r_1(\theta)}\Big) + \frac{\partial}{\partial \theta}\Big(\frac{GM_2 m}{r_2(\theta)}\Big). \tag{2.12}$$

We can then solve these partial derivatives using the chain rule, where if we assume a constant $a = GMm$ this results in

$$\frac{\partial}{\partial \theta}\Big(ar_i(\theta)^{-1}\Big) = -a\Big(r_i(\theta)^{-2}\Big)\frac{dr_i}{d\theta} = -\frac{a}{r_i(\theta)^2}\frac{dr_i}{d\theta}, \tag{2.13}$$

$$\frac{\partial L}{\partial \theta} = -\frac{GM_1 m}{r_1(\theta)^2}\frac{dr_1}{d\theta} - \frac{GM_2 m}{r_2(\theta)^2}\frac{dr_2}{d\theta}. \tag{2.14}$$

We then write out the equation of motion from the Euler-Lagrange as follows

$$m\ddot{\theta} = -\frac{GM_1 m}{r_1(\theta)^2}\frac{dr_1}{d\theta} - \frac{GM_2 m}{r_2(\theta)^2}\frac{dr_2}{d\theta}. \tag{2.15}$$

Then dividing through by the spacecraft mass, we find

$$\ddot{\theta} = -\frac{GM_1}{r_1(\theta)^2}\frac{dr_1}{d\theta} - \frac{GM_2}{r_2(\theta)^2}\frac{dr_2}{d\theta}. \tag{2.16}$$

This shows that the equation of motion is independent of spacecraft mass, assuming the mass is a constant throughout the trajectory.

The next step is to get rid of the $dr_i/d\theta$ term so that the equation is fully explicit. To do this we choose a geometry for the point of origin at star 1 where they are at a fixed distance D apart and θ is the spacecraft distance from star 1, so that $r_1 = \theta$

and $r_2 = D - \theta$, and so $dr_1/d\theta = +1$ and $dr_2/d\theta = -1$. Therefore, the equation of motion for the *Interstellar Flight Lagrange* becomes

$$\ddot{\theta} = -\frac{GM_1}{\theta^2} + \frac{GM_2}{(D-\theta)^2}. \tag{2.17}$$

In this context θ is the launch point from the origin star and $(D-\theta)$ is the distance to the destination target star minus any remaining distance, e.g. if the spacecraft trajectory stops 1 AU from the target star. How can we use this to calculate problems in spacecraft missions? Let us assume we have a distance to the nearest stars of $D = 4.3\ ly = 4.068 \times 10^{16}$ m and we are travelling from the origin star 1 at Sol to the star 2 at Alpha Centauri. If we assume this mission is to be completed in a trip time of T = 100-years = 3.1536×10^9 s, we can estimate the cruise speed required for this from

$$\dot{\theta} \approx v_\infty = \frac{D}{T} = \frac{4.3 \times 9.46 \times 10^{15}\ \text{m}}{100 \times 3.1536 \times 10^7\ \text{s}} \approx 12{,}899.59\ \text{km/s}. \tag{2.18}$$

This is essentially the speed in deep space where the gravitational potential from both stars is negligible. Now we will see what we get from the Lagrangian. Firstly, in the equation of motion, since L has no explicit time t within it, we can say that the specific energy is conserved. This can be shown by multiplying the equation of motion by $\dot{\theta}$ and taking derivatives so that

$$\frac{d}{dt}\left(\frac{1}{2}\dot{\theta}^2\right) = \dot{\theta}\ddot{\theta}, \tag{2.19}$$

$$\frac{d}{dt}\left(\frac{1}{\theta}\right) = -\frac{\dot{\theta}}{\theta^2}, \tag{2.20}$$

$$\frac{d}{dt}\left(\frac{1}{D-\theta}\right) = \frac{\dot{\theta}}{(D-\theta)^2}, \tag{2.21}$$

$$\frac{d}{dt}\left(\frac{1}{2}\dot{\theta}^2\right) = GM_1\frac{d}{dt}\left(\frac{1}{\theta}\right) + GM_2\frac{d}{dt}\left(\frac{1}{D-\theta}\right). \tag{2.22}$$

Alternatively

$$\frac{d}{dt}\left(\frac{1}{2}\dot{\theta}^2 - \frac{GM_1}{\theta} - \frac{GM_2}{D-\theta}\right) = 0. \tag{2.23}$$

Integrating this in time, where all the terms in the brackets are a constant, this leads to the total energy over the mass which is the specific energy ϵ expressed as

$$\epsilon = \frac{E}{m} = \frac{1}{2}\dot{\theta}^2 - \frac{GM_1}{\theta} - \frac{GM_2}{(D-\theta)} = constant. \tag{2.24}$$

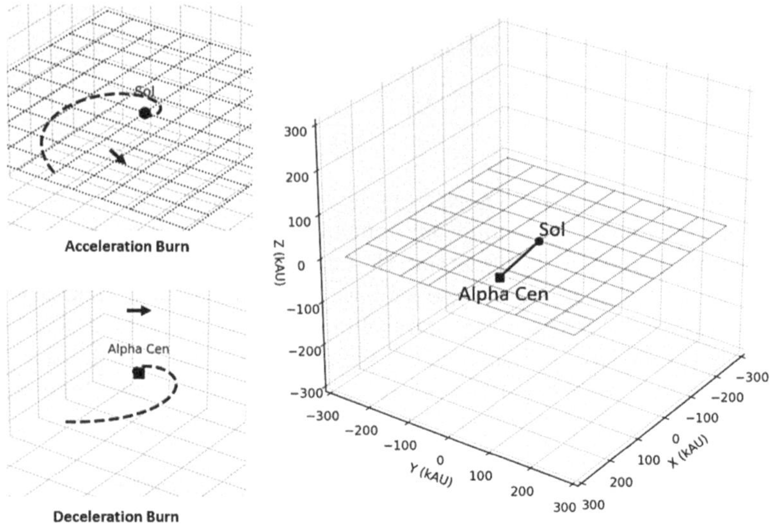

Fig. 2.2 Illustration of interstellar trajectory[KFL]

Now if we assume the mission is launched from Earth orbit, at $r_o = 1$ AU then we can calculate the velocity required at launch

$$\epsilon = \frac{1}{2}v_o^2 - \frac{GM_1}{r_o} - \frac{GM_2}{D - r_o} = \frac{1}{2}v_\infty^2. \tag{2.25}$$

Where v_∞ is the deep space potential which is negligible. For the two different stars we calculate gravitational parameters to be $\mu_1 = GM_1 = 1.3272 \times 10^{20}$ m^3/s^2 and $\mu_2 = GM_2 = 1.4317 \times 10^{20}$ m^3/s^2. Therefore

$$v_o = \sqrt{v_\infty^2 + \frac{2\mu_1}{r_o} + \frac{2\mu_2}{D - r_o}} \approx 12{,}899.66 \text{ km/s}. \tag{2.26}$$

We note that the difference between $v_o - v_\infty \approx 70$ m/s, which is the additional kinetic energy that is required near the target star and destination star, since v_∞ is the value calculated approximately mid-course in the mission and at a much lower gravitational potential, this becomes a correction. Because v_∞ is large compared to the characteristic escape speeds at 1 AU, the trajectory is well approximated by a straight-line segment in the line-of-centers model. The Stellar gravity perturbs the endpoints, adding only minor launch and arrival speed corrections.

What this shows is that $v_\infty = D/T$ gives a good approximation as the hyperbolic excess speed where the potential $V \sim 0$ in deep space. This is illustrated in Fig. 2.2, where although the acceleration and deceleration path may not be linear, most of the flight phase is. The conical part of the trajectory is likely to be present when the spacecraft must orbit into or out of the ecliptic plane. For example, the star Alpha Centauri has an ecliptic latitude of −42.6° South of the ecliptic. Otherwise, for the most part, due to the extremely high velocity of these sorts of vehicles we can neglect a consideration of the orbital elements.

Whereas when we use the Lagrange solution to calculate the value of $v(\theta)$ anywhere along the trajectory, specifically v_o near the origin it shows that a minor correction is required. So the $v \approx D/T$ solution appears to match the Lagrange solution very well, because the high cruise speed of the mission dominates over any gravitational effects of the stars. This would not be the case if the cruise velocity was much lower and especially so if the the origin and destination points were also much closer together such as in interplanetary flight, which is why for local trajectories a more sophisticated astrodynamics models is required. It is left as an exercise for the readers to play around with this equation and determine at which point the approximation might break down.

2.2 Relativistic Equation of Motion

Next, we briefly consider a relativistic correction to the interstellar Lagrange. This means that for speeds that become sufficiently high we must take account of the Special Theory of Relativity due to the constant speed of light, and we tend to think in terms of a velocity ratio $\beta = v/c \equiv \dot{\theta}/c$. To achieve this, we restate the potential

$$V(\theta) = -\frac{GM_1 m}{\theta} - \frac{GM_2 m}{D-\theta}. \tag{2.27}$$

Therefore

$$\frac{\partial V}{\partial \theta} = +\frac{GM_1 m}{\theta^2} - \frac{GM_2 m}{(D-\theta)^2}, \tag{2.28}$$

$$-\frac{\partial V}{\partial \theta} = -\frac{GM_1 m}{\theta^2} + \frac{GM_2 m}{(D-\theta)^2}. \tag{2.29}$$

In special relativity flat space-time is known as Minkowski space and we assume 1-dimensional motion along θ by defining a line element also known as a metric

$$ds^2 = -c^2 dt^2 + d\theta^2. \tag{2.30}$$

Proper time $d\tau$ is defined along a time-like metric as

$$ds^2 = -c^2 d\tau^2. \tag{2.31}$$

From here one can define the equation for the Lorentz factor γ given by

$$\gamma = \frac{1}{\sqrt{1 - \dot{\theta}^2/c^2}}. \tag{2.32}$$

In special relativity the kinetic energy is defined by

$$K = \gamma mc^2 - m_o c^2. \tag{2.33}$$

Where $m_o c^2$ is the rest mass energy. We neglect this term since this just zero point of energy and does not affect the equation of motion. Therefore

$$K_{rel} = \gamma mc^2 = \frac{mc^2}{\sqrt{1 - (\dot{\theta}^2/c^2)}}. \tag{2.34}$$

Then the modified relativistic Lagrangian becomes

$$L = \gamma mc^2 = \frac{mc^2}{\sqrt{1 - (\dot{\theta}^2/c^2)}} - V(\theta). \tag{2.35}$$

$$L = \gamma mc^2 = \frac{mc^2}{\sqrt{1 - (\dot{\theta}^2/c^2)}} + \frac{GM_1 m}{\theta} + \frac{GM_2 m}{D - \theta}. \tag{2.36}$$

We then go back to the Euler-Lagrange equation

$$\frac{d}{dt}\left(\frac{\partial L}{\partial \dot{\theta}}\right) = \frac{\partial L}{\partial \theta}, \tag{2.37}$$

$$\frac{\partial L}{\partial \dot{\theta}} = m\gamma\dot{\theta}, \tag{2.38}$$

$$\frac{d}{dt}\left(\frac{\partial L}{\partial \dot{\theta}}\right) = m\gamma^3\ddot{\theta}. \tag{2.39}$$

It is constructive to explain the γ^3 term given its importance, which comes from the product rule and the dependence on $\dot{\theta}$, where

$$\frac{d}{dt}(\gamma\dot{\theta}) = \gamma\ddot{\theta} + \dot{\theta}\dot{\gamma}, \tag{2.40}$$

$$\gamma(\beta) = (1 - \beta^2)^{-1/2}. \tag{2.41}$$

Which implies

$$\frac{d\gamma}{d\beta} = \frac{1}{2}(1-\beta^2)^{-3/2}(2\beta) = \beta(1-\beta^2)^{-3/2} = \gamma^3\beta. \tag{2.42}$$

Then since $\beta = \dot{\theta}/c$ and $\dot{\beta} = \ddot{\theta}/c$ we find

$$\dot{\gamma} = \frac{d\gamma}{d\beta}\dot{\beta} = \gamma^3\beta\Big(\frac{\ddot{\theta}}{c}\Big) = \gamma^3\frac{\dot{\theta}\ddot{\theta}}{c^2}. \tag{2.43}$$

Then substitute back in

$$\frac{d}{dt}(\gamma\dot{\theta}) = \gamma\ddot{\theta} + \dot{\theta}\Big(\gamma^3\frac{\dot{\theta}\ddot{\theta}}{c^2}\Big) = \Big(\gamma + \gamma^3\frac{\dot{\theta}^2}{c^2}\Big)\ddot{\theta}. \tag{2.44}$$

Using the identity

$$1 + \gamma^2\frac{\dot{\theta}^2}{c^2} = \gamma^2, \tag{2.45}$$

$$\gamma + \gamma^3\frac{\dot{\theta}^2}{c^2} = \gamma\Big(1 + \gamma^2\frac{\dot{\theta}^2}{c^2}\Big) = \gamma \times \gamma^2 = \gamma^3, \tag{2.46}$$

$$\frac{d}{dt}(\gamma\dot{\theta}) = \gamma^3\ddot{\theta}. \tag{2.47}$$

Then this leads to the potential

$$\frac{\partial L}{\partial \theta} = -\frac{\partial V}{\partial \theta} = -\Big(+\frac{GM_1m}{\theta^2} - \frac{GM_2m}{(D-\theta)^2}\Big) = -\frac{GM_1m}{\theta^2} + \frac{GM_2m}{(D-\theta)^2}. \tag{2.48}$$

If we then set them equal and cancel out the spacecraft mass m we are left with

$$\gamma^3\ddot{\theta} = -\frac{GM_1}{\theta^2} + \frac{GM_2}{(D-\theta)^2}. \tag{2.49}$$

This leads to the relativistic equation of motion as derived from the Lagrange or what we might term the *Relativistic Interstellar Flight Lagrange* as follows

$$\ddot{\theta} = \frac{-GM_1/\theta^2 + GM_2/(D-\theta)^2}{\gamma^3}. \tag{2.50}$$

We note that in the limit as $\dot{\theta} << c$, $\gamma \rightarrow 1$ and so the non-relativistic version is recovered. It will be clear then that $1/\gamma^3$ is the leading correction for the relativistic acceleration as derived from the Euler-Lagrange. It is useful to understand further how this approximates since $\beta^2 = \dot{\theta}^2/c^2$ and the Lorentz factor is $\gamma = (1-\beta^2)^{-1/2}$.

Therefore $\gamma^3 = (1-\beta^2)^{-3/2}$ and

$$\frac{1}{\gamma^3} = (1-\beta^2)^{3/2}. \tag{2.51}$$

We can then expand this using the binomial series for

$$(1+x)^n = 1 + nx + \frac{n(n-1)x^2}{2} + ... \tag{2.52}$$

If we assume $x = -\beta^2$ and $n = 3/2$ then we find

$$(1-\beta^2)^{3/2} = 1 + \frac{3}{2}(-\beta^2) + \frac{3/2(3/2-1)}{2}(-\beta^2)^2 + ... = 1 - \frac{3}{2}\beta^2 + \frac{3}{8}\beta^4 + ... \tag{2.53}$$

Therefore examining only the leading order correction

$$\frac{1}{\gamma^3} \approx 1 - \frac{3}{2}\frac{\dot{\theta}^2}{c^2}. \tag{2.54}$$

This shows that the acceleration is reduced by a factor $1/\gamma^3$, so for our 1-dimensional straight-line model, the parallel component of acceleration is given by $a = F/\gamma^3 m$, as the correction from special relativity. In this book we are mostly concerned with reaction engines with non-relativistic velocities so we will not examine this further. As the velocity approaches the speed of light the non-relativistic and the relativistic Euler-Lagrange solutions will deviate, and these corrections would be required in any comprehensive mission analysis. This would make a very constructive thesis project for an enthusiastic student.

2.3 Momentum Flux

A consideration from this is how do we get the rocket equation and thrust out of the Lagrange solution, and we examine the non-relativistic case only. To achieve this, we assume thrust as a non-conservative generalized force, which here is known as the momentum flux which is a generalised non-conservative force $F(\theta)$. The Lagrange equation is now modified to become the Lagrange-d'Alembert equation. Another way to express the Lagrange-d'Alembert principle is the sum of the differences between the forces acting on a system and the time derivatives of the momenta is equal to zero. For our derivation this now includes a momentum thrust term on the right-hand side as follows

$$\frac{d}{dt}\left(\frac{\partial L}{\partial \dot{\theta}}\right) - \frac{\partial L}{\partial \theta} = F(\theta). \tag{2.55}$$

Then the potential is

$$V(\theta) = -\frac{GM_1 m}{\theta} - \frac{GM_2 m}{D-\theta}. \tag{2.56}$$

We also must consider the mass to be time-dependent, since we are burning fuel, $m = m(t)$, and where $V = -m\Phi$, therefore

$$\Phi(\theta) = \frac{GM_1}{\theta} + \frac{GM_2}{D-\theta}, \tag{2.57}$$

$$\Phi'(\theta) = -\frac{GM_1}{\theta^2} + \frac{GM_2}{(D-\theta)^2}. \tag{2.58}$$

If we write out the Lagrangian we get the result

$$L = K - V = \frac{1}{2}m(t)\dot{\theta}^2 + m(t)\Phi(\theta). \tag{2.59}$$

Then solving this using the Lagrange-d'Alembert formalism we find

$$\frac{\partial L}{\partial \dot{\theta}} = m\dot{\theta}, \tag{2.60}$$

$$\frac{d}{dt}\left(\frac{\partial L}{\partial \dot{\theta}}\right) = m\ddot{\theta} + \dot{m}\dot{\theta}, \tag{2.61}$$

$$\frac{\partial L}{\partial \theta} = m\Phi'(\theta). \tag{2.62}$$

This leads to

$$m\ddot{\theta} + \dot{m}\dot{\theta} - m\Phi'(\theta) = F(\theta). \tag{2.63}$$

The spacecraft will have an exhaust velocity v_{ex} along the trajectory of the θ axis and so the momentum flux carried away by the vehicle per unit time will be $-v\dot{m}$ and the reaction of the rocket engine

$$F(\theta) = m\ddot{\theta} = -v\dot{m}. \tag{2.64}$$

For a rocket ejecting matter rearwards at the exhaust velocity relative to the vehicle in the positive (v_{ex}>0) rearward direction, the inertial exhaust speed is $v = \dot{\theta} - v_{ex}$, where the negative indicates the opposite direction of motion. Substituting this into the equation with the momentum flux (thrust) we find

$$m\ddot{\theta} + \dot{m}\dot{\theta} - m\Phi'(\theta) = (\dot{\theta} - v_{ex})\dot{m}. \tag{2.65}$$

Then cancelling the $\dot{m}\dot{\theta}$ terms

$$m\ddot{\theta} = -v_{ex}\dot{m} + m\Phi'(\theta). \tag{2.66}$$

If we neglect external forces so that $\Phi'(\theta) = 0$, and then for ease of visualisation let us rewrite this in derivative notation

$$m\frac{d\dot{\theta}}{dt} = -v_{ex}\frac{dm}{dt}, \tag{2.67}$$

$$d\dot{\theta} = -v_{ex}\frac{dm}{m}. \tag{2.68}$$

Then integrating between limits

$$\int_{\dot{\theta}_o}^{\dot{\theta}_f} d\dot{\theta} = -v_{ex}\int_{m_o}^{m_f}\frac{dm}{m}. \tag{2.69}$$

Which leads to the solution of the ideal rocket equation neglecting external forces such as gravity fields.

$$\dot{\theta}_f - \dot{\theta}_o = \Delta v = v_{ex} ln\left(\frac{m_o}{m_f}\right). \tag{2.70}$$

As we have seen from the Euler-Lagrange formulation you can get to the rocket equation, by substitution of the momentum flux term, which is the Lagrange-d'Alembert formulation. Throughout this book we will use the approximation $v = D/T$ as a back-of-envelope check on the velocity requirement for a mission. But as we have seen in this chapter, it gives a pretty accurate approximation to the actual velocity requirement when one considers the additional velocity required at the point of origin and destination. It is not necessary to completely understand the Lagrange approach to do rocket science, but it is important to have an appreciation for the sense of accuracy when undertaking starship engineering.

Problems of Chap. 2

2.1 In the derivation of the specific energy equation for the Lagrangian (i) show

$$\frac{d}{dt}\left(\frac{1}{2}\dot{\theta}^2\right) = \dot{\theta}\ddot{\theta}.$$

2.2 In the Project Daedalus study they designed a probe to go to Barnard's star at 5.9 ly distance in around 50 years. Calculate the value of (i) v_∞ and then also for (ii) v_o assuming the mission is launch from Jupiter orbit at 5 AU.

2.3 For a relativistic flight, (i) Using the line element and the definition for proper time, derive the equation for the Lorentz factor

$$\gamma = \frac{1}{\sqrt{1 - \dot{\theta}^2/c^2}}. \tag{2.71}$$

References

1. Mallove EF, Matloff GL (1989) The starflight handbook. A Pioneer's guide to interstellar travel. John Wiley & Sons Inc, Hoboken
2. Herrick S (1971) Astrodynamics: orbit determination. Perturbation theory, integration, vol 1. University of Wisconsin, Madison
3. Stroud KA (1995) Engineering mathematics, 4th edn. Macmillan, New York
4. Stroud KA (1986) Further engineering mathematics, 3rd edn. Macmillan, New York
5. Spiegel MR (1980) Theory and problems of theoretical mechanics. Schaum's Outline Series. McGraw-Hill Book Company, New York

Chapter 3
History of Research

For an interstellar flight, propulsion is the critical issue. The required velocity is so high that any conventional propulsion system in development today is inadequate, and the question of propulsion technology dominates all other technology considerations.

—Geoffrey A Landis, 2003 [1]

3.1 Early Design Studies (1952–1972)

In 1952 Les Shepherd [2] was the first to consider the requirements for interstellar flight. He determined that a spacecraft driven by fusion reactions with an energy release fraction of $\alpha = 0.0015$ might be able to reach the nearest stars using a cruise velocity of around ~20,000 km/s for a flyby trip time of ~65 years and a rendezvous trip time of ~130 years. This would require a minimum exhaust velocity of ~6,300 km/s, an acceleration of 0.003 m/stextsuperscript-2 and a vehicle mass ratio of $R = 4.9$. This would also correspond to an engine specific power of around 10 kW/kg. He concluded that: *"It would be possible... to construct a vehicle, deriving its power from known nuclear reactions, which would be capable of reaching the nearest stars in a time measured in centuries"*.

Various other authors had also written about the requirements for fusion propulsion systems such as Sanger in 1963 [3], Strong in 1965 [4], Hunter in 1966 [5]. This provided a valuable impetus for the concept developments that followed, and it was the starting point upon which the vehicle designs would evolve.

In the following, as the relevant propulsion concepts are presented, we will also state some of the key physics and engineering parameters as appropriate. This includes the spacecraft initial wet mass m_o, spacecraft initial dry mass m_f, payload mass m_{pay}, capsule mass m_{cap}, expellant mass m_{exp}, mass flow rate $\dot{m}$, burn-up fraction f_b, exhaust velocity v_{ex}, acceleration time t_b, total mission time t_{tot}, pulse detonation frequency f_{hz}, thrust T, jet power P_j, specific power P_{sp} and jet efficiency μ_j. We will also define the type and power level of the drive beamer used to initiate the ignition process where used.

K. F. Long, *Stellar Engines*, Springer Aerospace Technology,
https://doi.org/10.1007/978-3-032-18062-9_3

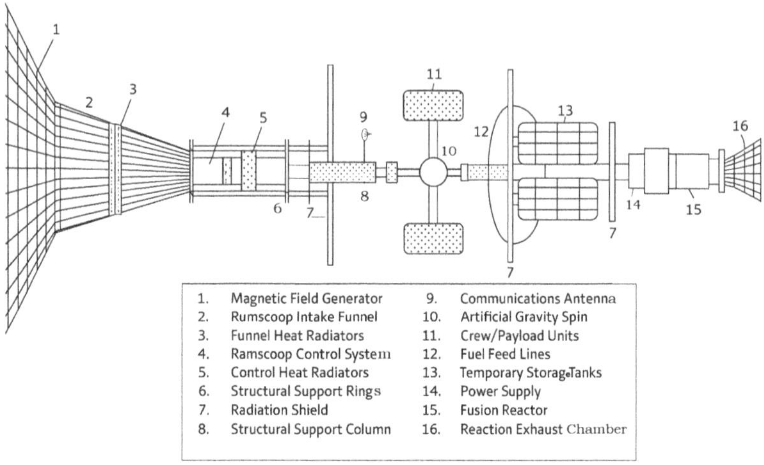

Fig. 3.1 Illustration of the Interstellar Ramjet Spacecraft Configuration[KFL]

Interstellar Ramjet, 1960

In 1960 Robert Bussard [6] published an original idea of mining hydrogen fuel from the interstellar medium and using this as the main fusion fuel source. The concept is illustrated in Figure 3.1 and is known as the interstellar ramjet. This had the benefit that large amounts of propellant mass would not be needed to be taken on the journey. However, proton-proton fusion is not easy to achieve without the presence of a large gravitational field like in the Sun, and in particular hydrogen has a very small cross section, where for a particle energy of say 1 MeV the reaction cross section would be around $\sim 10^{-49}$ cm^2. A proton has a mass of 1.6726×10^{-27} kg and at least two are required to form a reactant and give rise to the products via $p^+ + p^+ \rightarrow H_1^2 + e^+ + \nu_e$ and then decays to $H_1^2 + \nu_e + 1.442MeV$.

A critical component of the engineering of such a system is the ability to moderate the high energy protons down to a reasonable level and to a point where they can be confined inertially for a fraction of a second sufficient to implode them to ignition conditions such as by using lasers, which is problematic. Yet if the interstellar ramjet could be made to work it would revolutionise the possibilities for space exploration.

That said, if a way was found to make this work, then it would be theoretically possible for such a spacecraft to travel at speeds certainly exceeding 0.5c and likely approaching 0.99c. Indeed, taken to its limit, the spacecraft could travel closer and closer to the speed of light (although never reaching it), enabling trips to the centre of the Milky Way galaxy in a matter of decades ship time Δt as measured by clocks on-board the spacecraft, although much time t would have passed back on Earth preventing their viable return to anything they ever knew that was familiar [7]. This is because as $v \rightarrow c$, $1 - \beta^2 \rightarrow 0$ and $t = \Delta t / \sqrt(1 - \beta^2) \rightarrow \infty$, where $\beta = v/c$.

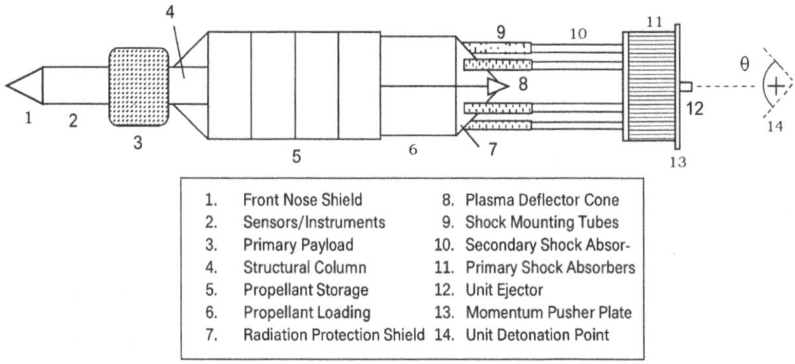

Fig. 3.2 Illustration of Orion Nuclear Pulse Engine and Spacecraft Systems[KFL]

This was the subject of a novel by Robert Heinlein called *Time for the Stars* [8] and includes considerations for special relativity. We discuss the interstellar ramjet further in chapter 13.

Orion, 1957–1965

The idea of using nuclear (fission) bombs to propel a spacecraft was first proposed by Everett and Ulam in 1955 as an application that followed the earlier Manhattan project [9]. Project Orion was a study conducted by the U.S government [10] and the vehicle is illustrated in Figure 3.2. The Orion proposed to use nuclear bomb technology to propel a spacecraft, firstly from the ground through the atmosphere and into orbit, and secondly through space to the outer planets. It would detonate small nuclear bombs at some distance from the rear of the vehicle, and when they detonated a certain fraction of the blast would be subtended by a rear 100 m diameter pusher plate and then transfer the momentum to the rest of the vehicle through a series of shock absorbers. This is known as a momentum limited Orion (as opposed to an energy-limited Orion).

Although Orion was originally conceived as an interplanetary study (mainly for military applications) after the project was brought to an end Freeman Dyson [11] looked at the interstellar potential of such a machine and found that versions of Orion were possible that could reach the nearest stars at 4.3 ly in around 133 years travelling at something like 10,000 km/s or $0.033c$ using fission bombs. If Hydrogen bombs were deployed, then it was possible that velocities on the order of $0.1c$ or 30,000 km/s could be achieved. The spacecraft would be massive, carrying a 50,000 tons payload, and 300,000 tons of propellant, each of which were 1 ton bomb units. The total mass of the vehicle would be around 400,000 tons. The acceleration profile involved detonating a unit every 3 s and sustaining this for around 10 days and involved an exhaust velocity of around 1,000 km/s. Since Orion would have a pusher plate, it does not make sense to talk in terms of a jet efficiency, although given the isothermal detonation the efficiency would be highly dependent upon the blast angle subtended and therefore proximity to the vehicle.

This technology was controversial, although there are no physics or engineering reasons why the design could not work. In principle it could have been built in the 1960s. The reason this study is so important to this work, is that if you shrink down the units of the externally driven nuclear pulse Orion system, and instead use laser beams to energy drive the capsules towards a state of ignition within an internal (or partially open) reaction chamber instead of using a pusher plate, you move to an ICF driven system. Therefore, the utilisation of ICF propulsion is Orion in the limit of small capsule masses and at lower driver energy.

Helios, 1964–1965

This was a study conducted in parallel with Project Orion that took the name Project Helios [12, 13], but instead of using an external pusher plate it focussed on using fission explosions detonated at 10 Hz but contained within an internal reaction chamber. The basis of the idea was that the mixing of a nuclear bomb detonation with a rocket propellant could produce a large enthalpy which when ejected rear from a spacecraft would lead to a large velocity addition on the vehicle. This can be considered a form of thrust augmentation.

They explored the propulsion of a 50 tons payload to 18 km/s using nuclear bombs that were 1–10 tons in mass. A particular emphasis of the work was the use of a conical nozzle with an angle of 20°. They also analysed the effect of the time between the detonations, where the longer pulsing periods would benefit from increased radiator cooling between pulses from the nozzle which therefore reduces the hydrogen coolant requirement, leading to a higher exhaust velocity. The time between detonations would depend on variables such as charge handling, pumping of the propellant and the time for the propellant and spacecraft material structure to reach thermodynamic equilibrium. Although they also concluded that there is a need to maintain an average acceleration profile and this favours the shortest possible pulse detonation time. Then, fixing the pulse detonation, engine mass and individual bomb mass they were able to maximise the exhaust velocity. These issues are a reference to optimisation of the mission profile and are also directly applicable to ICF propulsion, especially since the detonations are contained within a reaction chamber.

Enzmann, 1966

Robert Enzmann may have been one of the first to propose placing the small bomb units within an internal reaction chamber [14]. His principal interest was on powering space missions using a giant frozen sphere of Deuterium mined from the gas giants. The large propellant sphere would be positioned at the front of the spacecraft and any payload along a central column connecting to the rear engine and thrust chambers. His work was taken by others in the construction of a wider programme of interstellar exploration [15]. It is important to mention this concept because it was one of the first concepts to move towards detonations within a

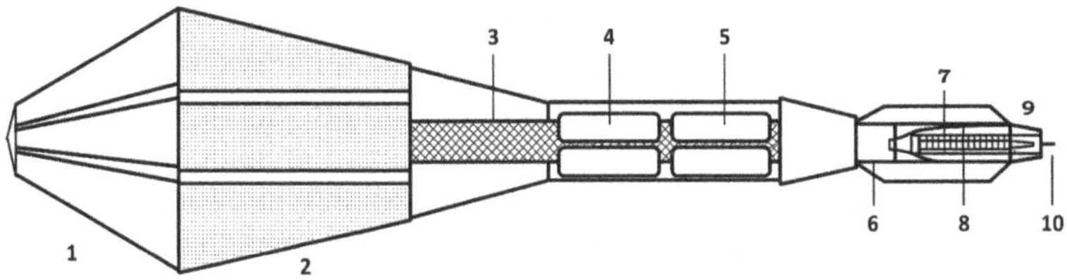

1	Payload	6.	Fusion Engine
2	Cryogenic Radiator	7	Bremsstrahlung Shield
3	Structural Column	8	Mirror Coils
4	Heat Exchangers	9	Magnetic Nozzzle
5	Propellant Tanks	10	Thrust Augmenter

Fig. 3.3 Illustration of a Magnetic Fusion Spacecraft Configuration[KFL]

reaction chamber. The concept lacked sufficient detail to enable a rigorous analysis of it at this time, but a comprehensive overview of this concept was earlier performed by this author with others [16]. It appears to have had a dry mass of around 30,000 tons, a propellant mass of 3 million tons. It assumed an exhaust velocity of around 11,700 km/s and using a mass flow rate of 5.02 kg/s it would achieve a cruise velocity of 27,000 km/s or 0.09c, which would permit it to arrive at its interstellar target destination of Alpha Centauri after around 60 years. Two versions of the concept were produced which had 8 or 24 engines. Assuming a pulse frequency of around 100 Hz, this would require ICF capsule masses of around 1 g or 0.4 g for the two different engine designs respectively. The concept known as the Enzmann Starship is discussed further in Chapter 11.

Hilton, Luce & Thompson, 1963

In 1963 an attempt was made to design a hypothetical fusion propulsion spacecraft utilising D^3He fusion reactions for the application of interplanetary missions [17]. This was termed a controlled thermonuclear reaction (CTR) and is illustrated in Figure 3.3. The system was to use a magnetic trap mechanism produced by superconducting magnets which would confine any plasma for 0.15 - 3 s within a magnetic mirror geometry. Yet it was a comprehensive study of a fusion propulsion design and helped to set the standard for what was to follow. The engine would have a mass of 11.4 tons and would generate a magnetic of around 50 kG ($\sim$5 T). The total wet spacecraft mass on departing low Earth orbit would be around 38.1 tons.

For the DT and DD reactions they release large numbers of neutrons and this would suggest a need for massive neutron shielding. In addition, the thermal heat loads would also require massive radiators to remove the heat which would further increase the system mass. This heating would especially be a problem since the superconducting magnet coils require cryogenic cooling. For the DD reaction the neutrons would be 2.45 MeV and 3.02 MeV, whereas for the DT reactions the neutrons would be 14.1 MeV. The spacecraft would make use of a radiation shield to absorb neutrons and x-rays and gyromagnetic radiation and reflect much of the radiation back into the plasma but would absorb around $\sim$10-30% of the energy.

So instead, considerations were made using D^3He fuel although a low number of neutrons from DD self-burn daughter reactions were to be expected. The authors estimated that this would be on the order of $\sim$1.5% of the total system power in neutrons which may have been overly optimistic. This resulted in an exhaust velocity in the range 785 - 4,611 km/s.

Any power loss in the system was mostly attributed to neutrons which resulted in a jet power of 25 - 100 MW. Operating at the high exhaust velocity it would exhibit a mass flow rate of around 1 kg/day or 0.0000116 kg/s. This would result in specific powers in the range 1,100 - 3,143 W/kg. This would lead to rapid missions to Mars or even to Jupiter within months.

Spencer, 1963–1964

In the 1960s Spencer wrote a couple of excellent papers laying out some of the requirements for an interstellar probe. In the first paper he discussed the broad issues pertaining to the feasibility of interstellar flight and in particular relating to the requirements of a fusion propelled vehicle [18]. Although this was mostly in consideration of a continuous thrust engine. He concluded that the minimum flight time for a probe to 5 ly would be around $\sim$50 years using a multi-stage vehicle provided the burn-up fraction exceeded 1% and acceleration exceeded 0.01 m/s^2. The work also suggested a requirement for fuel concentrations on the order of $10^{15} - 10^{16}$ particles/cm^3 and a confinement time for the average fuel ion of around $\sim$0.1 s.

In the 1963 study Spencer looked at the required mass ratio for an interstellar mission using fusion propulsion and assuming an energy release fraction of $\alpha =$ 0.004. It was found that for a 0.1c mission as a one-way trip without deceleration a mass ratio of $R = 3.0$ would be required. For the same mission but with deceleration to orbital velocity, the mass ratio would be squared to a value of $R = 9.0$. For a two-way trip with deceleration at both ends the mass ratio requirement would be $R = 81$. He also looked at up to 5-structural mass stage and showed that there was a limit to the fraction of light velocity attainable from different numbers of stages and this dependent on the payload fraction, which for a value of 0.1 which result in a fraction of light speed range $\sim 0.14 - 0.18c$ at five stages. The study assumed a fixed payload fraction and constant exhaust velocity.

This work continued in a second paper which examined fusion propulsion requirements [19]. A critical issue for mission success was determined to be the mass of any heat radiators which are required to dissipate any heat that enters the walls of the reaction chamber once the plasma achieves thermonuclear ignition. The authors developed equations for the fractional power loss due to radiation from the different contributions.

The studies emphasised the importance of minimising the structural dead weight fraction since it will have a significant effect on the overall performance. The study

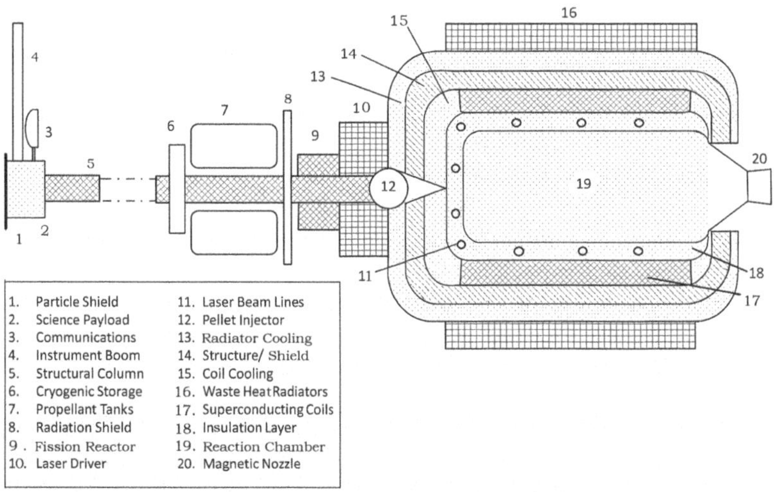

Fig. 3.4 Illustration of a Laser Fusion Engine and Spacecraft Systems[KFL]

also concluded that it should be expected that at least 20% of the fractional energy release will escape the fuel and must be dumped by a thermal radiator. Finally, the work concluded on a possible design point for a 5-stage vehicle to 5 ly distance in a 61-year trip time carrying a ~4.5 tons payload.

Figure 3.4 illustrates a basic spacecraft conceptual layout with emphasis on the propulsion elements. This is a modified drawing from one presented by Spencer but then significantly modified by this author and to include a laser-fusion element. The fusion material is ignited within a reaction chamber by a set of laser beam lines which facilitate a spherical implosion symmetry. The plasma is controlled within a magnetic field generated by superconducting coils and magnetically directed through the nozzle for thrust. At initiation the laser firing would require the use of a fission reactor. Any propellant is stored in cryogenic tanks.

The reaction chamber is surrounded by an insulation layer to capture high energy neutrons and x-rays. Any waste heat is ejected through the radiators. The field coils and the radiators require some form of refrigeration cooling. The cooling of the coils is required mainly due to the deposition of neutrons. The cooling of the reaction chamber is required mainly due to neutron heating but also x-rays produced by Bremsstrahlung and cyclotron radiation.

Sirius, 1970

Building on from the original work of Project Orion, members of the Los Alamos National Laboratory (LANL) considered a laser-fusion rocket under Project Sirius [20]. This was not an attempt to develop a vehicle concept description but more of a general discussion of the physics and engineering issues. The concept retained the idea of a pusher plate as was proposed for the Orion study but this time small capsules would hit the plate to transfer momentum. This may have been the first paper to suggest using laser-induced thermonuclear reaction in a space propulsion vehicle and so was among the first to suggest a detachment of the ignition system from the fusion fuel. The authors cited several key advantages to using lasers compared to the external nuclear pulse system of Orion. This included high energy density of the pulse units, short pulse durations and controllability. It also assumed a mercury-vapor laser.

The paper discussed milligram quantities of DT fuel with an input laser energy of ~10 kJ - 1 MJ delivered in $\sim 0.1ns - 0.1\mu s$. The pulse units are positioned at a target point where they can be initiated, and the resulting nuclear explosion and the heated fuel expands as a high temperature plasma interacting with the material structure as it expands within it. This leads to momentum transfer. If many of these are performed over a constant pulse frequency, then eventually the spacecraft velocity becomes large. Where Helios moved from an external nuclear pulse to an internal nuclear pulse and Sirius considered using lasers as the initiator of the fusion reaction, the two key components were in place for the design of nuclear pulse vehicles driven by laser beams. This is then the basis of the ICF propulsion concepts that will follow.

HWN, 1971–1972

One of the first concepts to consider the possibility of removing a pusher plate and moving to a magnetic field system to direct the thrust from thermonuclear explosions rearwards was by Hyde, Wood and Nuckolls (HWN) [21]. The design would use 0.015 g capsule masses detonated at a rate of 500 Hz pulse frequency, which would pose severe thermal and optical loading challenges. It would use a single 1 MJ mercury laser, reflecting off mirrors onto the target to produce 8 symmetrically convergent laser beams, which means that there would be around 125 kJ per beam line impinging on the target. The propulsion system mass would be 300 tons, and it would carry a 100 tons payload

The capsule design was to be mostly DD to produce a high flux of ions in the output to direct for thrust, but also with a small DT trigger to help initiate the burn at low ignition temperature. To direct the charged particles large superconducting coils would be required around the thrust chamber generating large magnetic fields. The exhaust velocity for the design was assumed to be 3,260 km/s with a thrust of 24.4 kN. Such a vehicle would be intended for interplanetary missions to any point in the Solar System within ~1 year rather than interstellar. Although the authors did imply that exhaust velocities approaching 10,000 km/s seemed feasible and this would lead to missions to the nearest stars in trip times of less than a human lifetime (<100 years). The spacecraft design was highly innovative for its time. However, it is likely that the laser energy would be insufficient to initiate fusion within the capsules, and the design would have to be scaled up to a larger system. The specification of kJ class beam lines likely relates to the thinking at the time in terms of what was required to generate thermonuclear energy within an ICF capsule. The high repetition rate of the lasers would also be a challenge to the engineering systems.

3.2 Golden Age Design Studies (1973–1991)

Daedalus, 1973–1978

Friedwardt Winterberg [22] was one of the earlier researchers to first suggest the idea of using thermonuclear micro-explosions for the propulsion of a spacecraft. This then informed the design considerations for the Project Daedalus study by members of the British Interplanetary Society in the United Kingdom [23–25]. The vehicle is illustrated in Figure 3.5.

It was one of the first efforts to create a theoretical spacecraft design on paper that had the potential to be a blueprint for an interstellar mission at some point in the future. The project was led by Alan Bond, Anthony Martin and Robert Parkinson. It was a flyby only probe carrying a 450 tons payload to 5.9 *ly* at Barnard's Star, believed to have a planetary system at the time [26]. The vehicle configuration was made up of two separate engine stages and so there were also two separate engine burns in the mission profile. As well as being an evolution from the Project Orion study and borrowing some ideas from Winterberg, the study also leaned on research of other work coming out at that time to help inform the study on the vehicle design

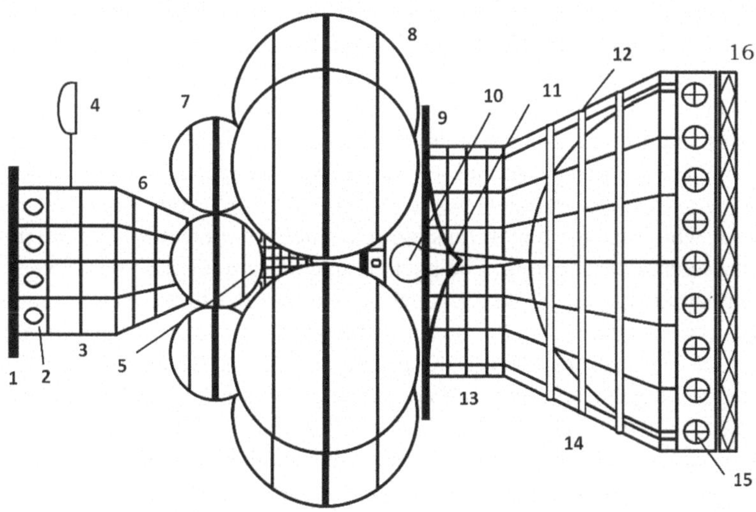

1. Particle Bombardment Shield
2. Sub-Probe Bays
3. Payload/Instruments/Wardens
4. Communications Antenna
5. Second Engine Stage
6. Computing/AI System
7. Second Stage Propellant Tanks
8. First Stage Propellant Tanks
9. Radiation Shield/Reflector
10. LH_2 Storage Tank
11. Pellet Injection System
12. Superconducting Colls
13. Thrust Support Structure
14. First Stage Reaction Chamber
15. Electron Gun Drivers
16. Power Induction Loop

Fig. 3.5 Illustration of a Daedalus Spacecraft Configuration[KFL]

and propulsion system. This was in relation to the Project Helios [13], Project Sirius [20] and Project Blascon [27].

The Project Daedalus spacecraft would begin its mission by initiating the first stage engine and then accelerate for 2.05 years taking the vehicle up to a velocity of 21,900 km/s (0.073*c*). After dropping the used propellant tanks and the first engine stage, the second engine stage would be ignited and then it would accelerate for a further 1.76 years with a velocity addition of 14,700 km/s (0.049*c*). The final cruise velocity would be 36,600 km/s (0.122*c*) and it would cruise for around 45 years. The total mission would be achieved in just less than half a century.

Both the first and second engine stage utilised an engine pulse frequency of 250 Hz. The first stage capsule mass was 2.84 g and the second stage capsule mass was 0.288 g driven by electron beams with energies 2.7 GJ and 0.4 GJ per pulse respectively. The capsule implosion and ignition would be initiated by a bank of 19 electron gun particle beam drivers. The first stage would be characterised by a mass flow rate, thrust, jet power and specific power of 0.713 kg/s, 7.54 MN, 40.056 TW, 40.54 MW/kg. Similarly, the second stage would be characterised with the parameters 0.072 kg/s, 0.663 MN, 3.054 TW, 9.6 MW/kg. The capsule gains were said to be around $G \sim 65$ and $G \sim 33$ for the first and second stage respectively, assuming burn-up fractions of 0.175 and 0.133. The jet efficiency for both stages was optimistically assumed to be around 91%. The exhaust velocities assumed were 10,600 km/s and 9,210 km/s for the first and second stages respectively.

The fuel adopted would be D^3He with 50:50 mixture. The Helium-3 was to be mined from the gas giant Jupiter [28], and although this may sound futuristic modern studies show that this was a reasonable proposal [29]. This would be made up of 46,000 tons for the first stage and 4,000 tons for the second stage. The total wet mass at the start of the mission would be around 52,670 tons and a dry mass of 2,670 tons, which included a 50 tons particle shield to protect it from the dust and ions of the interstellar medium. The payload mass fraction would be 0.85%.

It is apparent that the Project Daedalus is in fact the first ever mathematically rigorous attempt at the design of a system integrated interstellar spacecraft; although there were some contradictions in the design because the scientists involved were attempting to integrate current technology (1970s) with future projected technology (a few decades hence). This then prevented a full integration of the design. It goes beyond a concept design but not as complete as a detailed design and so it represents a preliminary design. The main design requirement was to demonstrate the flight of a flyby probe through Barnard's Star within approximately half a century, and it achieved this on paper. Although, the prolonged acceleration profile at high pulse frequency does suggest a large degree of risk and the potential for mission failure should such a vessel of this type ever be constructed. Such a vehicle would also require a high degree of fault tolerance, which is only viable with extensive autonomy, fault detection, isolation and recovery, and in-situ manufacturing of spare parts. This is why Daedalus was to be an artificial intelligence-based payload. Project Daedalus is discussed in more detail in chapter 9.

Soviet Icarus, 1976

The earlier Soviet Project Icarus was an initiative undertaken in Russia in parallel with the Project Project Daedalus study. It was presented at the 27th International Astronautical Congress in 1976 [30], but little information is known about the study other than what was detailed by others as a part of a review [31]. The concept apparently involved the use of multiple fuels (D, T, ^{3}He, ^{6}Li) in combination through a chain reaction, until sufficient ^{3}He had been built up through secondary daughter reactions. Unlike the Daedalus which was to be a two-staged vehicle design, the Icarus was to be a three-staged design, with the third stage intended for deceleration.

The authors assumed an exhaust velocity of 5,000 km/s and 300 tons mass for the ignition system, carrying a 10 tons payload (much less than the 450 tons for Daedalus). For the three stages the propellant mass would be 5,500 tons, 10,400 tons and 150 tons. The total wet mass at the start of the mission would be around 17,160 tons. No information on the nature of the capsules or pulse detonation is known but the driver was believed to be a $\sim$1 MJ laser. A key difference of this study from the Daedalus one is that it included propellant fuel for deceleration into its target system. The probe would accelerate to 21,000 km/s (0.07c) and then further to 84,000 km/s or 0.28c. No cruise phase would be included in the flight and after the acceleration phase was over the deceleration phase would begin, finally arriving at the destination target of Barnard's star $\sim$50-60 years after launch.

Due to the complexity of the fuel assumptions and the lack of information on the vehicle concept it is difficult to analyse the design. In addition, the propellant allocation does not seem right to this author and neither does the mission time given the claim that deceleration was included. It is also worth noting that the claimed velocity of 0.28c would appear to be well beyond practical fusion-rocket envelopes and so this study may not be credible. Indeed, an analysis by this author found that the maximum cruise velocity attained given the assumptions made was on the order of $\sim$0.05c. That said, little information is known about the study and the nature of the ignition system with its unique chain reaction mechanism so a proper assessment is left for another time until the paper can be fully accessed.

Voyager Probes, 1977 - Present

The two Voyager probes were launched in opposite directions out of the Solar System aiming to go beyond the solar heliosphere and to explore the interstellar medium [32]. They were both launched on a rocket from the ground onto a heliocentric trajectory. To pick up sufficient velocity to escape from the solar system with a velocity of around 17 km/s, and they also completed a sequence of gravity assists around other planets enabling a "Grand Tour". Thus, they did not really have a main engine although they did have a set of small hydrazine thrusters for trajectory corrections and attitude control. The power for the spacecraft came from the use of radioisotope generators (RTGs) on board using ^{238}Pu fuel. Since the power is running out, they are expected to be non-operational by the early 2030s and then will simply drift in interstellar space. These spacecrafts remain the only ones to date that have left our solar system and so whilst they are not relevant to the propulsion theme of this book, they are important in the overall story of our quest to build machines that can traverse the vast distances of space. At the time of publishing this book, Voyager 1 was at $\sim$170 AU and Voyager 2 at $\sim$141 AU and it is hoped their power supply will last until 2030 at least.

Halite-Centurion, 1978–1988

Although not a propulsion study, Halite-Centurion was a set of underground nuclear test experiments carried out by the US Department of Energy that have a direct relevance to fusion propulsion [33]. The concept was to place small pellets of ICF targets within the test chambers of nuclear warheads so that the radiation generated from the nuclear detonation would drive the capsules towards a state of thermonuclear ignition. The capsules would then be diagnosed using instruments to measure their performance. Spherical capsules of different sizes and masses were tested and all driven at different energies to inform on the requirements for thermonuclear ignition. Since most of the information pertaining to these tests remains classified we cannot comment much more on them.

It is important to explicitly highlight them as a direct experimental validation point for the idea of ICF ignition within a reactor, and therefore also for ICF propulsion. Statements by Lindl indicated that tests of up to $\sim$1 - 10 MJ and 5 mg fuel capsules were tested to ignition giving outputs in the range 100 - 1,000 MJ [34, 35]. Indeed, the tests were originally motivated by results from the Nova laser facility and then the results from the tests ultimately led to the proposal for a Micro Fusion Laboratory and then the design of the NIF which eventually achieved ignition with $\sim$2 MJ in 2022/2023 [36–38].

Hyde, 1983

The Inertial Fusion Rocket (IFR) [39] was initiated as a potential mission to the dwarf planet Pluto at just under 50 AU distance. The concept is illustrated in Figure 3.6. The concept was to use a 2 MJ Krypton Fluoride laser. The engine would be initiated by a 1 MW nuclear reactor. The spacecraft would detonate 15 mg capsules at a rate of 100 Hz, using 3.72 tons DD fusion fuel, augmented with 4,100 tons of expellant fuel. The expellant augmentation means that chemical fuel is ejected simultaneously with the plasma debris from the exhaust products which leads to a modification of the mass flow rate. The capsules would be seeded with 360 micromoles of a Tritium trigger at the centre. The gain of the capsules was said to be around $G \sim 1{,}000$ and the engine was characterised by a jet efficiency of 42%.

One of the key differences between the IFR and Daedalus was the use of indirect drive ignition. Whereas for the Daedalus design the energy drive would directly impinge onto the surface of the capsule, in an indirect drive scheme the energy is beamed into a cavity known as a 'Hohlraum' which has the function of creating a smooth radiation bath surrounding the capsule contained within. This then allows for a hydrodynamic implosion on a much lower adiabat and creates the conditions for minimal turbulence and hydrodynamic instabilities.

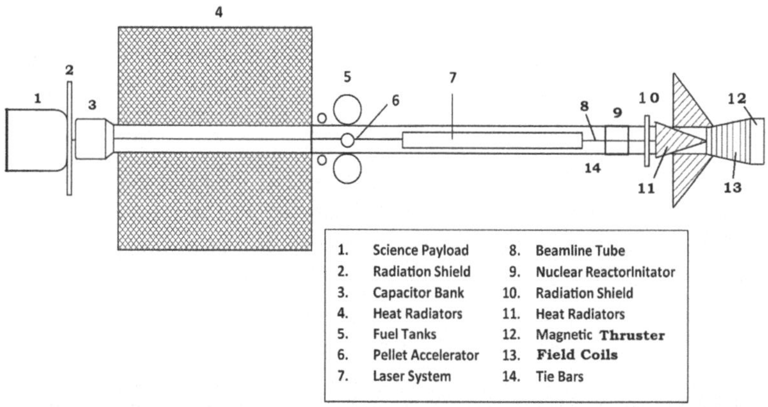

Fig. 3.6 Illustration of the Hyde Interstellar Fusion Rocket[KFL]

The IFR would carry a 1,500 tons cargo payload or a 50 tons payload for special VIP missions. The total dry mass of the spacecraft was on the order of 486 tons. It would utilise 2,700 tons wet mass. It would achieve a cruise velocity of 284 km/s, an effective mass flow rate, not just fusion, of 0.015 kg/s, thrust of 40 kN, specific power of 110 kW/kg.

World Ships, 1984

Building on from the earlier work of Enzmann [14, 15], Aan Bond and Anthony Martin took the study of interstellar slow boats to a whole new level in the development of vast world ship constructions [40, 41]. There are many issues with the idea of a world ship, such as the viability of population survival and genetic make-up, the sourcing and management of appropriate soils for plant growth and oxygen production, the potential for the break-out of a virus which must be contained, and the policing of a social system that must balance requirements for democracy with the need for strong determined leadership with a clear chain of authority. We do not discuss any of these issues here. Instead, the focus is on the external propulsion of the world ship using ICF engines and to describe what the mission profile and propulsion requirements look like given this assumption.

These were essentially large habitats capable of sustaining a massive population of people over multiple generations and have their origins in the research of O'Neill cylinder colonies such as proposed for the Lagrange 5 point by Gerard K O'Neill [42]. One of the models developed in the study was known as the Mark 2A dry worldship design concept (they also produced an oceanic model). This was a giant cylinder 87.5 km in length and with a radius of 7.24 km. It would have a wall thickness of 2.88 m and a surface area of 3,980 km^2.

Various other parameters for the Mark 2A design were defined including a wall mass of 1.04×10^{11} tons, regolith mass of 4.975×10^{10} tons, atmosphere mass of 1.403×10^{10} tons, making up the total habitat mass of 1.68×10^{11} tons. The design also specified a tankage mass of 3.09×10^{10} tons, propellant mass of 7.797×10^{11} tons, nuclear fuel mass of 3.696×10^{10} tons. This was obviously utilising nuclear pulse propulsion and so the nuclear fuel refers to the material associated with nuclear units.

In terms of the mission profile for the Mark 2A, it would accelerate at 0.001 m/s^2 and reach a cruise velocity of 1,500 km/s or 0.5%. The total time spent on the acceleration and deceleration period was to be 47.56 years. The exhaust velocity would have been 1,880 km/s. The engine power was to be 2.6×10^6 TW and the propellant mass flow 5.205×10^5 kg/s. The total time to complete the mission would be around 1,000 years. We discuss the world ships further in chapter 12.

Starwisp, 1984

Although not a reaction engine it is worth mentioning that in this period we also saw emerge the ideas from the physicist Robert Forward on the use of laser beams and microwave beams to propel spacecraft on deep space missions. These seminal concepts culminated in his Project Starwisp [43, 44] which involved a 560,000 tons Fresnel lens in Earth orbit to collimate a microwave beam onto a 1 ton spacecraft and accelerating it to around 34,000 km/s or $0.11c$. The spacecraft would require a beam power of 65 GW and arrive at the destination target around 40 years after launch.

An important lesson from this is that we can compare this to the Project Daedalus which had a similar flight profile but carried a 450 tons payload using 50,000 tons of fusion fuel. Thus, this suggests something about interstellar flight, that whichever way we do it, it requires substantial amounts of energy and suggests a principle of least action is at work in terms of a minimum requirement. As we saw in Chapter 2 the solution to the Euler-Lagrange equation gave us a minimum flight solution which was an adequate approximation provided the speeds were high. We might find however, that different propulsion systems become more competitive at much lower velocity and where corrections due to the gravitational potential of each star are required. Similarly for very high velocity in the relativistic limit.

TAU Probe, 1987–1990

This was a project ran out of the Jet Propulsion Laboratory (JPL) in California [45]. The goal was to develop a robotic probe that could travel one thousand units (hence the name) and it was abbreviated to the TAU mission. It was not a full interstellar probe but was not an interplanetary one either, so it is referred to as an interstellar precursor probe. It would take around 50 years to complete its journey. It was not an advanced fusion engine but instead would use a nuclear-electric propulsion engine with 10,000 kg of xenon gas and a 1 MW fission reactor to feed the propulsion system. It would accelerate for around 10 years until it reached its cruise velocity of around 100 km/s. It would perform a flyby of Pluto on its way out the Solar System.

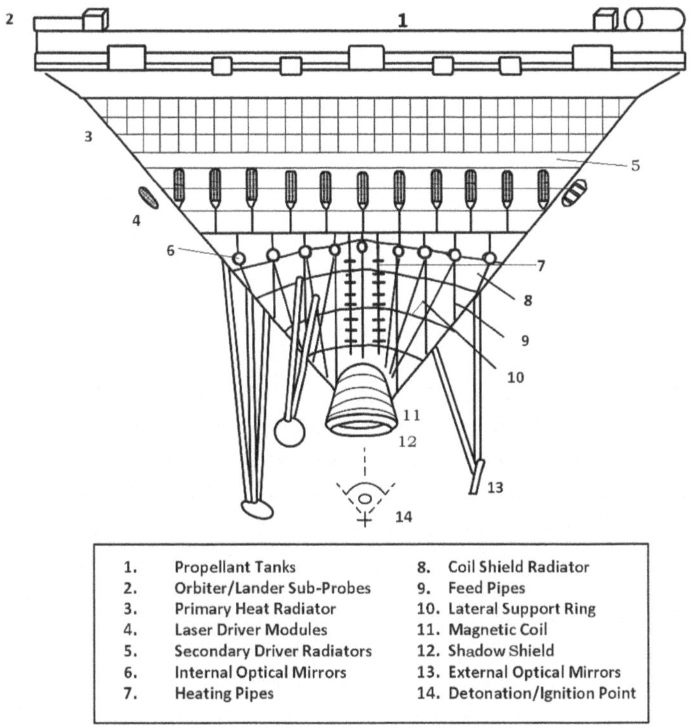

Fig. 3.7 Illustration of a VISTA Spacecraft Configuration[KFL]

The concept would perform important science experiments as a post-Voyager mission such as studying stellar parallax and the outer heliosphere and local interstellar medium. It would also performed refined tests of Einstein's General Theory of Relativity to improve our understanding of gravity. It would also carry a large 1.5 m astrometric telescope and communicate back to Earth using a 20 kb/s 1 m laser communications system. The TAU mission never left the concept planning phase, but it was sure to be an important step in designing vehicles of this type which will someday reach technological maturation and in service flight.

VISTA, 1987–2003

The vehicle for interplanetary transport applications or VISTA was a study initiated out of the Lawrence Livermore National Laboratory (LLNL) in the United States mainly by the physicist Charles Orth. Although the work was first initiated in 1987 [46], work continued until at least 2003 [47, 48]. The main motivation for the work was to develop a fusion concept for missions to Mars at 2.62 AU, with a return mission to be achieved in around 100 days. The vehicle concept is illustrated in Figure 3.7.

The designers of VISTA wanted to address a specific issue with previous fusion propulsion concepts and that was the large neutron and x-ray radiation flux emitted by the fusion reactions which irradiated the spacecraft structure. To address this VISTA was designed with an unusual conical shaped spacecraft to minimise the exposure of the vehicle structure to the emitted neutrons and x-rays. In addition, the magnetic nozzle and associated superconducting magnets were placed further back to minimise the radiation absorption.

The main energy driver was to be a 5 MJ excimer laser driver. The total jet efficiency was assumed to be around 34% which is much less than the previous concepts. The engine would operate with a pulse frequency of between 12.5 - 30 Hz (or even less) depending on whether the design utilised an induction bootstrap to power the next cycle or if the thermodynamic power cycle was of a Rankine type. It would utilise 0.046 g capsules containing DT fuel mixture. It was claimed that the capsules would be characterised by a gain in the range $G \sim$200-1,500, although this seems rather optimistic. The assumed burn-up fraction was 0.476 which is much higher than the assumptions for the Daedalus study and would be consistent with the expected high gain. Then again, the designers were close to actual experiments on ICF targets so were well informed as to what was possible. The basis of the capsule designs was via an indirect drive method and so smoother implosion symmetry may give rise to a higher burn-up fraction.

The spacecraft would carry a 100 tons payload and would have a payload mass fraction of around 1.5%. It had a 6,000 tons wet mass, 1,860 tons dry mass, with 3.72 tons of fusion fuel, but also with 4,100 tons of augmented expellant fuel (hydrogen) to push up the thrust performance. The engine performance would be characterised by a mass flow rate of 1.5 kg/s, thrust of 0.24 MN, jet power of 20 GW and specific power of 20 kW/kg. The spacecraft would also carry a 189 tons shield to protect the payload.

The spacecraft would accelerate for around 25 days, then cruise for around 20 days at a speed of around 120 km/s. Then it would decelerate for around 6 days into Martian orbit. The total mission time to Mars would be around 50 days, hence a return trip would be approximately twice this. It is worth noting that the VISTA study also went on later to look at utilising other fusion reactions for the Mars missions. This resulted in roundtrip times to Mars of 145 days (DT), 261 days (DD) and 241 days (D^3H) [47].

Although the Daedalus design was comprehensive in its consideration of all the spacecraft systems, as a propulsion study the VISTA is likely the most reliable. This is due to the fact it was able to build on the earlier studies (particularly Hyde [39]), it is more recent and so contains more up to date information on scientific discoveries, but it is also well informed by actual experiments using lasers on ICF designs, and presumably modern numerical codes. The innovative conical spacecraft design also makes for a creative way to minimise the radiation load on the structure. One must question the energy gain of G = 1,500 however, which would appear to have little basis in experimental validation and is highly optimistic for a theoretical study. VISTA is discussed more in Chapter 8.

Longshot, 1988

Project Longshot was a student study at the US Naval academy [49]. It was an attempt to design a robotic spacecraft that could reach the Centauri A/B system. The project was apparently a response to a 1986 report published by the National Commission on Space titled *Pioneering the Space Frontier* [50]. This study was conducted on the back of the Space Shuttle Challenger disaster where several astronauts lives were lost. The commission referred to the need for a programme to study solar and space physics which might include "*a long-life high velocity spacecraft to be sent out of the Solar System on a trajectory to the nearest star*". So, this was the motivation behind this design study.

The main goal was to design a spacecraft that could study the properties of the interstellar medium, the characteristics of the star system and conduct astrometry distance studies to stars using parallax measurements. It would feature a series of burn acceleration phases, followed by a cruise phase and then a deceleration phase. The mission was to be achieved in 100 years trip time travelling at around 0.05c. It would carry a 30 tons payload to the destination target.

The report specified that the propulsion system was to be based on an ICF engine using D^3He fuel using an exhaust velocity of 10,006 km/s. The report gave a range of capsule mass and engine pulse frequency values which were to be utilised. This included from 0.005 g mass at 250 Hz, up to 0.085 g mass using 14 Hz.

Unfortunately, independent analysis by this author has found that insufficient propellant was allocated for the mission to permit deceleration. Indeed, even with its allocated 264 tons of propellant in flyby mode, it would still take well over a century to reach its destination. So as a technical design study it is less useful. All the same, the report does give a good description of the various issues relevant to a robotic probe of this type and so it is worth mentioning and is a good introduction to the design of an advanced spacecraft of this type. Although, perhaps identifying the various deficiencies in the design serves as more of a good student exercise.

Schulze, 1991

An excellent study by Norman Schulze[51] was conducted by the US space agency NASA which although did not feature the development of a specific spacecraft concept, it contained a set of studies that nicely compliments this research. The purpose of the study was to examine 100-year missions to the Alpha Centauri system driven by fusion propulsion and to examine the different operational performance parameters to be expected from such missions.

The assumptions of the study were examining DT and D^3He fuel combinations, with assumed burn-up fractions of 0.57 and 0.31, driven by driver energies of 3 MJ and 7.3 MJ respectively. The capsule masses were not defined but gains of $G \sim 10$ were assumed for the capsule output performance, when compared to the driver input energies. Although not stated, it is possible to infer from this that the capsule masses would have been on the order of 1.54 g and 6.72 g for the DT and D^3He designs respectively. However, the authors clearly indicate a preference towards steady state propulsion instead of pulse detonation. To use ICF would require gains on the order of $G\ \sim$1,500 in their view and presumably this was based on an examination of the Hyde [39] and VISTA [46] work discussed earlier.

The study looked at both flyby and rendezvous mission scenarios, but also for single stage and multiple staged (1–4) engine configurations for a parallel thrust profile. The vehicle was modelled to carry a 10 tons payload using a typical exhaust velocity of around 9,810 km/s and if ICF was the adopted fusion method then a pulse frequency of 30 Hz was assumed, and these were to be manufactured in the vehicle as it proceeded through the flight.

The amount of propellant carried depended on the number of engine stages, but was 4,000 tons (1–2 stage concepts), 65,400 tons (4 stage concept) and 1,308,000 tons (4-stage concept). The study then examined the vehicle performance for three values of specific power, to include 1 kW/kg, 10 kW/kg and 100 kW/kg. For the higher value of specific power, the trip time was reduced, since the specific impulse, cruise velocity, thrust and jet power was higher. A higher specific power allows a vehicle to go much faster and so reduce the trip time. For example, for a single stage system, as the specific power was increased by an order of magnitude, the cruise velocity went from 4,184 km/s, to 9,015 km/s to 19,422 km/s, resulting in a trip time of 460 years, 213 years and 99 years respectively. The corresponding thrust went from 1.8 kN, 8.44 kN to 39.2 kN resulting in a jet power of 195 MW, 1.95 GW to 19.5 GW respectively. Overall, the effect of staging was noticeable but not dramatic. Going from a single stage to a 4-engine stage resulted in trip times of 99 years (1 stage), 98 years (2 stage), 79 years (3 stage) to 71 years (4 stage). Overall, this was an outstanding study, and these few paragraphs barely do it justice.

3.3 Modern Design Studies (1992–2025)

Halyard, 1997

In the studies by Halyard [52] the Project Daedalus was revised as DD but also using a light ion beam as the 1.4 GJ energy driver system. An additional change was adoption of a Lithium-6 blanket which had the function of breeding sufficient Tritium for the use in fusion triggers to be used during the deceleration part of an interstellar mission. The spacecraft would use an exhaust velocity of 8,300 km/s and would carry a payload of 100 tons. The total wet mass was to be 258,534 tons and the total dry mass 2,534 tons, so there would be 256,000 tons of propellant. The modified Daedalus concept would use capsule masses of 23 g and these would be detonated at 15 Hz pulse frequency. The spacecraft performance would be characterised by a mass flow rate of 0.3446 kg/s, thrust 2.87 MN, jet power 11.91 TW and specific power 6.473 MW/kg (based on a propulsion mass of 1,840 tons). The total travel time to its target of Barnard's Star at 5.9 ly would be 102 years.

Comparing this concept to the original Daedalus concept, it is noticeable that the use of DD fusion instead of D^3He leads to a much larger requirement for the quantity of propellant, although this is also due to the allocation of propellant for the deceleration phase. It is also observed that the travel time is twice that of the Project Daedalus. This is because the concept is using a reduced exhaust velocity and much reduced pulse frequency compared to Daedalus. Another factor is the time for deceleration since this was for a rendezvous mission compared to the flyby mission only of the Daedalus. So, because DD reactions are neutron rich, shield and radiator mass penalties grow substantially versus D^3He, which contributes to higher propellant demands and longer trip times.

Halyard also looked at modified versions of the VISTA concept. Again, these were deceleration missions to Barnard's Star and so these were for an interstellar VISTA variant. Two versions of the interstellar VISTA were produced utilising DD based fuels. In the first version it would be driven with an energy of 30 MJ and use an exhaust velocity of 4,540 km/s, carrying a payload mass of 100 tons using 23,300 tons of propellant. In the second version it would be driven with a driver energy of 120 MJ and use an exhaust velocity of 5,720 km/s, carrying a payload mass of 50 tons using 60,500 tons of propellant. The first version of the interstellar VISTA would use capsule masses of 0.351 g driven at a pulse frequency of 13.7 Hz. The second version of the interstellar VISTA would use capsule masses of 1.404 g also driven at a pulse frequency of 13.7 Hz.

In terms of the performance, the first version of the interstellar VISTA would be characterised by a mass flow rate of 0.0048 kg/s, thrust 21.8 kN, jet power 49.486 GW, specific power 28.117 kW/kg (based on a propulsion mass of 1,760 tons). The spacecraft would take 395 years to complete the mission. The second version of the interstellar VISTA would be characterised by a mass flow rate of 0.0192 kg/s, thrust 110 kN, jet power 0.314 TW, specific power 0.118 MW/kg (based on a propulsion mass of 2,660 tons). The spacecraft would take 1,404 years to complete the mission.

An initial examination of the assumptions of the study shows inconsistences in the values. For example, the author claims that the concepts had an unrealistic jet efficiency of 98% but with the burn-up fraction, mass defect and exhaust velocity values assumed, one in fact finds that the jet efficiency would have to be characterised by a value much greater than 100% which is not possible. It is likely therefore that there is an erroneous assumption here, and in any case even a jet efficiency of 98% is a near perfect machine and exceeds even the performance of the non-energy loss-based Daedalus which doesn't seem realistic. However, the paper was well done and the calculations rigorous, even if based on incorrect assumptions.

Clearly, these are missions that take centuries to a millennium and the survivability of the spacecraft over such long-time scales is doubtful. However, if the spacecraft were fully autonomous, had a high degree of computational intelligence and decision making, and an ability to self-reproduce vital replacement parts as they broke down (such as through additive manufacturing) then such missions begin to appear plausible. Reproductive star probes have been studied by Robert Freitas [53].

Discovery-II, 2004

Although this study was not ICF propulsion but instead MCF, as a fusion propulsion study it is so important and relevant to the field since it was conducted by the US space agency NASA under the leadership of Craig Williams and Stan Borowski. This concept was named the Discovery-II [54]. The study was in relation to the earlier concept of the Discovery-I as appeared in the book and movie of the same name '*2001: A Space Odyssey* [55] which was written by Arthur C Clarke. In the book the author utilised a nuclear propulsion concept although the details lacked specifics. Clarke's thinking had always been that a deep space interplanetary vehicle would encompass an spherical-torus reactor and a crew payload, both separated by a long boom to minimise the radiation dosage to the crew. A similar concept known as the Ares spacecraft had appeared in his earlier novel '*The Sands of Mars*" [56]. So it was that passing through the year 2001 that the NASA team decided to design a spacecraft concept that was close to Clarke's original vision except that it would be driven by a magnetic fusion engine.

It was a crewed vehicle designed to accommodate between 6–12 astronauts. Its key design requirement was a rendezvous mission to Jupiter or Saturn with a design that was characterised by a payload mass fraction exceeding 5% and a trip time not greater than one year. The 240 m long spacecraft would carry a 172 tons artificial gravity payload. This would be on the end of a long structural boom, from which would also be attached the heat radiators and propellant cryo-tanks. On the rear end of the structure would be the spherical torus nuclear fusion reactor and magnetic nozzle. The total structural mass would be 646 tons.

The spacecraft would be propelled by 11 tons of D^3He fuel, which was then augmented by Hydrogen expellant propellant. From the total expellant of 861 tons, around 807 tons would be used for the main fusion engines, whilst the remainder would be used for reaction control and flight performance reserve. The use of the expellant increased the mass flow rate to a total of 0.079 kg/s for the Jupiter mission and 0.045 kg/s for the Saturn mission. The increase in mass flow rate would also increase the vehicle thrust.

In addition, the spacecraft would carry a 10 tons nuclear fission reactor for start-up of the main engine with a power output of 2 MWe. There would be 4 propellant tanks with a total mass of 6 tons. The main fusion engine would have a mass of around 310 tons and would exhibit a nozzle jet efficiency of 80%. The total wet mass at launch would be around 1,690 tons and the dry mass would be around 872 tons, with a total mass ratio of 1.9.

The 118 days mission to Jupiter at 4.7 *au* would require a jet power of 4,830 MW and a specific power of 8.62 kW/kg. It would use an exhaust velocity of 347 km/s and achieve a cruise velocity after engine stage burn of 223 km/s, although it would not enter a cruise phase but begin to decelerate towards its rendezvous.

For the Saturn mission to 9.57 *au* the Discovery-II spacecraft would take 212 days to complete the rendezvous. It would be characterised by a mass flow rate of 0.044 kg/s. A Saturn mission would be more in line with the original Arthur C Clarke story, as opposed to Jupiter in the movie. Overall, this was an outstanding study that helped to paint a vision about what a fusion powered spacecraft to the outer planets would look like. We will revisit this spacecraft design in Chapter 8 as we attempt to modify the outstanding Discovery-II design as an ICF vehicle.

Icarus, 2009

Project Icarus was a design project set up (primarily by this author and the physicist Richard Obousy) to re-design the original Project Daedalus concept run mostly by British, American and European designers. The idea was to take the original design, gradually perturb it towards a new design based on scientific and technological developments in the proceeding decades. Although the project was initiated in 2009 [57, 58] it continued until 2023. The team were not obligated to remain with ICF based fusion, although the requirement was to use mostly fusion energy for the propulsion system which was interpreted to mean that fusion reactions were responsible for at least 80–90% of the thrust generation. The mission requirement also called for a 100-year trip time carrying a 150 tons payload and to fully decelerate into the orbital system of the target star. Whereas for Daedalus the target was Barnard's Star at 5.9 ly, for Icarus the target was the Alpha Centauri system at 4.3 ly. Overall, there was a desire to reduce the environmental constraints, reduce the vehicle mass and relax the mission to make it more plausible.

Some of the concepts examined were non-ICF based and so these will not be discussed here, the most notable being the concept Firefly which utilised a Z-pinch type effect to achieve fusion ignition [59, 60]. Others were ICF based but did not achieve a full level of theoretical development past a certain point. Icarus Ghost which was a DD fast ignition concept using 148,000 tons of propellant, and Icarus UDD (Ultra-Dense-Deuterium) which was a plasma jet driven magneto-inertial fusion concept using 75,468 tons of propellant [61]. The main vehicle design put forward by this author was the laser-ICF driven Pegasus which had a capsule mass between 28.8 - 72 mg and a pulse frequency in the range 1,000 - 2,500 Hz. Some of the designs from this project are described more in Chapter 9.

Project Starshot, 2016+

Again, not a reaction engines system but we would be remiss not to give a mention to the Breakthrough Initiatives Project Starshot; an initiative of the physicist Yuri Milner. This investigated the possibility of using a 100-GW class laser beam on the ground to propel a $\sim$3-4 m sail in Earth orbit carrying a gram-scale payload to the stars at 60,000 km/s (0.2c) [62, 63]. The project was also strongly related to Project Starlight, and both had their origin with the physicist Professor Philip Lubin and his innovative concept work at the University of California, Santa Barbara. These ideas also built on the earlier work of Robert Forward. The project received considerable funding on the order of millions of dollars.

The goal was to launch a vehicle on a trip time on the order of 20 years, although the payload would have been so small that it is difficult to see how such a weakly transmitted communications signal could be received over light years distance, and especially if it was many Megabytes of information sufficient to form images. Yet Project Starshot led to the publication of dozens of peer reviewed papers and extensive research into lasers and materials science which is sure to have tremendous value in future decades when such a concept becomes more realisable. A useful read at the close-out of Project Starshot was given by Scientific American [64]. This author served on the Starshot advisory committee for its duration, and although disappointed not to see it go further, a personal observation is that it was a thrilling adventure to even be considering such ambitious plans surrounded by other optimistic scientists. This was the experience of being paid to design a starship, not something many others get to do, and I will always be grateful for the Breakthrough Initiates team for that.

Interstellar Probe, 2018+
Building on from the earlier work of the TAU mission was the Interstellar Probe, an initiative of the Johns Hopkins Applied Physics Laboratory (JHAPL) [65] under the leadership of Pontus Brandt and Ralph McNutt. Its stated aim was to propel a spacecraft to 1,000 AU at a speed on the order of 100 km/s which means it would arrive at its destination in around 50 years trip time. It would likely be propelled by either a SolarSail or a Nuclear Electric Engine.

SunVoyager, 2023–2024
The SunVoyager mission also built on from TAU and the Interstellar probe but proposed to use an ICF engine [66, 67]. It would be a mission to 500 - 1,000 AU and it would achieve this in a trip time of less than a decade carrying a 100 tons payload. The propulsion system would use a 5.76 mg pellet detonated at 10 Hz. The spacecraft was developed as a part of discussions relating to the development of an interstellar roadmap. The SunVoyager is discussed in more detail in chapter 8.

Discovery-III, 2020
This was a design to build on from the earlier Discovery-II design but instead using an ICF engine [68]. Although it also involved missions to Jupiter and Saturn and was intended as a form of technology comparison for the same mission. Interestingly, the calculations seem to suggest that only half the thermonuclear fuel would be required when compared to a magnetic fusion engine. The Discovery-III would use a 2.88 mg ICF capsule detonated at 100 Hz pulse frequency. Discovery-III is discussed in more detail in Chapter 8.

World Ship Ra, 2024
Building on from the earlier work of Bond and Martin [40, 41] using a nuclear pulse driven world ship, this involved the use of 1,024 ICF engines to propel the vehicle at a speed of 0.05c and therefore reach the nearest stars in a trip time on the order of 850 years [69, 70]. The design would utilise a 230 g pellet mass detonated at 100 Hz. Like the world ships developed by Bond and Martin these were massive vessels that were tens of km in length and represent small worlds with a capacity to carry millions of people. In the future such vessels might represent a part of a colonisation

strategy or even a part of an exodus from our solar system in the event of a severe existential threat to our existence. The world ship is discussed further in chapter 12.

Long-Enzmann, 2024
This project built on from the earlier calculations [16] of the Enzmann starship but investigated the possibility of an augmented thrust profile using expellant propellant [71]. Missions that were consistent with the original design concept were derived but also further details about the layout of the vehicle to include its 24 ICF driven engines. This would require a 2.09 g pellet mass detonated at 100 Hz. They are smaller vessels than the world ship but still massive undertakings compared to what we can do in space today. Such missions may represent our first adventures as humans to the planets and beyond. This is described more in Chapter 11.

Summary of Concepts
This author is very fortunate to have worked on several of the above projects. This includes Project Icarus, Project Starshot and the Interstellar Probe. For someone that merely dreamed of designing ships that could travel to the stars such participation represented the fulfilment of an ambition. Given this, it is a privilege to call oneself a starship designer, since I have worked on many actual starship projects that may be a part of our plausible future. Taking all the concept designs mentioned above, these are collated in Table 3.1.

The original idea for a pulsed propulsion system takes its origins with Orion in the 1950s. A realisation that such a machine could be developed on a much smaller scale lead to the concepts for fusion powered spacecraft that literally create mini-suns within a reaction chamber for a fraction of a second; using the power of the stars to travel between them sounds serendipitous at least. The gradual developments of laser technology over the decades have inched us closer to the possibility of achieving ignition within a laboratory environment. Once this is achieved, next comes gain and then it is a question of shrinking the technology down and applying it to the extreme applications of spacecraft propulsion. This process may take many more decades or even into the next century, but there is every reason to be optimistic that it will someday be achieved. Provided physicists continue to develop vehicle concepts on a theoretical basis, there will come a time when these are proven by experiments and then the stars will be open for the exploration and reconnaissance of interstellar probes from Earth.

Problems of Chapter 3
3.1 Reviewing the concepts discussed in this chapter (i) List the range of unit mass and pulse frequencies of detonation for the different systems. Provide your answers as a minimum to maximum range. The term unit refers to ICF pellets or nuclear bombs in the case of external nuclear pulse propulsion.

Table 3.1 List of Concept Developments Cited in this Research

Concept	Year	Propulsion
Shepherd	1952	Fusion and other, interstellar
Bussard ramjet	1960	ISM fed, interstellar
Orion	1957–1965	External nuclear pulse, interplanetary
Spencer	1963–1964	Internal fusion, interstellar
HLT	1963–1965	CTR, interplanetary
Helios	1964–1965	Internal nuclear pulse, interplanetary
Enzmann	1966	Internal nuclear pulse, interstellar
Sirius	1970	External nuclear pulse, lasers, interplanetary
HWN	1971–1972	ICF, pusher plate removed, interplanetary
Project Daedalus	1973–1978	ICF, electron guns, hybrid, interstellar
Soviet Icarus	1976	ICF, 3-stage Daedalus, interstellar
Voyager	1977	Gravity assist, precursor
Halite-Centurion	1978–1988	nuclear tests of ICF capsules, ground
Hyde-IFR	1983	ICF, KrF laser, interplanetary
World Ship	1984	External nuclear pulse, interstellar
Project Starwisp	1984	Microwave driven sail, interstellar
TAU	1987–1990	Nuclear-electric, precursor
VISTA	1987–2003	ICF, excimer laser, interplanetary
Longshot	1988	ICF, interstellar
Schulze	1991	MCF/ICF, interstellar
Halyard	1997	ICF, light ion beam, interstellar
Discovery-II	2004	MCF, interplanetary
Project Icarus	2009–2023	ICF/Z-pinch, interstellar
Project Starshot	2016–2024	laser-sail, interstellar
Interstellar Probe	2018–2024	Nuclear-electric sail, precursor
SunVoyager	2023–2024	ICF-laser, precursor
Discovery-III	2020	ICF-laser, interplanetary
World Ship Ra	2024	ICF-laser, interstellar
Long-Enzmann	2024	ICF laser, interstellar

3.2 Look up the NASA Technology Readiness Level (TRL) definitions which were originally published in a white paper. (i) Reviewing the various spacecraft concepts of this chapter what TRL would you assign to the propulsion systems in general?

3.3 A spacecraft accelerates for 6 months at $0.01g_o$ from rest. (i) What is the cruise velocity achieved by the probe at the end of stage burn? What is the distance reached by the probe in the units of km and AU? (ii) If the probe is to continue to 500 AU, how many years after the original launch would it have taken to reach this distance? (iii) At the defined cruise speed how many hours does it take for the spacecraft to pass through 1 AU, equivalent to the Sun-Earth mean orbit distance?

References

1. Y. Kondo, F. C. Bruhweiler, J. Moore, C. Sheffield (editors), Interstellar Travel and Multi-Generation Space Ships, Apogee Books Publication, 1971.
2. L. R. Shepherd, Interstellar Flight, JBIS, 11(4), 149–167, July 1952.
3. E. Sanger, Space Flight, Countdown for the Future, *Econ-Verlag GmbH*, Germany, 1963.
4. J. Strong, Flight to the Stars, *Temple Press Books*, 1965.
5. M. W. Hunter, Thrust into Space, Holt, *Rinehart and Winston Inc*, 1966.
6. R. W. Bussard, Galactic Matter and Interstellar Flight, *Astronautica Acta*, 6, pp.170–194, Fasc.4, 1960.
7. C. Sagan, "Direct Contact Among Galactic Civilisations by Relativistic Interstellar Spaceflight", Planetary and Space Science, 11, 1963.
8. R. Heinlein, Time for the Stars, Scribners, 1956.
9. C. J. Everett, S. M. Ulam, On a Method of Propulsion of Projectiles by Means of External Nuclear Explosions, LAMS, 1955.
10. G. Dyson, Project Orion - The Atomic Spaceship 1957–1965, *The Penguin Press*, 2002.
11. F. J. Dyson, Interstellar Transport, *Physics Today*, 21(10), 41–45, October 1968.
12. E. A. Platt, D. W. Hanner, The Effective Specific Impulse of a Pulsed Rocket Engine. In Proceedings of the AIAA Propulsion Joint Specialist Conference, UCRL-12296, Colorado Springs, CO, USA, 14–18 June 1965.
13. J. W. Hadley, T. F. Stubbs, M. A. Janssen, L. A. Simons, The Helios Pulsed Nuclear Propulsion Concept. UCRL-14238. In Proceedings of the AIAA Propulsion Joint Specialist Conference, Salt Lake City, UT, USA, 14–18 June 1965.
14. R. Enzmann, Rocket Science and Fact-Based Science Fiction, *Planetology and Space Mission Planning Conference*, New York Academy of Sciences, 1966.
15. G. H. Stine, A Program for Star Flight, *Analog*, October 1973.
16. A. Crowl, K. F. Long, R. Obousy, The Enzmann Starship: History & Engineering Appraisal, JBIS, 65(6), 185–199, June 2012.
17. J. L. Hilton, J. D. Luce, A. S. Thompson, Hypothetical Fusion Propulsion Rocket Vehicle. *J.Spacecraft*, 1, 276–282, May-June 1964.
18. D. F. Spencer, L. J. Jaffe, Feasibility of Interstellar Travel, *Acta Astronautica*, 9, 49–58, 1963.
19. D. F. Spencer, Fusion Propulsion System Requirements for an Interstellar Probe. Technical Report No.32-397, NASA CR-50964, Jet Propulsion Laboratory, 15 May 1963.
20. K. Boyer, D. J. Balcomb, System Studies of Fusion Powered Pulsed Propulsion Systems. AIAA 71-636, Presented at the 7th Propulsion Joint Specialist Conference (LA-DC-12520), 14–18 June 1971.
21. R. Hyde, L. Wood, J. Nuckolls, Prospects for Rocket Propulsion with Laser-Induced Fusion Microexplosions, AIAA, 72-1063, Presented at the 8th Propulsion Joint Specialist Conference, 29 November - 01 December 1972.
22. F. Winterberg, Rocket Propulsion by Thermonuclear Microbombs ignited with Intense Relativistic Electron Beams, *Raumfahrtforschung*, 15, 208–217, 1971.
23. A. Bond, A. R. Martin, Project Daedalus: The Mission Profile, Final Study Report, JBIS, Special Supplement, S37-S42, 1978.
24. A. R. Martin, A. Bond, Project Daedalus: The Propulsion System Part 1: Theoretical Considerations and Calculations, JBIS, Special Supplement, S44-S62, 1978.
25. A. Bond, A. R. Martin, Project Daedalus: The Propulsion System Part 2: Engineering Design Considerations and Calculations, JBIS, Special Supplement, S63-S82, 1978.
26. A. R. Martin, A. Bond, Project Daedalus: An Analysis of the Photometric Data on Barnard's Star and its Implications. JBIS, Special Supplement, S24-S32, 1978.
27. A. P. Fraas, The Blascon - An Exploding Pellet Fusion Reactor, ORNL-TM-3231, 1971.
28. R. C. Parkinson, Project Daedalus: Propellant Acquisition Techniques, JBIS, Special Supplement, S83-S89, 1978.

29. B. Palaszewski, Atmospheric Mining in the Outer Solar System: Resource Capturing, Storage and Utilization, AIAA 2012-3742, *Presented 48th AIAA/ASME/SAE/ASEE Joint Propulsion Conference & Exhibit*, Atlanta, Georgia, 30 July–01 August 2012.
30. A. A. Filyukov, U. N. Zakirov, On the Possibility of Creating the Project of a Subrelativistic Probe to Search for the Forms of Extraterrestrial Life. Presented at the 27th IAF Congress, Anaheim, California, 10–16 October 1976.
31. A. R. Martin, A. Bond, Nuclear Pulse Propulsion: A Historical Review of an Advanced Propulsion Concept, JBIS, 32(8), 283–310, August 1979.
32. S. J. Pyne, Voyager, Exploration, Space, and the Third Great Age of Discovery, Penguin Books, 2010.
33. J. D. Lindl, Development of the Indirect-Drive Approach to Inertial Confinement Fusion and the Target Physics Basis for Ignition and Gain, *Physics of Plasmas*, 2(11), 3933, 1995.
34. J. D. Lindl, Progress on Achieving the ICF Conditions Needed for High Gain, UCRL-98669, 1988.
35. W. Happer (Chairman), Review of the Department of Energy's Inertial Confinement Fusion Program, National Academies Press, 1986.
36. A. B. Zylstra, O. A. Hurricane, D. A. Callahan, et al., Burning Plasma Achieved in Inertial Fusion, *Nature*, 601(7894), 542–548, 26 January 2022.
37. A. B. Zylstra, A. L. Kritcher, O. A. Hurricane, et al., Experimental Achievement and Signatures of Ignition at the National Ignition Facility, *Phys. Rev E*, 106, 025202, 8 August 2022.
38. A. R. Christopherson, O. A. Hurricane, C. Weber et al., Alpha-Heating Analysis of Burning Plasma and Ignition Experiments on the National Ignition Facility, *Physics of Plasmas*, 30(6), 062705, June 2023.
39. R. A. Hyde, A Laser Fusion Rocket for Interplanetary Propulsion, UCRL-88857, University of California, Livermore, CA, Presented 34th IAC, IAF-83-396, October 1983.
40. A. R. Martin, World Ships - Concept, Cause, Cost, Construction and Colonisation, JBIS, 37(6), 243–253, June 1984.
41. A. Bond, A. R. Martin., World Ships - An Assessment of the Engineering Feasibility, JBIS, 37(6), 254–266, June 1984.
42. G. K. O'Neill, The High Frontier: Human Colonies in Space, Space Studies Institute, Space Frontier Foundation, Apogee Books, 2000 (originally published 1976).
43. R. L. Forward, Roundtrip Interstellar Travel using Laser-Pushed Lightsails, *J. Spacecraft*, 21, 2, 187–195, March/April 1984.
44. R. L. Forward, Starwisp: An Ultra-Light Interstellar Probe, *J. Spacecraft*, 22(3), 345–350, May/June 1985.
45. K. T. Nock, TAU - A Mission to a Thousand Astronomical Units, AIAA-87-1049, Presented at the AIAA/DGLR/JSASS International Electric Propulsion Conference, Colorado Springs, 11–13 May 1987.
46. C. D. Orth, G. Klein, J. Sercel, N. Hoffman, M. Murray, F. Chang-Diaz, The VISTA Spacecraft - Advantages of ICF for Interplanetary Fusion Propulsion Applications, UCRL-96676, *Presented IEEE 12th Symposium on Fusion Engineering*, Monterey, California, 12–16 October 1987.
47. C. D. Orth, Parameter Studies for the VISTA Spacecraft Concept, LLNL, UCRL-JC-141513, Presented at NASA Advanced Propulsion Workshop in Fusion Propulsion, Huntsville, Alabama, 8–10 November, 2000.
48. C. D. Orth, VISTA: A Vehicle for Interplanetary Space Transport Application Powered by Inertial Confinement Fusion, LLNL, UCRL-TR-110500, 16 May 2003.
49. K. A. Beals, M. Beaulieu, F. J. Dembia, J. Kerstiens, D. L. Kramer, J. R. West, J. A. Zito, Project Longshot, An Unmanned Probe to Alpha Centauri, US Naval Academy, N89-16904, NASA-CR-184718, 1988.
50. Paine, Pioneering the Space Frontier, 1986.
51. N. R. Schulze, Fusion Energy for Space Missions in the 21st Century, NASA TM-4298, August 1991.

52. R. J. Halyard, Comparison of Proposed Inertial Confinement Fusion Propulsion Systems for Interstellar Missions, JBIS, 50(4), 129–136, April 1997.
53. R. A. Freitas Jr, A Self-Reproducing Interstellar Probe. *J.British Interplanetary Society*, 33(7), 251–264, July 1980.
54. C. H. Williams, L. A. Dudzinski, S. K. Borowski, A. J. Juhasz, Realizing 2001: A Space Odyssey: Piloted Spherical Torus Nuclear Fusion Propulsion, NASA/TM-2005-213559, AIAA-2001-3805, March 2005.
55. A. C. Clarke, 2001: A Space Odyssey, *Hutchinson*, 1968.
56. A. C. Clarke, The Sands of Mars, *Sidgwick & Jackson*, 1951.
57. K. F. Long, R. K. Obousy, A. Hein, Project Icarus: Optimisation of Nuclear Fusion Propulsion for Interstellar Missions, *Acta Astronautica*, 68(11–12), 1820–1829, June-July 2011.
58. K. F. Long, M. Fogg, R. Obousy, A. Tziolas, A. Mann, R. Osborne, A. Presby, Project Icarus: Son of Daedalus - Flying Closer to Another Star, JBIS, 62(11/12), 403–414, November/December 2009.
59. R. M. Freeland II, and M. Lamontagne, Firefly Icarus: An Unmanned Interstellar Probe using Z-Pinch Fusion Propulsion, JBIS, 68(3/4), 68–80, March/April 2015.
60. R. M. Freeland, Plasma Dynamics in Firefly's Z-Pinch Fusion Engine, JBIS, 71(8), 288–293, August 2018.
61. M. Stanic, J. T. Cassibry, R. B. Adams, Project Icarus: Analysis of Plasma Jet Driven Magneto-Inertial Fusion as Potential Primary Propulsion Driver for the Icarus Probe. *Acta Astronautica*, 86, 47–54, May-June 2013.
62. P. A. Lubin, A Roadmap to Interstellar Flight, JBIS, 69(2/3), 40–72, February/March 2016.
63. K. L. G. Parkin, The Breakthrough Starshot System Model, *Acta Astronautica*, 152, 370–384, November 2018.
64. S. Scoles, Voyage to Nowhere, How a Billionaire's Plan to Reach Another Star Fell Apart, Scientific American, October 2025.
65. Interstellar Probe, Humanity's Journey to Interstellar Space, NASA Solar And Space Physics Mission Concept Study for the Solar and Space Physics 2023–2032 Decadal Survey, led by Ralph L McNutt Jr, Michael V Paul, Pontus C Brandt, Jim D Kinnison et al., 13 December 2021.
66. K. F. Long, Sunvoyager: Interstellar Precursor Probe Mission Concept Driven by Inertial Confinement Fusion Propulsion, *J. Spacecraft & Rockets*, 60(3), May 2023.
67. K. F. Long, Development of the SunVoyager Interstellar Precursor Probe Driven by Inertial Confinement Fusion Propulsion, *Journal of Spacecraft & Rockets*, 62(1), January 2025.
68. K. F. Long, Discovery III: Missions to the Outer Planets using Inertial Confinement Fusion Propulsion, Journal of Spacecraft & Rockets, 62(3), May 2025.
69. K. F. Long, Population Demographics and other Issues for the Massive Ra World Ship Model - Part 1, JBIS, 76(11), 262–272, November 2023.
70. K. F. Long, Inertial Confinement Fusion Propulsion for the Massive Ra World Ship Model - Part 2, JBIS, 77(4), April 2024.
71. K. F. Long, Mission Architecture Calculations for the Enzmann Interstellar Spacecraft Concept, JBIS, 77(1), 226–232, October 2024.

Chapter 4
Laser Fusion Physics

In the last century, we have discovered this new energy source....which gives us the ability to destroy ourselves with advanced weaponry. But oddly enough, the energy available from the elementary use of this new, potentially destructive energy source allows us to build rocket engines of such performance capabilities that we can leap the energy barrier that stops chemistry.

—Robert Bussard, 2007 [1]

4.1 Fusion Reactions

Typical fusion reactions consist of Hydrogen and Helium isotopes. Figure 4.1 shows a plot of the reaction average cross section (reaction rate) as a function of plasma temperature and shows that DT reactions will ignite at the lowest temperature, followed by DD and D^3He. DT ignites at the lowest temperatures because it has the smallest effective Coulomb barrier and the largest Maxwellian-averaged reactivity at 5–30 keV.

The reaction cross section is a measure of the probability of a fusion reaction occurring as a function of the relative velocities of the two reacting nuclei. The nuclei will have a thermal distribution of velocities, so we represent this by use of an average of the two velocity distributions and cross section products $<\sigma_c v>$. We then express the number of fusions which occur per unit volume per unit time as a function of this average. The following describe the reactivities (cm^3/s) as a function of temperature (keV) for the four main reactions used in this work which were derived by others [2–4]:

$$< \sigma_c v >_{DT} = 9.10 \times 10^{-16} \exp\left(-0.572 \left| \ln\left(\frac{T}{64.2}\right) \right|^{2.13} \right), \tag{4.1}$$

$$< \sigma_c v >_{D^3He} = 4.98 \times 10^{-16} \exp\left(-0.152 \left| \ln\left(\frac{T}{802.6}\right) \right|^{2.65} \right), \tag{4.2}$$

K. F. Long, *Stellar Engines*, Springer Aerospace Technology,
https://doi.org/10.1007/978-3-032-18062-9_4

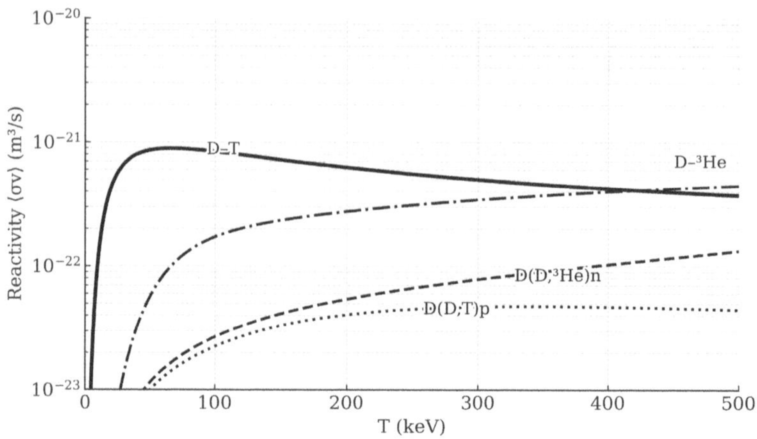

Fig. 4.1 Nuclear reaction rate data

$$< \sigma_c v >_{DD_p} = 2 \times 10^{-14} \frac{1 + 0.00577T^{0.949}}{T^{2/3}} \exp\left(- \frac{19.31}{T^{1/3}} \right), \tag{4.3}$$

$$< \sigma_c v >_{DD_n} = 2.72 \times 10^{-14} \frac{1 + 0.00539T^{0.917}}{T^{2/3}} \exp\left(- \frac{19.80}{T^{1/3}} \right). \tag{4.4}$$

Where they are accurate to 10% in the range 3–100 keV, except for the D^3He which has the same accuracy but in the range 0.5–100 keV. The DT is accurate to 20% in the range 0.3–3 keV. These power law equations are used as one of the options to model fuels to get a more accurate reactivity for a given input temperature within the fuel. The reaction rate rises sharply, peaks, then slowly declines due to the v^{-1} weighting in $< \sigma_c v >$. Deuterium-Tritium reacts at the lowest temperature and is therefore the obvious choice for laboratory experimentation but also the next generation of commercial reactors. The type of fuel defined will be important in calculating the total energy generation possible. The most important parameters here are the mass of fuel m_{cap}, the mass defect α, the specific energy of the fuel E_{fuel} (J/g) and the burn up fraction ϕ, which are related to the energy output by

$$E = \alpha m_{cap} \phi c^2. \tag{4.5}$$

The burn-up fraction is design dependent and involves calculations for the implosion, hot-spot ignition and thermonuclear burn of the fusion fuel. It depends on many factors but is a measure of the efficiency of the implosion engine in converting the hydrodynamic work to useful energy output. It will be a function of how well the laser driver couples to the capsule, the interaction of the materials within the capsule and the geometrical design of the capsule itself. It is a critical parameter

Table 4.1 Specific energy release for various fusion reactions, where charged = energy to charged products useful for direct thrust (no neutrons)

A_r (amu)	Reaction	Products (MeV)	Tot (MeV)	Energy $(J/g)_{total}$ $\times 10^{11}$	Energy $(J/g)_{charged}$ $\times 10^{11}$
2	H+H	$D+e^+ + \nu_e$	1.442	0.451	0.451
5	D+T	^{4}He(3.52) + n^0(14.06)	17.58	3.408	0.684
4	D+D	^{3}He(0.82) + n^0(2.45)	3.27	0.792	0.198
4	D+D	T(1.01) + p^+(3.02)	4.033	0.977	0.243
5	D+^{3}He	^{4}He (3.67)+p(14.67)	18.34	3.556	0.714
6	T+T	^{4}He+2n^0	11.32	1.829	1.829
6	^{3}He+^{3}He	^{4}He+2p^+	12.9	2.077	2.077
6	^{3}He+T	^{4}He+$p^+ + n^0$	12.1	1.953	1.953
6	^{3}He+T	^{4}He+D	14.3	2.312	2.312
7	H+^{6}Li	^{4}He+^{3}He	4.019	0.768	0.768
7	^{6}Li+n	^{4}He+T	4.8	0.874	0.874
12	H+^{11}B	3^{4}He	8.682	0.907	0.907

to determine the overall performance of the capsule in any engine design, since the energy output will also affect the spacecraft mission profile.

The mass defect for the four main fusion reactions are D(T,^{4}He)n, $\alpha = 0.0038$; D(D,T)p, $\alpha = 0.0011$; D(D,^{3}He)n, $\alpha = 0.0009$; and D(^{3}He,^{4}He)p, $\alpha = 0.0039$. The specific energy of different fusion reactions is shown in Table 4.1, to include without neutrons (charged only). When the fuel reactions are assessed the energy output and particle energy deposition are based on the Spitzer stopping power coefficients [5] and Maxwell averaged cross section at a given temperature [6]. The choice of any of these fuels will imply a value for the energy deposition, from all reacting nuclides, including neutrons and protons. The reaction rate dn/dt is dependent upon the reacting fuel adopted and associated reactivity $< \sigma_c v >$ (defined as the probability of reaction per unit time per unit density of target) multiplied by the reactant number densities n_1 and n_2 as

$$\frac{dn}{dt} = n_1 n_2 < \sigma_c v > . \tag{4.6}$$

Although for species that are reacting with itself (i.e. DD) the product is modified so that the reaction rate becomes

$$\frac{dn}{dt} = \frac{1}{2} n^2 < \sigma_c v > . \tag{4.7}$$

The number density $n = \rho_{hs} N_A / M_A$ where M_A is the molar mass of a nuclide. The rate of fusion energy production dE/dt is then given by multiplying the energy output per reaction by the reaction rate and the volume for the reacting zone otherwise known as the hot spot

$$\frac{dE}{dt} = \frac{1}{2}n^2 < \sigma_c v > \frac{4\pi}{3} r_{hs}^3 \alpha m_{hs} c^2. \tag{4.8}$$

Multiplying the instantaneous fusion power per reaction by the burn-up fraction ϕ and mass of the fuel capsule m_{cap} gives an estimate for integrated energy yield

$$E_{yield} = \phi \alpha m_{cap} c^2. \tag{4.9}$$

An actual ICF pellet design will consist of many different material layers, to include the thermonuclear fuel and an ablative shell layer on the outside which facilitates the inertial confinement through material pressure on the imploding gas. However, it is currently an assumption in the equations used for this work that where a capsule mass is defined m_{cap} that this is assumed to be all thermonuclear fuel and any mass due to these additional layers have been neglected—so this will overestimate the yield slightly for a given capsule.

It is assumed that the target is compressed to a smaller volume with a high number density and temperature, particularly within a central hot spot region. The hydrodynamic compression will lead to the generation of mechanical work through a PdV process adding energy to the hot spot. Once the central hot spot region has ignited it will generate ^{4}He alpha particles and then deposit them into the surrounding fuel. This will also heat up the material and add energy. Whilst these energy addition processes are occurring, several energy loss mechanisms will be in effect. This includes the loss of energy due to electron thermal conduction away from the central hot spot, as well as radiation loss through the process of Bremsstrahlung emission. The race for the capsule is to inertially confine the central hot spot sufficiently long enough for the energy deposition (and subsequent Thermonuclear burn wave) to propagate and fully ignite all the fuel, before the energy loss mechanisms and capsule disassembly can detract energy from it.

One of the useful forms of fusion for space propulsion is the utilisation of aneutronic reactions. This is where the neutrons carry $\lesssim 1\%$ of the total energy release, compared to the typical 80% nuclear fraction release associated with conventional fusion reactions being investigated for the next generation of nuclear reactors. Aneutronic fusion is potentially good because it reduces the amount of radiation shielding necessary to mitigate ionising damage and neutron activation. However, the physics requirements to achieve aneutronic fusion are much more demanding.

Daedalus had used D^{3}He fuel since it was aneutronic and would present a reduced radiation environment with little or no neutrons produced. However, Hyde [7] pointed out that self-burn daughter reactions within a D^{3}He fuel will account for around 15% of the reactions with half of these going to produce a neutron [7]. Hilton, Luce and Thompson stated that side reactions produce less than $\sim 1.5\%$ of the total system power in neutrons [8]. Williams argued that the use of D^{3}He offers up to a 95% reduction in high energy neutrons [9]. In earlier work Linlor and

Clauser state that the power represented by the neutron flux from secondary D(D,^{3})n chain is around 2% [10]. This is also discussed by Hilton [11].

To investigate, we examine the D(^{3}He, ^{3}He)p chain, where the ^{3}He gives a total energy of 3.67 MeV and the proton an energy of 14.67 MeV. This is a total energy of 18.3 MeV, but if we assume this only reacts at 95% then it is reduced to 17.38 MeV, where 5% of the Deuterium goes into DD side reactions. There are then two DD chains to consider. The first is the neutron producing chain of D(D,^{3}He)n, where the ^{3}He is 0.82 MeV and the neutron is 2.45 MeV, with a total energy of 3.27 MeV. The second is the proton producing chain of D(D,T)p, where the Tritium is 1.01 MeV and the proton is 3.02 MeV, with a total energy of 4.03 MeV. However, if we assume 2.5% split between the two DD chains. The other half will produce Tritium which then reacts with Deuterium to produce another neutron.

Consider that all the side reaction energy goes into the DD_n chain which produces ^{3}He (0.82 MeV) and n (2.45 MeV) giving rise to a total energy of 3.27 MeV. Therefore, adding this to the primary chain energy of 17.385 MeV leads to a total energy output of 20.655 MeV. This means that the DD_n side reaction chain would have contributed 3.27/20.655 = 15.83% of the total energy to the system.

We can do the same calculation but for the DD_p chain which produces T (1.01 MeV) and p (3.02 MeV) giving a total energy output of 4.03 MeV, so that 4.03 / (4.03 + 17.385) = 18.82%. We might expect around $\sim$2.5% to be split between the DD_n and DD_p chain so that the total energy contribution from each would be 3.27/2 = 1.635 MeV and 4.03/2 = 2.015 MeV respectively. This would lead to energy contributions of 1.635/21.035 = 7.77% and 2.015/21.035 = 9.58% bringing the total possible energy contribution of the two DD side reactions to $\sim$17.35%.

A third level of tertiary reactions may occur, so that the total contribution from side reactions may be as high as $\sim$20%. In any spacecraft design that utilises fusion reactions this contribution should really be considered since the additional energy may contribute to the thrust generation but also produce enhanced radiation via neutrons and x-rays thereby requiring additional shielding. However, in the models presented in this book side reactions are not explicitly included. Instead, based on the arguments of Hyde [12] a scaling factor of $\times$1.15 (15%) will be applied to the neutron flux measurements for the power cycles of Chap. 11.

4.2 Laser Drivers

4.2.1 Experimental Facilities

The essence of the Laser (Light Amplification by Stimulated Emission of Radiation) is that it is distinguished from other radiation sources by the properties of being monochromatic, coherent (spatially and temporally) and of high brightness or intensity. With a low input pulse of duration $\sim$ 1 ms the laser energy is greatly

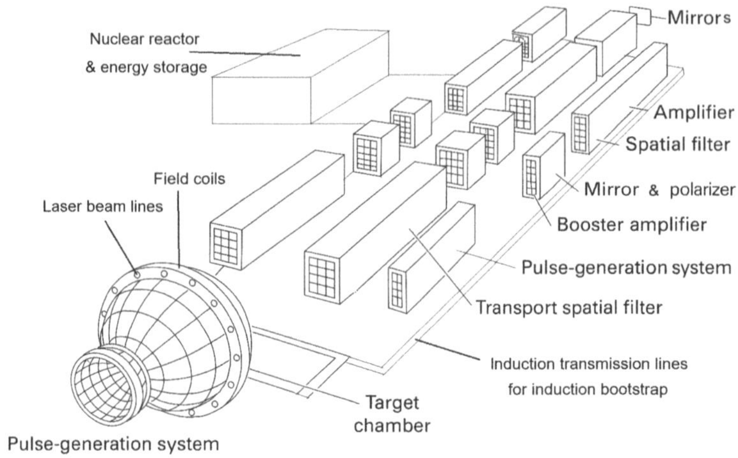

Fig. 4.2 Illustration of laser system behind an open engine reaction chamber[KFL]

amplified by the short release duration ~ 1 ns. This means that the peak power is many orders of magnitude greater than the average input power.

Important properties for the laser driver are the intensity and wavelength, and it is desirable to have a high intensity and shorter wavelength. However, there is a practical physical limit due to mirror damage at high intensity, and the shorter wavelength (higher frequency) reduces inverse-bremsstrahlung absorption length and many parametric instabilities; practical limits then come from optics damage at high fluence. Ideal values for the intensity are around $\sim 10^{14} - 10^{15}$ W/cm^2 and for the wavelength is around 0.3–0.5 μm [13]. At lower wavelength resonance absorption and Raman scattering effects are minimised.

The emitted laser energy can consist of several different transitions. The one mainly used is 1.06 μm light (infrared, 1ω) with either a double 0.5 μm (green visible, 2ω) or triple 0.35 μm (ultraviolet, 3ω) frequency multiplication. The choice of higher or lower frequency is a trade-off between higher energy and laser-plasma interaction effects, where a higher frequency corresponds to shorter wavelength. These trade-offs can also be mitigated by improving the quality of the beam uniformity and minimising hydrodynamic instabilities.

Figure 4.2 shows an illustration of the sort of system envisaged with a fusion engine reaction chamber coupled to a large laser system. A large technology leap must be made before it is possible for the scaling of large laser facilities within a laboratory to be demonstrated on Earth can be applied in space. Whilst in space, envelope size is less constraining, more critical is the mass which influences the launch and staging to Low Earth Orbit.

The most difficult part of a spacecraft design would be the energy drivers since it is not well understood how plausible it would be to shrink a laser system down and apply it within these extreme conditions. However, there are many types of energy drivers that could in principle be applied to such a machine. A more comprehensive discussion is given by others [12, 14]. In a review of inertial fusion Reupke [15] defines the main requirement of a laser driver as "*the ability to deliver nanosecond energy pulses and to sustain high average power, operation at high electrical efficiency, and good target coupling*".

A Carbon Dioxide laser can generate energy at high efficiency perhaps up to $\sim$20% in infrared light but long wavelength (10.6 μm). This can enhance laser-plasma interactions and increase absorption length, making ignition harder. Since they operate at high wavelengths it can be difficult to generate high energy for an implosion. Yet they are apparently sensitive and require regular realignment. A Krypton Fluoride laser will generate light in ultraviolet and is capable of high repetition rate, although the efficiency is limited up to perhaps $\sim$6% and developments are required to increase the beam brightness.

A Free electron laser has the advantage of being tuneable for a broad range of frequencies to include at microwave, infrared, visible, ultraviolet and even x-ray and may achieve efficiencies up to $\sim$20%. However, there are potential issues with the transport of charged particles near strong magnetic fields and the ability to focus the beam which may reduce the implosion pressure of a target. They are mass heavy and there are issues with steering the beam over a large distance.

Heavy Ion Beams have an efficiency up to $\sim$ 30% and involve accelerating a large ion mass to high velocity, so it needs to be long and massive. Light Ion beams have an efficiency up to $\sim$15% but due to its short focusing distance there is a risk of beam divergence from the target. These are accelerator-based drivers.

Perhaps of special interest may be Diode Pumped Solid State Lasers (DPSSL) and this was the laser of choice for the VISTA design. These have a driver efficiency exceeding $\sim$6–12% and may approach $\sim$20% with petawatt-class laser power. They may also be tailored for a high repetition rate of $\sim$10s Hz. They use doped crystals such as Yb^{3+}- doped fluorapatite $S_{r5}(PO_4)_3F$. Light is generated at a wavelength of 1047 nm and can then be frequency shifted to either 300 or 500 nm with minimal laser plasma instabilities. DPSSLs also have the advantage that they tend to be compact. They do not have a flash lamp as the light source and so can reach high laser powers whilst maintaining good beam quality. They are sensitive to temperature however and so may only operate within a narrow range of environmental conditions if stability issues are to be avoided. The wall-plug efficiency depends strongly on the diode efficiency and thermal management.

In this book we assume Nd:glass lasers as the ones for which our civilisation has the most experimental physics validation. Experimental facilities with laser energy and peak power output are shown in Fig. 4.3. These were chosen due to the recent experimental achievement of ignition. They are easy to focus on a target across a broad range of frequencies but also come with significant laser plasma instability issues particularly at high laser power greater than $\sim 10^{15}$ W/cm^2.

Statements by Lindl [16] indicate that around 10 MJ would be required to drive a pellet to ignition. Early laser concepts to explore the physics of laser impingement onto a capsule include the Janus, Cyclops, Argus, Novette and Shiva facilities in the U.S. These experiments eventually lead to the Nova facility which used a Neodymium glass laser and operated between 1984 and 1999. Nova delivered 100 kJ

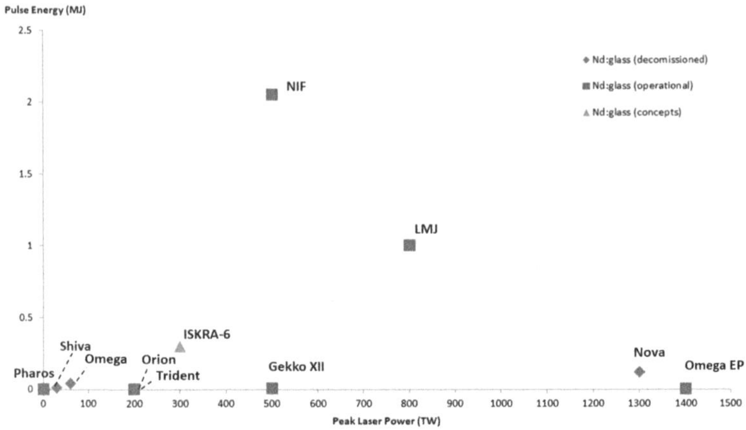

Fig. 4.3 Historical performance data for Nd:glass lasers to include legacy and modern facilities; decommissioned, operational and near-term concepts [KFL]

at 1054 nm or 40–45 kJ at 351 nm. It would deliver around 16 TW of power in ~2–4 ns using 10 separate beam lines for indirect drive.

Another facility in the U.S was Trident which operated between 1988–2017. It was also a Nd: glass laser with a long pulse beam of 527 nm (2ω) operating over 100 ps up to 5 ns. It also had a single long pulse beam at 1045 nm (1ω) operating from 100 ps to 10 μs. Finally, it had a short pulse beam delivering 100 J/200 TW in 600 fs. The intensity of the short pulse beam would be $I_L \sim 10^{20}$ W/cm^2.

Probably the most successful facility to date is the US National Ignition Facility (NIF) [17]. Construction on the project began in 1997 and was completed in 2009. The first experiments began in mid-2009 and the ignition experiments began in late 2010. NIF operates at 1ω light at 1053 nm infrared. Although, if the frequency is tripled to 350 nm wavelength, then it has 2 MJ maximum energy and 500 TW maximum power. It utilises 192 long pulse laser beam lines. This corresponds to around ~10 kJ/beam or a peak power of ~2.6 TW/beam.

In 2021 NIF achieved ~70% of the output energy from fusion reactions [18]. Then in 2022 the facility achieved ignition with a gain of G ~1.5 from an input energy of $E_{in} \sim 2.15$ MJ. The NIF may utilise ~mg capsule masses at $<<1$ Hz pulse frequency as an experimental laboratory. Future facilities which generate inertial fusion energy as a part of a national electrical supply can be expected to operate at 10s Hz but in principle could achieve a gain of $G > 10$.

The Laser Megajoule (LMJ) is in France. Construction began in 1999 and it was operational by late 2014. With 22 beam lines it can produce a peak power of 800 TW using a 1 MJ energy pulse. Although there are plans to upgrade it to a 1.8 MJ ultraviolet facility using 240 beams. LMJ is also an indirect drive scheme.

The Omega laser at the University of Rochester in the US has been operational since 1995 for experiments in both direct and indirect drive mode using 60 beams for a 40 kJ pulse at 60 TW in ~1 ns. Although it has been recently upgraded to include

4 short pulse laser beams with 2.6 kJ delivered to the target in 1–10 ps pulses with PW power levels. For the application to fast ignition experiments it can achieve a laser intensity of $I_L \sim 10^{20}$ W/cm^2.

The UK has the Vulcan laser at the Rutherford Appleton Laboratory which has been operational since 1977. It is an 8 beam (1ω, infrared) Nd:glass laser delivering 100 J to target at 200 TW peak power in a ps pulse. It can generate a laser intensity of $I_L \sim 10^{21}$ W/cm^2. More recently the UK has developed the Orion laser which became operational in 2010 using 10 long pulse beam lines at 351 nm (3ω) to deliver 200 J in ns, in addition to two Chirped Pulse Amplification (CPA) arms at 5 kJ in 0.5 ps (1ω). The target chamber size for Omega is 4 m.

An advanced proposal was the High Power Laser for Energy Research (HiPER) [19–21]. The project planned to use the fast ignition technique to achieve a high gain of $G > 100$. Fast ignition is a method of separating the compression and ignition phases of the ICF capsule implosion using a combination of long pulse and short pulse laser beams. HiPER would have used a combination of a long pulse 250 kJ ns beam at 3ω with a short 70–100 kJ 15 ps pulse (2ω or 3ω).

The Japanese also operate a Nd: glass laser facility at Osaka University called Gekko XII which has been operational since 1983. It has been designed for direct drive experiments. It uses a wavelength of 532 nm in green light (2ω) using 12 beam lines. It uses a 10 kJ/10–20 TW pulse at 1–2 ns in addition to a 0.4 kJ/0.5 PW ultra-short pulse. It is currently being upgraded with an additional laser to deliver 10 kJ in 10 ps for fast ignition experiments.

Prior to the construction of NIF the proposal was for a Laboratory Microfusion Facility (LMF), but this was scaled back by a recommendation from the National Academy of Sciences [22]. The LMF would have been a 10 MJ facility and may have delivered gains of $G \sim$10–100s, depending on the capsule designs.

If a facility that is even half of the capability of the LMF was constructed in the next two decades, this immediately presents the real possibility that spacecraft concepts like the 5 MJ driven VISTA become feasible in principle. That said, as will be discussed later, the claimed gain $G = 1500$ for VISTA was not really justified in any of the reports and appears to require a yet unproven physics performance.

There are exciting times ahead in ICF energy research and it is quite possible that the long-standing ambition of harnessing the power of the Sun in a laboratory on earth will be achieved within years using this method. A key question for all of this is when such technology would reach maturation to the point of full space mission utilisation. In a recent study by this author attempts were made to project the timescales for the development of ICF propulsion, and it was concluded that they would likely occur in the timeframe $\sim$2100-2150 [23]. For some this may sound optimistic, but it is the view of this author that continued technology developments justify these projections. For others this may sound pessimistic, and they may desire such technology much sooner. However, one must consider how long it takes to design new technologies which integrate new physics. Even if the design blueprints for a fusion powered spacecraft were available, it would likely take at least a decade to refine the design to a sub-system level, and then several decades to construct it. In between this all the sub-systems must be independently tested in the appropriate environment. The timescale for development likely exceeds that of say a typical air force jet fighter which may take on the order of two decades.

In this book we mostly focus on inertial fusion using solid state neodymium glass laser facilities and no other facilities such as magnetic fusion [24], although it cannot be overstated how much of a contribution other types of fusion facilities have contributed towards our overall knowledge about fusion physics. We are not making a statement here about the practicality of utilising these laboratory facilities towards the application of an engineering propulsion machine; although we would hope that such a spin-off application may be possible from this or another type of laser technology. The aim is to simply illustrate the technological potential.

We are also not suggesting that the development of propulsion technology of this type depends only on the evolution of lasers. There are other technologies which must be advanced in parallel including magnetic nozzles, efficient power and energy storage systems, cryogenic propellant storage containment systems, materials with thermal and radiation absorption properties. An interstellar probe will also depend on the maturation of artificial intelligence computing capability but this and the other issues are not discussed further here.

The emphasis on Nd:glass lasers is merely an illustrative one and instead we are mostly interested in what happens once the energy arrives, and a capsule has been successfully compressed to a state of thermonuclear ignition. That said, since they are topical and have recently achieved ignition within a laboratory environment, our working assumption is in the adoption of Nd:glass lasers. These systems also have an ease of focussing onto a target at a required power level. Although they are also low in efficiency, high cost and characterised by a low laser-coupling efficiency.

There are differences in the requirements for an ICF driver in a propulsion engine and those intended for an inertial fusion power reactor. For the propulsion system the primary concern is mass reduction of the driver, whilst for the ground reactor this is not an issue, but instead it is maximising the driver efficiency as the primary requirement. Whilst desirable this is not the case for a propulsion system. Whilst a large efficiency η is the issue for the ground power reactor, for the propulsion system it is a large ηT^4 which relates to the mass of any waste heat radiators where it is desirable to dissipate heat at the highest temperature to minimise radiator mass and surface area.

In terms of scaling a NIF-type laser system, there are fundamental requirements that must be considered for the application of advanced space propulsion. This includes good wall plug efficiency, a reasonable burn-up fraction, high capsule gain and high pulse frequency. It may also include a high automated capsule production rate and advanced target fabrication methods at competitive cost. There is a requirement for neutron absorbing materials and radiation resistance materials around the driver. The laser system mass must also be minimised. An ability to alter capsule mass and/or pulse frequency for effective thrust alteration.

There is a need for a pellet injection system that operates so undetonated pellets are protected from the radiation of existing detonations. There is a requirement for an efficient decoupling of the plasma from the magnetic field generated in the exhaust plume to minimise radiation of charged particles onto the spacecraft structure. Smart ICF pellet designs aim for enhanced confinement and maximise energy release and reliable storage systems. The use of a dedicated radiator heat rejection system and/or cooling system for the driver. The ability of optical materials performance to remain stable under extreme thermal environmental conditions. Reliable beam line sub-division system with redundancy to maintain adequate beam smoothness on capsule target. Use of a reaction chamber where its size is a function of the plasma expansion time between detonations. It should be obvious then that post-NIF any next generation experimental facility, or even a commercial nuclear fusion reactor, will be a critical stepping stone towards the maturation of this technology for space, but it will also be a long development path.

Laser Efficiency

Lasers are defined by the input energy E_{in} (MJ), frequency conversion efficiency ϵ_1 and pulse duration τ_p. The peak power P_{peak} (TW) is then given by

$$P_{peak} = \frac{\epsilon_1 E_{in}}{\tau_p}. \tag{4.10}$$

Then with a set number of individual beam lines N_{beam} the energy per beam E_{beam} (kJ) and power per beam P_{beam} (TW) is given by

$$E_{beam} = \frac{\epsilon_1 E_{in}}{N_{beam}}, \tag{4.11}$$

$$P_{beam} = \frac{E_{beam}}{\tau_p}. \tag{4.12}$$

Then for a pellet outer radius R_{pell} with surface area $A_{pell} = 4\pi R_{pell}^2$, the area of an individual beam line spot size A_{spot} (the area illuminated by a single laser beamline) assuming 100% coverage is given by

$$A_{spot} = \frac{A_{pell}}{N_{beam}}. \tag{4.13}$$

Then with the radius of the spot size $R_{spot} = (A_{spot}/\pi)^{1/2}$ the laser intensity I_L (W/cm^2) for a direct drive system with multiple beam lines is given by

$$I_L = \frac{P_{beam}}{A_{spot}}. \tag{4.14}$$

However, in the case of indirect drive the individual laser beams do not impinge directly onto the surface of the pellet, but instead they irradiate the internal surface of the Hohlraum cavity which then creates a radiation bath to surround the pellet. In which case we can represent the total power P_{tot} to the pellet as

$$P_{tot} = \frac{\epsilon_1 E_{in}}{\tau_p}. \tag{4.15}$$

It is a convention in ICF not to quote a laser intensity for indirect drive shots since there are no beam spots directly impinging onto the pellet surface. However, for the discussion of this book we will quote what we might refer to as an effective laser intensity which for indirect drive would be given by

$$I_L = \frac{P_{tot}}{A_{pell}}. \tag{4.16}$$

Alternatively, for either we can express the effective intensity requirement as

$$I_L = \frac{\epsilon_1 E_{in}}{4\pi \tau_p R_{pell}^2}. \tag{4.17}$$

The angular frequency of a laser pulse is given as a function of the wavelength

$$\omega_L = \frac{2\pi c}{\lambda}. \tag{4.18}$$

The laser intensity can also be related to the electric field amplitude E_o via

$$I_L(W/cm^2) = c\rho = \frac{c\epsilon_o \mid E_o^2 \mid}{2}. \tag{4.19}$$

Where ρ is the energy density (J/cm^3). Alternatively, the Amplitude can be described in appropriate units [13] as follows

$$E_o(V/cm) = 2.75 \times 10^9 \left(\frac{I_L}{10^{16}\ W/cm^2}\right)^{1/2}. \tag{4.20}$$

Critical to an analysis of ICF pellet performance is the total energy budget and losses as illustrated in Fig. 4.4. This begins with the wall plug electrical energy with a loss of ϵ_1 = 99.32%, so that for NIF a E_{wp} = 300 MJ supply becomes 2.05 MJ. Then there is the frequency conversion loss from 1ω to 3ω which for NIF is around ϵ_2 = 50%. For an indirect drive target there is a loss to the Hohlraum on the order of ϵ_3 ~10-15%, followed by loss due to the energy absorption within the pellet outer layers of order ϵ_4 ~10-15%. Finally, there is energy loss due to the capsule implosion physics on the order of ϵ_5 ~90-99%. Collectively this leads to a deposited energy within the fuel given by these losses

$$E_{dep} = E_{wp} \times \Big[(1-\epsilon_1)(1-\epsilon_2)(1-\epsilon_3)(1-\epsilon_4)(1-\epsilon_5)\Big]. \tag{4.21}$$

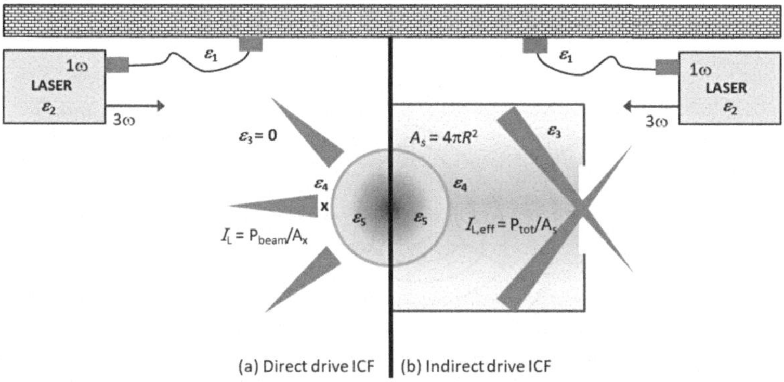

Fig. 4.4 Illustration of laser driver energy efficiencies for direct drive and indirect drive hot-spot ignition scenarios[KFL]

Although this is for indirect drive where Hohlraum losses are present. In the case of direct drive this equation reduces to the following form

$$E_{dep} = E_{wp} \times \Big[(1-\epsilon_1)(1-\epsilon_2)(1-\epsilon_4)(1-\epsilon_5)\Big]. \tag{4.22}$$

4.3 Laser-Plasma Interactions

The interaction of a laser beam with any solid matter leads to the generation of an ionised state described as a plasma; quasi-neutral gas of charged particles which exhibit a collective behaviour. It differs from a gas in that a gas will be thermalised to an equilibrium state by direct particle collisions, whereas for a plasma this is enabled by Coulomb collisions of the charges.

Laser-plasma interactions are described by an electromagnetic wave (the laser) propagating through a material medium (the plasma). When a laser arrives at a pellet it will ablate the surface and create a coronal plasma layer. Most of the energy deposition occurs just below the critical surface and the reflection peaks at the surface. Thereafter, the energy must be absorbed into the electrons, by inverse Bremsstrahlung absorption (collisional), resonance absorption (collisionless) and parametric instabilities which is the excitation of plasma and electromagnetic waves which correspond to stimulated scattering (Brillouin, Raman) and decay instabilities. Figure 4.5 illustrates the various laser-plasma interaction mechanisms.

In physics, wave phenomena are described by dispersion relations. These are functions which describe a particle's position at any time $y = f(x,t)$. The wave velocity can be described as a function of the wavelength λ and frequency f. The dispersion relation for electromagnetic waves can be found by solving the wave equation in a vacuum. Using the Maxwell's equations, we begin with Gauss's Law

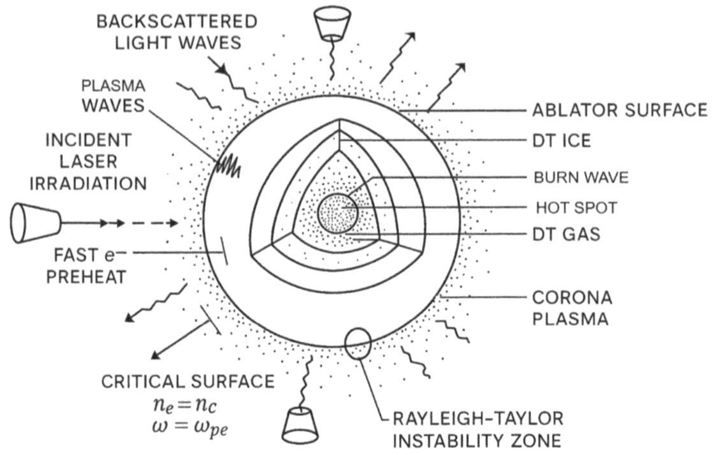

Fig. 4.5 Illustration of ICF laser-plasma interactions[KFL]

for the divergence of the electric field E and magnetic field B as follows

$$\nabla \cdot E = 0, \tag{4.23}$$

$$\nabla \cdot B = 0. \tag{4.24}$$

Then Faraday's law of induction and Amperes current law are defined by

$$\nabla \times E = -\frac{\partial B}{\partial t}, \tag{4.25}$$

$$\nabla \times B = \frac{1}{c^2}\frac{\partial E}{\partial t}. \tag{4.26}$$

The phase velocity is equal to the wave speed or the speed of light in free space $v_{ph} = c = 1/\sqrt{\epsilon_o \mu_o}$, and μ_o is the vacuum permeability or magnetic constant and ϵ_o is the permittivity of free space. In these equations the magnetic field B and electric field E will have the units of tesla (T) and V/m respectively. Taking the curl of the electric field and substituting $\nabla \times B$ it is then found

$$\nabla \times \nabla \times E = -\nabla \times \frac{\partial B}{\partial t}, \tag{4.27}$$

$$\nabla \times \nabla \times E = -\frac{1}{c^2}\frac{\partial^2 E}{\partial t^2}. \tag{4.28}$$

Then using the vector identity $\nabla \times (\nabla \times E) = \nabla(\nabla \cdot E) - \nabla^2 E$, and using $\nabla(\nabla \cdot E) = 0$ in a vacuum—the vacuum Poisson equation. This gives the wave equation for an electromagnetic wave propagating at the speed of light.

$$\nabla^2 E_y \equiv \frac{\partial^2 E_y}{\partial x^2} = \frac{1}{c^2}\frac{\partial^2 E_y}{\partial t^2}. \tag{4.29}$$

This has a plane wave solution of the form

$$E_y = E_o e^{i(\omega t - kx)} = E_o \sin(\omega t - kx). \tag{4.30}$$

That is to say that it satisfies the differential equation. In this equation E_o is the complex amplitude measured in V/m which represents the longest period in the wave under study, and it is the coefficient of the plane wave solution. The wave will propagate sinusoidally with respect to position x and time t. The phase angle is given by $\phi = \omega t - kx$. The phase velocity of the wave is given by

$$v_\phi = \frac{dx}{dt} = \frac{\omega}{k}. \tag{4.31}$$

The interaction of an electromagnetic wave with a plasma should then satisfy the following dispersion relation which is expressed as a function of the wave number $k = 2\pi/\lambda$ and angular frequency $\omega = 2\pi f = 2\pi c/\lambda$ so that the dispersion relation for a plane light wave in a vacuum is

$$\omega^2 = k^2 c^2. \tag{4.32}$$

Alternatively, in a non-vacuum the plasma will have a natural frequency of oscillation of the non-propagating charge density in a conducting medium given by the electrons oscillating about the ions at a certain frequency.

$$\omega^2 = \omega_{pe}^2 + c^2 k^2. \tag{4.33}$$

Where the electron plasma frequency ω_{pe} (rad/s) is given by

$$\omega_{pe} = \sqrt{\frac{n_e e^2}{\epsilon_o m_e}} \approx 56.349\sqrt{n_e (m^{-3})}. \tag{4.34}$$

Where m_e is the electron mass, n_e is the electron number density and e is the electron charge. Any plasma where the charge density oscillates at this frequency is known as a 'cold plasma' and the ions can be considered as a fixed background or with infinite mass. At some point the frequency of the laser ω will become equal to the frequency of the oscillating electrons ω_{pe} within the plasma and this will lead to reflection and scatter of the laser beam. This creates a problem for the propagation of an electromagnetic wave since the implosion of an ICF capsule will depend on the pressure profile generated. In a plasma when the density is underdense $n_e <$

n_c, propagation occurs and when overdense $n_e > n_c$ there is reflection, where the critical density is given by

$$n_c = \frac{\epsilon_o m_e \omega^2}{e^2} \approx 1.115 \times 10^{21} cm^{-3} \left(\frac{1 \mu m}{\lambda}\right)^2. \tag{4.35}$$

An electromagnetic wave will only propagate through a cold plasma provided it has a frequency greater than the electron plasma frequency, i.e. $\omega^2 \geq \omega_{pe}$. If the laser frequency is less then it will not propagate. From the EM dispersion relation given above reflection occurs at $\omega = \omega_{pe}$. Within the context of a laser beam interacting with the surface of an ICF pellet this translates from a situation of energy propagation to one of energy absorption and reflection.

We also need to know about electron plasma waves which are propagating plasma oscillations due to the thermal motion of the electrons, and these are described by electrostatic waves. Essentially, where electrons are oscillating whilst streaming into an adjacent region the average of these propagating oscillations is a wave of charge. This results in a modified dispersion relation, from the electron plasma frequency. It is an electrostatic wave but without a transverse magnetic component. It is a longitudinal electric wave only and the thermal velocity is an addition of a thermal motion term known as the Bohm-Gross dispersion relation

$$\omega^2 = \omega_{pe}^2 + 3k^2 v_{th}^2. \tag{4.36}$$

Where v_t is the thermal velocity due to the motion of electrons and the pressure acts as a restoring force on deviations from charge neutrality. To propagate an electrostatic wave requires a temperature $T > 0$. The thermal velocity is given by

$$v_{th} = \left(\frac{k_B T_e}{m_e}\right)^{1/2}. \tag{4.37}$$

Meanwhile, the heat from this process will cause mass ablation of the pellet and the inwards movement of the sphere towards a state of spherical convergence and fuel compression by the mechanism of momentum transport. Provided this state is maintained for a continued period, the fuel will be confined by the material above it and this will lead to the thermonuclear ignition of the fuel via the hydrogen nuclei overcoming their own Coulomb force.

The creation of a plasma cloud (corona) around the ICF pellet due to the interaction of the hot laser source sets the conditions for a reduction in the efficiency of the implosion engine. This is because any further laser light must propagate through an ionised layer of ions and electrons and the physics processes at work in that laser-plasma interaction are not all for the benefit of the shell collapse.

As an example of a laser interacting with the plasma of an ICF pellet surface, a calculation was conducted using an open source 1-dimensional Particle-In-Cell (PIC) code [25], where charged particles are represented by pseudo particles and

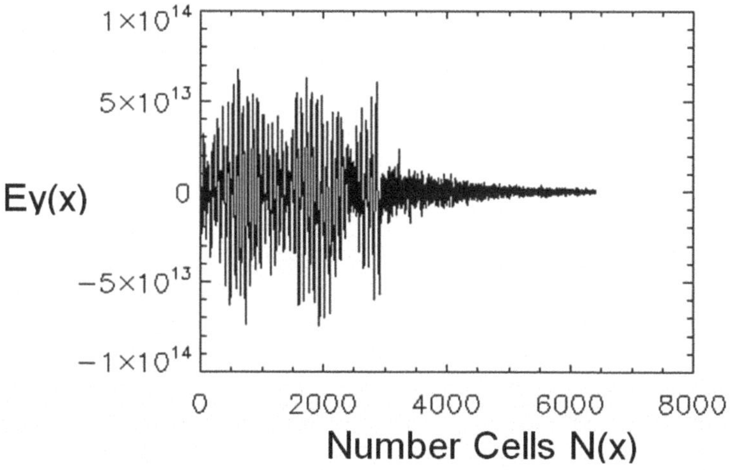

Fig. 4.6 PIC code calculation of laser-plasma interaction with a uniform plasma of ions and electrons. Where $I_L = 9.99 \times 10^{14}$ W/cm^2 and $\lambda_o = 350$ nm[KFL]

solves the Maxwell's equations for the electric and magnetic field on a finite difference domain to represent a uniform plasma. The particles in the simulation are pushed around via the Lorentz force.

Efforts were also made to ensure that the numerical resolution was consistent with the Nyquist sampling theorem. The numerical results are expected to diverge away from the analytical at some value of wave number k consistent with deviations from the Nyquist sampling theorem. This states that the shortest frequency (or wavelength) that can be sampled is given by $2\times$ the sampling interval, which will be dt for obtaining frequency (or dx for obtaining wave number). The Nyquist frequency (and wave number) will depend upon the cell size used in the problem and thereby the resolution for a given number of wavelengths.

As shown in Fig. 4.6 a left-hand boundary condition was imposed as a laser intensity within the domain. Although the material medium was modelled as planar with no density ramp, it was assumed this was representative of a section of a spherical shell of material that might surround a fusion fuel as is typical in ICF ignition physics. The relevant equations are obtained by combining the Maxwell equation for the curl of the electric field with that for a plane wave solution

$$\frac{\partial E_y}{\partial x} = -\frac{\partial B}{\partial t}, \tag{4.38}$$

$$E_y = E_o \sin(\omega t - kx). \tag{4.39}$$

Then integrating one is left with an equation for the magnetic field component as well as the electric field at time $t = 0$ to represent the initial laser condition

$$B_z = \frac{E_o}{\omega} k \sin(kx), \tag{4.40}$$

$$E_y = E_o \sin(kx). \tag{4.41}$$

This was for a wavelength of $\lambda_o = 0.35 \times 10^{-6}$ m, a domain size of $x = 1 \times 10^{-3}$ m, number of cells $N_{cells} = 6400$ cells, with 32,000 particles of species electrons and species ions, so a total of 64,000 particles in a run time of 10^{-12} s. The right boundary condition describes the plasma outflow as a distribution function. The laser was input as a left boundary condition and the pulse profile as a representative Gaussian line shape as a single laser pulse. The laser intensity was $I_L = 9.99 \times 10^{14}$ W/cm^2 corresponding to an electric field amplitude of $E_o = 8.679 \times 10^{10}$ V/m. This had an angular frequency of $\omega_L = 5.382 \times 10^{15}$ s^{-1} and this was verified by fast fourier transform analysis by then taking the FFT of $\boldsymbol{E}y(x, t)$ to produce $\boldsymbol{E}y(k, \omega)$. Since the frequency of the laser is greater than the electron plasma frequency, $\omega \geq \omega_{pe}$ (or $n_e < n_c$), the laser will continue to propagate into the plasma of the ICF capsule surface.

In this chapter we introduced the technology of laser drivers and the interaction with a plasma. It is necessary to have some understanding of this subject since in later chapters we apply this technology to spacecraft propulsion systems.

Problems of Chap. 4

4.1 Look up the reactants and products for a proton-Boron-11 fusion reaction. (i) Estimate the value of α for this reaction and therefore the maximum likely limit for the exhaust velocity. How does it compare to the DT, DD and D^{3}He reactions?

4.2 An x-ray laser with a wavelength of 1 nm is used to cause the break-down of the quantum electrodynamic vacuum so that electron and positron pairs can be used for the purpose of thrust in a novel type of electric engine. The Schwinger critical electric field amplitude for pair production is given by $E_s = m_e^2 c^3 / e\hbar \approx 1.3 \times 10^{18}$ V/m. (i) Calculate the angular frequency and normal frequency (ii) Calculate the laser intensity and power required to achieve this?

4.3 For a diffuse plasma with density $n_e = 10^7$ cm^{-3} (i) Calculate the angular electron plasma frequency ω_{pe}. Convert to an ordinary frequency and express in Hz. (ii) What wavelength of a laser would be reflected from this plasma density?

References

1. Dewar J, Bussard R (2009) The nuclear rocket, making our planet green, peaceful and prosperous. Apogee Books, Burlington
2. Hively LM (1977) Convenient computational forms for maxwellian reactivities. Nucl Fusion 17(4):873
3. Hively LM (1983) A simple computational form for maxwellian reactivities. Nucl Technol Fusion 3(2):199–200
4. Atzeni S, Meyer-ter-Vehn J (2004) Nuclear fusion reactions. Chapter in the physics of inertial fusion, international series of monographs on physics, vol 125. Oxford Science Publications, Oxford
5. Spitzer L (2006) Physics of fully ionized gases. Dover Publication, Garden City. Originally published 1956
6. Rand McNally Jr J, Rothe KE, Sharp RD (1979) Fusion reactivity graphs and tables for charged particle reactions. Oak Ridge National Laboratory, ORNL/TM-6914
7. Hyde RA (1983) A laser fusion rocket for interplanetary propulsion, UCRL-88857. Presented 34th IAC, IAF-83-396. University of California, Livermore
8. Hilton JL, Luce JD, Thompson AS (1964) Hypothetical fusion propulsion rocket vehicle. J Spacecraft 1:276-282
9. Williams CH, Borowski SK (1995/1996) Commercially driven human interplanetary propulsion systems: rationale, concept, technology and performance requirements. 13th symposium on space nuclear power and propulsion, Albuquerque, New Mexico, 7–11 January
10. Linlor WI, Clauser MU (1962) Fusion plasma propulsion system. Hughes Research Laboratories, Malibu
11. Hilton JL (1963) Plasma and engineering parameters for a fusion powered rocket. Aerojet-General Nucleonics, San Ramon
12. Hyde R, Wood L, Nuckolls J (1972) Prospects for rocket propulsion with laser-induced fusion microexplosions. AIAA, 72-1063, presented at the 8th propulsion joint specialist conference, 29 November–01 December
13. Pfalzner SP (2006) An introduction to inertial confinement fusion. Taylor & Francis, Milton Park
14. Orth CD (2003) VISTA: a vehicle for interplanetary space transport application powered by inertial confinement fusion. LLNL, UCRL-TR-110500, 16 May
15. Reupke WA (1985) Inertial fusion systems studies and nuclear pulse propulsion. J Br Interplanetary Soc 38(11):483-493
16. Lindl JD (1995) Development of the indirect-drive approach to inertial confinement fusion and the target physics basis for ignition and gain. Phys Plasmas 2(11):3933
17. Dewald E (2016) Status of indirect drive ICF experiments on the national ignition facility. LLNL, TR-686482
18. Zylstra AB, Hurricane OA, Callahan DA et al (2022) Burning plasma achieved in inertial fusion. Nature 601(7894):542–548
19. Dunne M (2006) A high-power laser fusion facility for Europe. Nature Phys 2(2–5)
20. Atzeni S (2008) Fast ignitor target studies for the HiPER project. Phys Plasmas 15:056311
21. Garrec BJL, Hernandez-Gomez C, Winstone T, Collier J (2010) HiPER laser architecture principles. J Phys 244:032020
22. Hackerman N, Bickel PJ, Carrier GF et al (1990) Review of the department of energy's inertial confinement fusion program. Final report, committee on physical sciences, mathematics and applications, national research council. National Academy Press.
23. Long KF (2022) On the application of pulse propulsion frequency in inertial fusion space mission design. J Br Interplanetary Soc 75(4):127–141

24. Litaudon X, Abduallev S, Abhangi M, Abreu P, Afzal M, Aggarwal KM, Ahlgren T, Ahn JH, Aho-Mantila L, Aiba N (2017) Overview of the JET results in support to ITER. Nucl Fusion 57(10)
25. Brady C (2014) Epoch, An open source PIC code developed for high energy density physics. University of Warwick, Coventry

Chapter 5
Inertial Confinement Fusion

> *It thus appears likely, that within the next 20–30 years, we can achieve a transport capability which allows us complete and rapid transit throughout interplanetary space.*
>
> *—Roderick A. Hyde [1]*

5.1 Methods of Ignition

In 1955 John Lawson [2] showed the minimum physical conditions necessary for a fusion reaction to attain a state known as ignition. This occurs when the heating of the plasma products due to alpha particle deposition is sufficient to maintain the temperature of the plasma against any other losses (i.e. radiation losses such as bremsstrahlung), without any further energy inputs. It was shown that there are three fundamental components to achieving ignition (1) the energy confinement time τ in units of s (2) the particle number density of particle species n in units of m^{-3} (3) plasma temperature T in units of keV.

The basic problem is that when a fuel is compressed it will tend to heat up and expand under its own internal pressure, and this is significantly enhanced if the heating source is due to fusion reactions. Yet, it is required to inertially confine the capsule for a minimum time duration to allow the fusion reactions to take place at a sufficient rate of production. A star can achieve this because it uses its enormous gravity to confine the plasma but in ICF we must use the inertia of the material itself to provide the containment so that it can be hot enough, for long enough at high compression ρr. It must be sufficiently hot since the Coulomb barrier of the positively charged protons will not allow them to fuse at lower temperature.

Lawson considered a hot plasma with a Gaussian velocity distribution function such that the energy release per unit volume is

$$P_e = fE = n_1 n_2 < \sigma_c v > E_{ch}. \tag{5.1}$$

Where n_1 and n_2 is the species number density, σ_c is the reaction cross section as a function of temperature, v is the average velocity of the particles and E_{ch} is

K. F. Long, *Stellar Engines*, Springer Aerospace Technology,
https://doi.org/10.1007/978-3-032-18062-9_5

the energy output per fusion reaction for the charged particle products which at 3.52 MeV (for a DT reaction) amounts to a specific energy release of 3.3×10^{11} J/g. Since the reaction cross section for DT is two orders of magnitude larger than for the other fusion reactions it is the most accessible and so is typically the one considered for commercial reactors.

Following on from the pioneering work of Lawson we derive the minimum 'two-product' requirement by considering an equi-molar mixture of DT fuel where the ion density is assumed equal to the electron density. It is known as the Lawson Criteria and we neglect losses due to conduction but assumed that any energy losses per unit volume from the "hot cloud" were due to light radiation, meaning electromagnetic emissions (i.e. Bremsstrahlung radiation), approximated by

$$P_{loss} = 1.4 \times 10^{-34} n^2 T^{1/2} \text{ W/cm}^3. \tag{5.2}$$

Then equating the energy release and energy loss term, we get an expression for the plasma temperature

$$T = \left(\frac{< \sigma_c v > E_{ch}}{1.4 \times 10^{-34}} \right)^2. \tag{5.3}$$

Lawson showed that for a DT plasma $T \sim 2.6$ keV (~ 30 million K), where E_{ch} is the energy of the charged particle products in a reaction minus any neutrons which don't contribute to heating the plasma. For the DT reaction this would be E_{ch} = 3.52 MeV (^{4}He), for the D^{3}He reaction E_{ch} = 18.34 MeV (^{4}He + p), for the DD$_n$ reaction E_{ch} = 0.82 MeV (^{3}He), for the DD$_p$ reaction E_{ch} = 4.03 MeV (T + p). The average energy per particle for a plasma is given by

$$E = \frac{3}{2} k_B T. \tag{5.4}$$

Lawson then defined the combined energy density or the energy that is stored in a plasma per unit volume by

$$P_e = (n_e + n_i) \frac{3}{2} k_B T \sim 3 n k_B T. \tag{5.5}$$

Where $n = n_i + n_e$ is the number density of ions and electrons combined, the factor of 2 cancels and k_B is the Boltzmann constant. If the reaction rate for the mixture is defined as 50:50 D:T, the heating per unit volume per unit time or rate of fusion reactions is

$$f = n_D n_T < \sigma_c v >= 1/4 n^2 < \sigma_c v > . \tag{5.6}$$

The volume rate of heating is then given from the Lawson criteria requiring that the fusion heating exceeds the losses

$$f \times E_{ch} \geq P_{loss}, \tag{5.7}$$

$$\frac{1}{4}n^2 < \sigma_c v > E_{ch} \geq P_{loss}. \tag{5.8}$$

The inertial confinement time is measured as the ratio of the energy density per unit volume generated over the power loss density per unit volume

$$\tau_c = \frac{P_e}{P_{loss}}, \tag{5.9}$$

$$P_{loss} = \frac{P_e}{\tau_c} = \frac{3nk_BT}{\tau_c}, \tag{5.10}$$

$$\frac{1}{4}n^2 < \sigma_c v > E_{ch} \geq \frac{3nk_BT}{\tau_c}. \tag{5.11}$$

Therefore

$$n\tau_c \geq \frac{12k_BT}{< \sigma_c v > E_{ch}}. \tag{5.12}$$

For a DT plasma $E_{ch} = 3.52\,\text{MeV}$ and a 2.6 keV temperature then $n\tau_c \geq 2.5 \times 10^{20}\,\text{s/m}^3$. For a temperature of $\sim$5 keV, $n\tau_c \geq 1.2 \times 10^{20}\,\text{s/m}^3$ and for $\sim$10 keV, $n\tau_c \geq 6.4 \times 10^{19}\,\text{s/m}^3$. Multiplying the equation by T we can also write the Lawson criteriaLawson Criteria in terms of a fusion triple product which includes the temperature

$$n\tau_c T \geq \frac{12k_BT^2}{< \sigma_c v > E_{ch}}. \tag{5.13}$$

For the temperature of 2.6 keV for example with DT, this gives the minimum conditions of $n\tau_c T \geq 6.4 \times 10^{20}\,\text{s keV/m}^3$. The reaction cross section for all the reactions is lowered if the plasma is reacted at a higher temperature. For DT the minimum is typically in the $\sim$10–20 keV range and the reactivity continues to rise beyond that. For DD around $\sim$ 50 keV and for D^3He it is at around $\sim$ 60 keV. Ultimately, for ICF ignition it boils down to a competition between the rate of energy production through fusion reactions and the energy losses.

When considering fusion research in commercial industry this is usually associated with magnetic confinement fusion (MCF), which utilises a low number density ($n \sim 10^{14}\,\text{cm}^{-3}$) of fusion particles at high temperature ($T \sim$10 keV) for a relatively long ($\tau_c \sim$s) confinement time. MCF research has made substantial progress with the attainment of 16 MW of fusion power using DT fuel on the Joint European Torus

(JET) [3] and led to the design of the International Experimental Thermonuclear Reactor (ITER) which should become operational in 2025 and aims to achieve 500 MW of fusion power for 400–600 seconds. JET began operation in 1983 and attained a gain of $G = 0.7$. ITER aims to achieve a gain of $G > 10$. In contrast, in this book we have focussed on the ICF approach to igniting fusion fuels, which requires the use of high-power intensity lasers.

For ICF to be effective, this requires a high plasma density ($\sim 10^{23}\,\text{cm}^{-3}$), in which a small sub-gram capsule of low mass isotopes (typically Hydrogen, Deuterium, Tritium and Helium) is compressed to extremely high density (ρ/ρ_o $\sim$1000s) and temperature (T $\sim$10s keV), whilst being inertially confined in that state for a sufficiently short period (τ_c $\sim$ns) so that fusion ignition can be initiated within a hot spot region, then propagating outwards as a thermonuclear burn wave.

The confinement is complicated by the interaction of the laser beam where mass is initially lost to the surface via ablation and sets up a coronal plasma layer limiting penetration to a critical density surface, and instead it is absorbed into the electrons, mainly by collisional inverse Bremsstrahlung, collisionless resonance absorption and parametric instabilities. The electrons transport the energy inwards to transfer their energy into the thermal energy of the compression. Convergent shocks with an implosion velocity of $\sim 10^7$cm/s then compress the fusion fuel up to 100 times that of lead producing a high temperature $\sim 10^8$ K hot spot at the centre of the capsule so the fusion reactions can begin depositing ^{4}He alpha particles into the surrounding high density but cooler fuel, defining the point of ignition. Provided this burn-up occurs faster than the time for the capsule to expand under its own pressure, then 10s MJ's of energy are liberated in principle.

The objective is to achieve energy gain; i.e. more energy would be released from the fusion reaction than the driver energy required to achieve the compression and burn. To achieve this in a spacecraft is no small challenge. The spacecraft must have the apparatus of a driver beam (i.e. laser) and some way of accelerating the pellets into the target chambers [4]. Energy driver beams would then confine each pellet for a fraction of a second until ignition conditions were achieved.

In ICF it is more useful to express an areal density ρr condition for the central fuel after compression where for a 10 keV plasma there is a breakeven requirement that $\rho r > 0.1\,\text{g/cm}^2$ where breakeven refers to the energy of the fuel being equal to the energy that is released, and this would correspond to a burn-up fraction of around 1.6%. Any losses beyond this would be less than breakeven. However, in order to produce more energy than went into the fuel, account must be taken of the inefficiencies in the laser driver and so this immediately suggests a higher ρr requirement and as stated earlier for a DT plasma it is desirable to have the condition $\rho r > 3\,\text{g/cm}^2$ for a 33% burn-up fraction.

We can estimate the minimum energy required to heat up a fuel to a given temperature by multiplying the energy stored per unit volume by the quantity of hot spot within the capsule. The minimum heat energy for the hot spot is then given by the number of particles, which for a thermalized plasma it is the convention to

Table 5.1 Estimates of minimum energy for heating the hot-spot mass of a 1 mg ICF fuel capsule based on average thermal energy at different temperatures

m_{hs} (%)	T = 1 keV (MJ)	T = 5 keV (MJ)	T = 10 keV (MJ)
1.0	0.0011	0.0058	0.0116
3.0	0.0035	0.0174	0.0347
5.0	0.0058	0.0289	0.0579
10.0	0.0116	0.0579	0.1158
100.0	0.1158	0.5790	1.1581

count both the ions and electrons

$$E_{heat} = \frac{3}{2} N_{particles} k_B T m_{hs}. \tag{5.14}$$

Where the number of particles assumes the number of ions and electrons are equal so that $N_i + N_e = N_i + N_i = 2N_i$, therefore $N_i = m_{hs}/m_i$. The mass of an ion depends on the fuel but for DT will be multiplied by the average atomic mass $(2+3)/2) = 2.5$ amu multiplied by the average of the proton and neutron rest mass $m_p = 1.66 \times 10^{-27}$ kg. So the minimum energy becomes

$$E_{heat} = \frac{3}{2} \frac{m_{hs}}{m_i} k_B T m_{hs}. \tag{5.15}$$

In these equations the temperature T(keV) must be converted $1eV/k_B = (1.602 \times 10^{-19})$ J/ $(1.3806 \times 10^{-23}$J/K). As an example, we have a DT fuel of mass 1 mg and we want to estimate how much energy it will take to heat 3% (0.03 mg) to a temperature of 10 keV. Calculating this gives a minimum energy of E ~0.035 MJ. The results of estimating this for other hot spot masses at different temperatures are shown in Table 5.1.

What if we wanted to heat the entire fuel. Then calculating this for our problem we find the energy required to heat the entire fuel to 10 keV is E ~1.15 MJ. Allowing for losses this suggests we need around > 1 MJ, and so it is easy to see why laser facilities have focussed on ~MJ class experiments as what is required. Although the actual amount of losses apparent in such facilities is a lot greater than perhaps was envisaged and is one of the reasons why fusion ignition in a laboratory has taken so long. Although the assumed ion + electron degrees of freedom used means that this is a minimum and does not take account the further energy requirements for compression, conduction and radiation and so the energy required is going to be larger than this minimum value.

The driver energy is likely going to be inefficient and so this suggests an even higher energy input is required, but this is also made up from the compression process and the propagation of the alpha particles away from the hot spot which rapidly heats up the surrounding fuel. This partly explains the ~2 MJ shots fired at facilities like NIF where ignition has been achieved. Such targets will require a main fuel $\rho = 10^3$ g/cm^3 and $\rho r = 1$ g/cm^2, whereas in the hot-spot $\rho = 10^2$ g/cm^3, ρr

= 0.3 g/cm^2 for a minimum temperature of $T \sim$ 5–10 keV in a DT fuel. This would hopefully then lead to an energy output on the order of 10s MJ.

As has been presented the ignition of a pellet to a state where thermonuclear fusion can occur is non-trivial and involves many stages. This is further complicated in the context of a space propulsion engine where many pellets must be imploded at a set repetition rate (also called pulse frequency). The process of physics demonstration towards demonstrating 'proof of concept' is typically measured by three achievements which is the attainment of (1) breakeven (2) ignition (3) gain.

The gain process is interesting because it really depends on how it is defined. Physicists have come up with several ways of defining gain in the context of lasers for example (i) the fusion energy release as a function of the electrical grid (wall plug) energy release (ii) the fusion energy release as a function of the laser input energy release to the target (iii) the fusion energy release as a function of the capsule absorption energy. In this book, as a definition of ignition we will generally be consistent with definition (iii) which is really the ratio between the thermonuclear energy release and the laser energy absorbed into the fuel capsule.

One of the first authors to propose using laser beams to compress matter for fusion ignition was by Nuckolls [5], where it was estimated at $\sim$1 kJ energy requirement for DT ignition, although $\sim$ 1 MJ would be required for high gain. He estimated for a laser intensity of 10^{17} W/cm^2 implosion velocities of $10^7 - 10^8$ cm/s and pressures of 10^{12} atmospheres would be possible, compressing a fuel by $10,000$ times. He identified symmetric implosion pressures and the control of hydrodynamic instabilities to be critical. Provided these factors were mitigated, a fusion energy output of $10^7 - 10^8$ J was credible for commercial power production, attaining $\sim$1 GW of electric power burning 100 pellets per second in 10 chambers. Fortunately, we have since invented several methods by which this may be achieved.

According to conventional direct drive ICF theory, the burn-fraction ϕ of a DT target is given by a relation that is dependent upon the areal density ρR, where ρ is the fuel density and R is the external fuel radius. The quantities are related by

$$\phi = \frac{\rho R}{\Phi_i + \rho R}. \tag{5.16}$$

Here Φ_i is the specific column density and is fuel dependent, but for DT it is $\sim$6–10 g/cm^2. Standard theory predicts that an areal density of $>$ 3 g/cm^2 [6] is required to obtain a state of ignition, defined to be when alpha particle deposition from self-sustaining fusion reactions begins to compensate for losses. This would suggest a burn-up fraction of $\phi \sim 0.33$ (33%) is required to obtain ignition. However, the value is dependent upon the types of reactants being fused and therefore the ignition temperature.

Direct Drive ICF Method

Typically, an ICF pellet is a DT fuel surrounded by an ablator shell with a total outer diameter on the order of $\sim$1 mm. The confinement of a fusion pellet is complicated by the interaction of the laser beam and the implosion process. Mass is initially lost

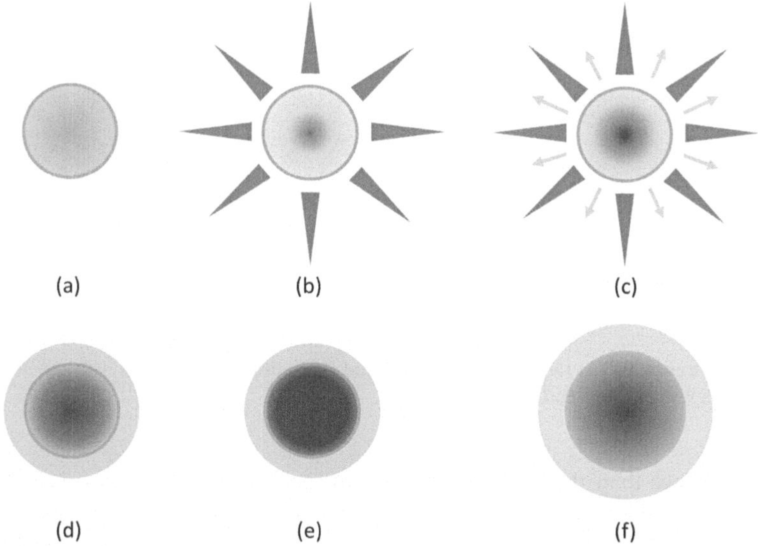

Fig. 5.1 ICF process (**a**) Position of target (**b**) Laser impingement (**c**) Laser plasma interaction & mass ablation (**d**) Symmetrical compression of fuel to high T and high ρ with TN ignition (**e**) α-particle energy deposition and TN burn wave (**f**) Disassembly & plasma expansion with neutron/radiation loss[KFL]

to the surface via ablation and sets up a coronal plasma layer. The laser beam can then only penetrate up to a critical surface, as discussed in the previous chapter. This is determined by the electron plasma frequency which is proportional to the square root of the number density $\omega_{pe} \propto \sqrt{n_e}$ and the frequency of the laser ω, and instead it is absorbed into the electrons. Then electrons transport the energy inwards to transfer their energy into the thermal energy of the compression. The Direct Drive method process is illustrated in Fig. 5.1

Prior to full compression high energy electrons can be accelerated by plasma waves into the fuel ahead of the ablation front, heating it which makes it less compressible. This is known as electron preheat and is dependent upon the laser intensity and wavelength. Once sufficient energy has been absorbed into the capsule, convergent shocks then compress the fuel producing a high temperature $\sim 10^8$ K hot spot at the centre of the capsule. When the conditions are right, the fusion reactions will begin depositing alpha particles into the surrounding high density but cooler fuel. However, the alpha particles do not penetrate very far into the surrounding fuel so come to a stop and instead deposit their energy into the surrounding ions and electrons to produce further reactions in a process known as self-heating of the central hotspot and leads to a supersonic fusion burn wave moving radially outwards.

Ignition is defined as the point where the energy production and alpha particle deposition are sufficient to initiate a self-sustaining burn wave. Provided this burn

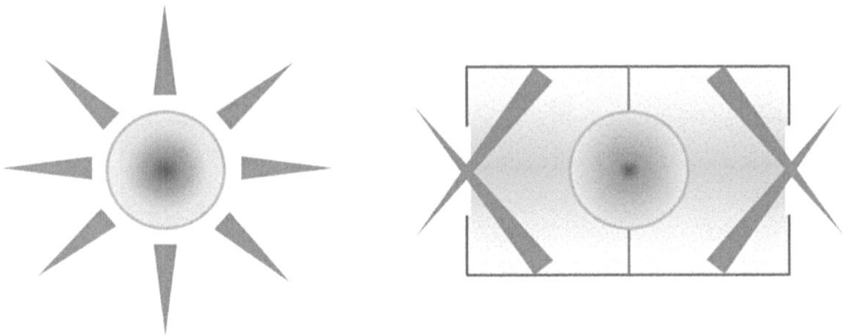

Fig. 5.2 Illustration of (**a**) Direct drive method and (**b**) Indirect drive method for generation of radiation bath within 'hohlraum'cavity[KFL]

up occurs faster than the time for the capsule to expand under its own pressure, ∼10s MJ of energy will be liberated [7]. An advantage of the hot spot ignition method is that the outward moving burn wave smoothes out any hydrodynamic instabilities originating from the compression phase.

The above description is for direct drive ignition. However, one of the problems with direct drive is the growth of instabilities due to the beam interaction but also due to perturbations in material surface finish. Further research over the years has shown that the requirements for ignition are more stringent, mainly due to these losses and inefficiencies in the implosion and laser beam-pellet energy coupling.

InDirect Drive ICF Method

Another approach to avoiding surface instabilities is to surround the capsule with a cavity known as a 'Hohlraum'. This is the route currently being pursued by NIF [8, 9] and LMJ [10]. Laser beams then only enter through two holes either side of the cavity and irradiate the cavity surface. This leads to the production of hot x-ray emission which then surrounds the capsule in a bath of radiation, giving rise to a much smoother energy driver. The cavity wall does absorb some of the radiation with losses up to ∼10–15%, but most of the energy is reemitted. Only a few micrometers of the cavity wall are heated and evaporate so the physics of the radiation-capsule interaction is not complicated by ablation products. The Indirect drive method is less sensitive to preheat. The method is illustrated in Fig. 5.2.

Fast Ignition ICF Method

Fast ignition [11, 12] is an attempt to separate out the compression phase due to the hydrodynamic implosion, from the thermonuclear ignition phase. The method is illustrated in Fig. 5.3. It has been argued that the gain from a fast ignition model may be much higher than for a conventional direct drive or indirect drive hot spot approach. It proposes to use a long pulse laser for the compression and just prior to the hot spot ignition condition being met a short pulse ∼ 1–100 ps petawatt laser ($\sim 10^{19} - 10^{20}$ W/cm^2) is then fired in to complete the process. The short pulse laser may either enhance the earlier pulse through further heating or alternatively generate a hot electron beam which can penetrate the fuel.

A problem with this approach however is that the electrons must penetrate through all the high-density plasma without being stopped. A way to solve this is to use a long pulse laser to compress the fuel but to have a gold cone in part of the capsule that tapers off at an angle from the axis. A short pulse laser beam is fired down this channel impinging on the gold cone tip and causing the generation of electrons at the tip which go straight into the central hot spot to ignite the fuel. The problem with this method is that the generated electrons do not leave the tip as a collimated beam but instead diverge widely so that only a fraction of the electrons produced will reach the central hot spot. Another problem with this approach is that the act of passing the laser beam down the channel causes plasma infilling, which is an obstacle to the continued laser transport. Some of the laser energy will be scattered reducing the laser-capsule coupling efficiency.

Shock Assisted Ignition ICF Method

Shock assisted ignition [13–15] also moves away from the idea of hot spot ignition and decouples the compression and ignition phase. The method is also illustrated in Fig. 5.3. Instead, the pellet is designed with an unusually thick ablator shell which makes for a slow implosion velocity and on a low adiabat using a low laser intensity ($\sim 5 \times 10^{14}$ W/cm^2). This helps to minimise the hydrodynamic instabilities during the implosion phase. The capsule will be compressed up, and the shock waves will centrally reflect, but on their way out they will collide with a second incoming shock, and it is at this point that ignition conditions will be achieved. Since the capsule is imploded on a low adiabat, in principle it can also be imploded with a lower laser energy. This is possible because the strength of the ignitor shock launched by a laser spike ($\sim 5 \times 10^{15} - 10^{16}$ W/cm^2) is significantly amplified as it passes through the thick shell.

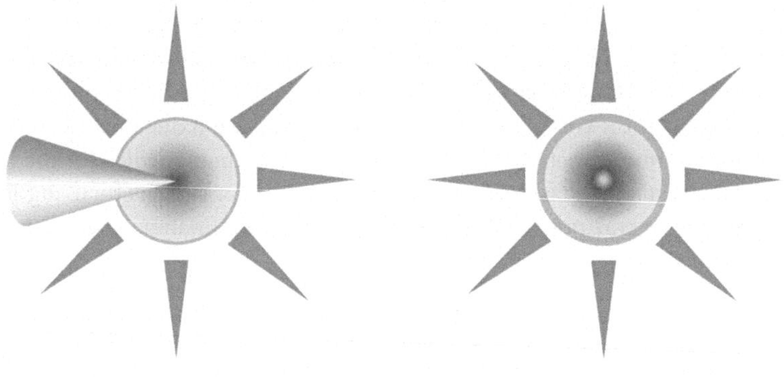

Fig. 5.3 Illustration of fast ignition and shock ignition ICF methods[KFL]

Whereas conventional ICF ignition depends upon a single hot spot ignition at the centre of the fuel, in contrast shock ignition depends upon the dynamics and interaction of three shock waves: (1) initial-return shock (2) ignitor shock (3) collision shock. This is a simple scheme and does not require the use of any short pulse lasers, which means laser-plasma instabilities are minimised.

In conventional designs instabilities are a problem with suprathermal electrons preheating the pre-compressed fuel. For shock ignition, the high intensity laser pulse is not applied until late time when the fuel approaches stagnation, so the dense shell absorbs any hot generated electrons up to high energies and shields the fuel from any electron preheat. Another important effect in shock ignition is heat conduction from free-streaming electrons as they enter the ablation front, as caused by the penetration of heated thermal coronal electrons during the power spike, with the electrons augmenting the driver shock by increased heat transfer.

The reason that the gain is higher for shock ignition than conventional ICF is due to the low implosion velocity, so that more fuel mass can be assembled for the same kinetic energy in the shell. Calculations [15] suggest targets at NIF would be driven by a 1 MJ laser input, produce a fusion yield output of 100 MJ at 30% burn-up fraction, with 2.5g/cm^3 peak areal densities of fuel, a mean aspect ratio of $\sim$2.5. Then potential thermonuclear yields on NIF under shock ignition would be 9.1 MJ for the smallest target driven at 0.17 MJ with a gain of G $\sim$53. Alternatively, 261 MJ for the largest target driven at 1.59 MJ with a gain of G $\sim$164. This shows energy gains 5–10 times higher than conventional ICF schemes. Experiments have been proposed to test shock ignition within a laboratory setting [16–18].

5.1.1 Antimatter Catalysed ICF Method

Finally, there have been proposals for the use of antimatter particles to catalyse a fusion reaction, which is known as Antimatter Catalysed Fusion (ACF). Such as with the mission concepts ICAN-II and AIMSTAR [19]. The collision of a matter-antimatter particle pair results in the annihilation of both into energy proportioned approximately as 1/3 gamma-rays and 2/3 charged pions. The pions move at $\sim$0.94c and exist for a very short time so will only move a distance of order $\sim$20 m and yet in that time somehow, they must be directed magnetically for the purpose of thrust. There is also the issue of antimatter containment such as by using Penning traps. There are several potential methods by which antimatter catalysed ignition may be facilitated:

1. Direct injection of antiprotons into a hydrogen fuel stream which then reacts through annihilation in a method known as beam core or plasma core reactor.
2. Injection of protons or other elements into a cloud of magnetically contained antiprotons (using a Penning trap) from which annihilation will then occur.
3. A beam of antiprotons reacts with the inside wall of a DT capsule and so initiates the fusion reaction. Thus, annihilating the protons and creating the

plasma conditions for copious neutrons and heat. A metal shell surrounding the fuel produces a self-generated magnetic field that thermally insulates the metal from the hot plasma.

The physics of matter-antimatter annihilation reactions is immature, as well as the ability to conduct experiments. That said, it is briefly mentioned here since it is possible that in the future such a method could be used to facilitate ignition within a fusion fuel but also to produce extraordinarily high gains. This would also result in a very high exhaust velocity and would take spacecraft mission concepts to speeds beyond the currently estimated $\sim 0.15c$ limit for fusion reactions. For this reason, developments in this field should be kept in mind - especially since the potential specific energy may be of order $\sim$90 billion MJ/kg.

Summary of Methods

Table 5.2 shows a summary of the four main methods (we neglect ACF) discussed in this chapter which seem viable for the application of space propulsion, although we have excluded antimatter catalysed fusion for the reasons given. For simplicity, in the remainder of this book the focus will be on the hot spot ignition scenario since the inputs and outputs from this can be easily represented in propulsion calculations, but more complex ignition scenarios would require a more sophisticated approach to modelling them in-line with the calculations.

As mentioned, the successful ignition of the fusion fuel depends on achieving the fusion triple product, the generation of a sustained thermonuclear burn wave and then disassembly of the capsule as it explodes outwards in a high temperature plasma state [5–7]. Whilst the technology is still being investigated in Earth based laboratories for the application of national grid commercial power generation, here the extended application of this technology to the advanced propulsion of a

Table 5.2 Comparison of ICF ignition methods and minimum conditions

Property	Indirect Drive	Direct Drive	Shock Ignition	Fast Ignition
Fuel e^- preheat	high	high	low	high
Compression	high density	high density	low density	high density
Imp.Vel. (m/s)	$\sim 4 \times 10^7$	$\sim 4 \times 10^7$	$\sim 3 \times 10^7$	$\sim 3 \times 10^7$
Adiabat	high	high	low	high
Central hotspot	Yes	Yes	No	Yes
RT instabilities	Low	High	Low	High
Energy gain	$\sim$ 10–20	$\sim$ 10–20	$\sim$ 100–250	$\sim$ 10–100
Min driver energy	>1 MJ	>1 MJ	<1 MJ	1 MJ+100 kJ
No. driver pulses	1	1	1	2
Long/Short pulse	LP	LP	LP	LP+SP
Comp-Igni	Coupled	Coupled	De-coupled	De-coupled
Target design	Complex	Simple	Simple	Complex
Shell thickness	thin	thin	thick	thin

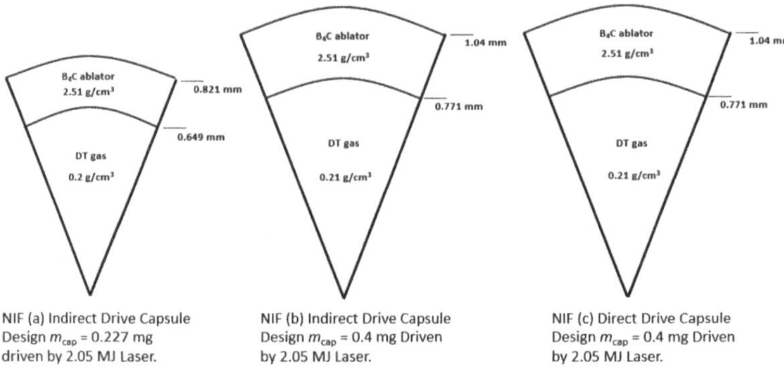

Fig. 5.4 NIF-like ICF capsule designs[KFL]

spacecraft on an interstellar trajectory is considered. In addition to correctly understanding the ignition physics of the capsule and its predicted output performance, one must also understand the interaction of the output effects onto the spacecraft vehicle structure - mainly through x-ray Bremsstrahlung radiation and high energy neutrons which will radiate the materials causing significant structural heating and fatigue.

For a realistic space propulsion engine it will be necessary to assume something like each laser fires a total of $\sim$0.5 MJ of energy, so that with 20 individual optical lines there will be a total of $\sim$10 MJ fired at the target. However, for current experimental reactors the situation is something like a total energy of 1.8 MJ delivered onto a DT gas target at 0.3 mg/cm^3 density up to a radius of 0.9 mm, surrounded by a layer of DT solid up to a radius of 1 mm, then finally an outer ablator layer of CH plastic up to a radius of around 1.1 mm.

For a set-up with many individual beam lines, these are fired onto the target simultaneously and that the distance between the beam source and the target is sufficiently small that the divergence of the beam can be neglected, so all the beam spots collectively fully cover the surface area of the capsule with no gaps. Once the laser arrives at the capsule surface it will ionize the atoms within the layers of the outer ablator shell, leading to a plasma cloud and in particular free electrons.

Direct drive is likely the most appropriate route for space propulsion if the instabilities can be overcome and a high symmetry of implosion maintained. This can be achieved by use of long pulse lasers. Figure 5.4 shows the designs for the NIF-like capsule designs that were studied by this author using the sorts of equations that will be described next.

A major issue with these sorts of designs is that the efficiency of the laser for these models is very low, in the range $\sim$1.3–2.9%, even though they are much higher than is used on NIF (<1%) but these parameters are what was required to give the desired performance in this model. The input for these models was the mass, radii and densities, then also the laser wavelength. From this the models were tuned to give a laser performance and fusion output performance like the shot experiments on NIF so that a baseline could be established for a credible model. The other two models were then scaled up for a larger mass. We note that the gain for model NIF(a) of $G = 1.5$ is consistent with the first successful NIF shot to achieve ignition. The other models NIF(b) and NIF(c) were speculative models constructed by this author as a part of a theoretical project, deliberately designed as higher mass systems, but they illustrate what may be built in the future.

5.2 Ignition Physics

Pressure and Velocity

In the model below we assume that the hot spot has a radius r (cm) and is imploded with a velocity v (cm/s), ρ is the hot-spot density (g/cm^3), T is the hot-spot temperature (keV). Electrons and ions are assumed to be in equilibrium. The total number of particles in the hot-spot is defined by n_0 (cm^{-3}).

In the following relations we will use a normalised laser intensity ($I_{15} = I/10^{15}$ W/cm^2) and normalised wavelength ($\lambda = \lambda_o/10^{-6}$ m). In the work by Lindl [6] it states that for direct drive scenarios the following analytic expressions have been found to be good approximations in calculating the pressure, mass flow rate of ablated surface material, implosion velocity and ablation velocity:

$$P(Mbar) = 40\left(\frac{I_{15}}{\lambda}\right)^{2/3}, \tag{5.17}$$

$$\dot{m}(g/cm^2 s) = 2.6 \times 10^5\left(\frac{I_{15}}{\lambda^4}\right)^{1/3}, \tag{5.18}$$

$$V_{imp}(cm/s) = 1.5 \times 10^8 (I\lambda^2)^{1/3} \ln\left(\frac{m_o}{m_f}\right) = V_{ex} \ln\left(\frac{m_o}{m_f}\right), \tag{5.19}$$

$$V_{abl}(cm/s) = 4.3 \times 10^4 \beta_f^{3/5} I_{15}^{-1/15} \lambda^{-14/15}. \tag{5.20}$$

Where β_f is the ratio of pressure P at a given density to the Fermi pressure P_{fd} at that density and $P = P_{fd}\beta_f\rho^{5/3}$ and λ is in the units of μm. However, for the implosion velocity the equation did not give a reasonable value and for this reason a modified version was used (also for v_{abl}) in all direct drive ignition models which approximated to within $\sim 1\%$ and where λ is in the units of m, along the lines:

$$V_{imp}(cm/s) = 3.5 \times 10^6 (I\lambda^2)^{1/3} \ln\left(\frac{m_o}{m_f}\right), \tag{5.21}$$

$$V_{abl}(cm/s) = 4.3 \times \beta_f^{3/5} I^{-1/15} \lambda^{-14/15}. \tag{5.22}$$

We now need to calculate the In Flight Aspect Ratio (IFAR), which is the ratio of the shell inner radius R to its thickness ΔR which is obtained by simple integration of the rocket equation under the assumption that most of the work done is when the shell is at a large fraction of its initial radius. The implosion velocity is less than the ablation exhaust velocity and for typical implosion velocities and typical laser intensities of $10^{14} - 10^{16}$ W/cm^2, the relation between the IFAR and the rocket parameters, for the range $V_{imp}/V_{ex} = 0.1$–0.8 is given by

$$\left(\frac{R}{\Delta R}\right) = 0.7 \times \frac{V_{imp}^2}{V_{abl} V_{ex}}. \tag{5.23}$$

Similarly, the following analytic expressions have been found [6] to be good approximations for indirect drive scenarios in calculating the pressure, mass flow rate of ablated surface material, implosion velocity, ablation velocity:

$$P(Mbar) = 170 I_{15}^{7/8}, \tag{5.24}$$

$$\dot{m}(g/cm^2 s) = 10^7 I_{15}^{3/4}, \tag{5.25}$$

$$V_{imp}(cm/s) = 1.8 \times 10^7 I_{15}^{1/8} \ln\left(\frac{m_o}{m_f}\right), \tag{5.26}$$

$$V_{abl}(cm/s) = 6.6 \times 10^5 \beta_f^{3/5} I_{15}^{9/40}. \tag{5.27}$$

Then calculating the IFAR for the indirect drive case, the implosion velocity is greater than the rocket exhaust velocity, with typical implosion velocities 2–4 $\times$ 10^7 cm/s and with radiation temperatures on the order of a few hundred electron volts, for rocket parameters in the range $V_{imp}/V_{ex} = 0.8$–4, we get

$$\left(\frac{R}{\Delta R}\right) = 0.56 \times \frac{V_{imp}}{V_{abl}}. \tag{5.28}$$

The pressure generated at the capsule external surface is then used to estimate the compressed volume of the fuel. This is done by using the combined ideal gas equation as an approximation for the volume reduction and where we approximate the compressed boundary pressure by the ablation pressure

$$\frac{P_o V_o}{T_o} = \frac{P_c V_c}{T_c}. \tag{5.29}$$

So that the compressed volume is given by

$$V_c = \frac{P_o}{P_c}\frac{T_c}{T_o} V_o = \frac{P_o}{P_{abl}}\frac{T_c}{T_o} V_o. \tag{5.30}$$

And then the compression ratio is given by the two volumes

$$\Gamma_c = \frac{V_o}{V_c}. \tag{5.31}$$

The compression should be carried out isentropically (constant thermodynamic entropy) and adiabatically (heat does not leave or enter the system) to minimise the losses from the ignition whilst the fuel is still undergoing compression. This is known as a Fermi degenerate state and the temperature of the compressed gas should be below the temperature at which a Fermi degenerate state occurs. If the compression is not isentropic then the energy due to the compression is dominated by the energy due to the ignition. In this text we define the state of the gas by the ratio of the pressure at a given density to the Fermi pressure

$$\beta_f = \frac{P}{\rho^{5/3} P_{fd}}. \tag{5.32}$$

The time for a sound wave to reach the pellet centre from an initial radius of r_o at the surface is given as a function of the Fermi degenerate pressure $P_{fd} = 7.516 \times 10^9$ N/m^2 (for D^{3}He) and the density of the compressed volume ρ_c as follows

$$\tau_s = r_o \left(\frac{\gamma P_{fd}}{\rho_c} \right)^{-1/2}. \tag{5.33}$$

Where γ is the ratio of specific heats.

PdV Work

To estimate the rate of mechanical work per unit volume P_w performed on the hot-spot region of the fuel we adopt the model of a piston compressing a gas with a uniform pressure. The work rate per unit volume is then given [6] by

$$P_w = \frac{P(\frac{dV}{dt})}{V} = \frac{PAv}{V} = \frac{3 \times Pv}{r}. \tag{5.34}$$

We can then use the fact that $P = nk_BT$ for an isothermal ideal gas equation of state and that $\rho = n_i m_i + n_e m_e$, which refers to the number density and mass of the ions and electrons. The solution to this equation is

$$P_w(W/cm^3) = \frac{2.3 \times 10^{15} \times \rho T}{r} \left(\frac{v_{imp}}{10^7} \right). \tag{5.35}$$

Alpha-Particle Deposition

The thermonuclear heating rate per unit volume P_α is found [6] by multiplying the burn-up rate $\dot{\phi}$ and the mass defect α (0.0038 for DT)

$$P_\alpha = \rho\dot{\phi}\epsilon_\alpha. \tag{5.36}$$

Where the burn rate is given by the following approximation

$$\dot{\phi} = \frac{1}{2}(1-\phi)n_0\langle\sigma_c v\rangle \approx 1.2 \times 10^{23}\rho\langle\sigma_c v\rangle. \tag{5.37}$$

The burn-up fraction is given by [20]

$$\phi = \frac{\rho r}{\rho r + 6(g/cm^2)}. \tag{5.38}$$

We can estimate the energy release per reaction ϵ_α by counting the atomic mass of the reacting species. We then multiply the energy release in units of MeV by 1.609×10^{-13} MeV and then multiplying this by the Avogadro number 6.022×10^{23} atoms^{-1}, as the number of constituent particles (atoms or molecules) in one mole of a given substance. We then divide the total by the atomic mass N (grams/mole) to arrive at the total energy release per gram. The earlier Table 4.1 showed the energy release data for various fusion reactions. Also shown is the energy release but not including the excess protons and neutrons, labelled fraction. For a DT plasma we assume an alpha-particle energy per gram of $\epsilon_i = 0.684 \times 10^{11}$ J/g.

Table 5.3 shows the fusion reactivities for various Maxwell averaged cross sections. For these three set of reactants they will have a peak of around 8.76×10^{-16} cm^3/s (DT), 1.48×10^{-22} cm^3/s (DD) and 2.6×10^{-16} cm^3/s (D^3He). We can then substitute these terms into the equation for P_α and then multiply the top and bottom by 10^{-17} so that the Maxwell averaged cross section can also be written in normalised units. The heating rate per unit volume is then given by

$$P_\alpha(W/cm^3) = 8.04 \times 10^{16}\rho^2\langle\frac{\sigma_c v}{10^{-17}}\rangle\alpha. \tag{5.39}$$

The reaction cross section $\langle\sigma_c v\rangle$ is averaged over a Maxwellian distribution of particles and for an equimolar mixture of DT or otherwise.

Table 5.3 Approximate Maxwellian-averaged reactivities (cm^3/s) as a function temperature (keV) for different reactants in a fusion plasma

Reaction	1 keV	2 keV	5 keV	10 keV	20 keV
DT	6.27×10^{-21}	2.83×10^{-19}	1.35×10^{-17}	1.13×10^{-16}	4.31×10^{-16}
DD	9.66×10^{-23}	3.05×10^{-21}	8.97×10^{-20}	5.99×10^{-19}	2.65×10^{-18}
D^3He	3.10×10^{-26}	1.41×10^{-23}	6.46×10^{-21}	1.97×10^{-19}	3.26×10^{-18}

Bremsstrahlung Emission

The radiation loss rate per unit volume P_r is assumed to be all via Bremsstrahlung emission which is escaping the hot-spot region of the fuel. Classically, in the theory of electrodynamics this is photons emitted when electrons accelerate around positive ions during Coulomb collisions. Quantum mechanically, it is due to electrons undergoing transitions between two states of the continuum in the field of an atom. It is a temperature and density dependent phenomena given [6] by

$$P_r(W/cm^3) = 3.0 \times 10^{16} \rho^2 T^{0.5}. \tag{5.40}$$

Electron Conduction

The conduction losses are approximated by a Spitzer conductivity κ [21] assuming a steady-state temperature profile, where the heating rate is $Q = -\kappa \nabla T$,

$$\kappa = \frac{9.4 \times 10^{12} S(z) T^{5/2}}{Z \ln} \Lambda, \tag{5.41}$$

$$Q = -\frac{9.4 \times 10^{12} S(z)}{Z \ln} \Lambda T^{5/2} \nabla T. \tag{5.42}$$

In these equations $\ln \Lambda$ is the Coulomb Logarithm which defines the ratio of the upper and lower cut-offs in Coulomb collisions. Z is the electric charge. The term $S(Z)$ is defined by Spitzer [21] and ranges from $S(Z) = 1$ at $Z = 1$ to $S(Z) = 4$ at $Z = \infty$. Differentiating, the heating rate becomes the following

$$Q = -\frac{9.4 \times 10^{12} S(z)}{Z \ln \Lambda} \frac{4}{7} \frac{T^{7/2}}{r}. \tag{5.43}$$

The electron conduction loss rate per unit volume is given by equating it to the volume heating as follows:

$$P_e = \frac{AQ}{V} = \frac{4\pi r^2 Q}{4\pi r^3/3} = \frac{3Q}{r}, \tag{5.44}$$

$$P_e = \frac{3 \times 9.4 \times 10^{12} S(z)}{Z \ln \Lambda} \frac{4}{7} \frac{T^{7/2}}{r^2}. \tag{5.45}$$

So that finally we get the equation as follows

$$P_e(W/cm^3) = \frac{8 \times 10^{12} T^{7/2}}{r^2}. \tag{5.46}$$

Energy Conservation

We approach the problem from the principle of total energy conservation in that the sum of the energy gains minus the sum of the energy losses should be greater than zero for ignition and gain. The energy gains are represented by the PdV mechanical work per unit volume P_w and the alpha particle deposition thermonuclear heating rate per unit volume P_α. The energy losses are represented by the radiation losses (due to Bremsstrahlung emission) per unit volume P_r and the electron conduction losses (approximated with a Spitzer conductivity) per unit volume P_e

$$P_w + P_\alpha - P_r - P_e > 0. \tag{5.47}$$

Another energy gain term should technically be added to account for the neutron energy deposition, but reference [7] makes the point that this is small and so can be neglected as an approximation. These equations describe a competition between the gains and losses. For the gains, the PdV work P_w will increase as a function of density ρ, temperature T and implosion velocity v_{imp}, but also the smaller the radius of the compressed volume since $\propto 1/r$. The alpha particle deposition P_α will increase with density squared ρ^2 and it's important to raise this to a high value since it's a dominant term reflecting a sustained thermonuclear burn wave in an igniting plasma. However, these must be balanced against the losses which includes the electron conduction P_e which will increase as a function temperature as $T^{7/2}$, but also the smaller the radius squared of the compressed volume since $\propto 1/r^2$. In addition, the losses due to the radiation P_r will increase with density squared ρ^2 and the square root of the temperature $T^{1/2}$.

Neutron Flux

Estimating the neutron flux from the fusion reaction will determine the radiation shielding required but also the magnitude of expected heat loads to the spacecraft structure. Reference [22] discusses radiation output from the Daedalus pellets. The report examines the difference between the reaction probability for non-neutron producing reactions of 3×10^{-22} m^3/s and for those producing neutrons of 6×10^{-23} m^3/s. The reaction cross section is a measure of the probability of a fusion reaction as a function of the relative velocity of two reactant nuclei. The reaction rate is the number of fusions occurring per unit volume per unit time and is defined by $<\sigma_c v>$ multiplied by the reactant number densities n_1 and n_2 so that $f = n_1 n_2 < \sigma_c v >$. For species that are reacting with itself such as in the DD reaction, then the product $n_1 n_2$ is modified so that the reaction rate becomes $f = \frac{1}{2} n^2 < \sigma_c v >$. The reaction rate for different species is then

$$f_{DT} = n_D n_T < \sigma_c v >_{DT}, \tag{5.48}$$

$$f_{DD} = \frac{1}{2} n_{DD}^2 < \sigma_c v >_{DD}, \tag{5.49}$$

$$f_{D^3He} = n_D n_{^3He} < \sigma_c v >_{DHe_3} . \tag{5.50}$$

If we take the ratio of the DD versus D^3He reaction we get

$$f_{DD} = \frac{1/2n_{DD}^2 < \sigma_c v >_{DD}}{n_D n_{^3He} < \sigma_c v >_{D^3He}} = \frac{1/2 \times 0.6}{3} = 0.1. \tag{5.51}$$

This equation shows that the number of neutrons produced will be $\sim$10% the number of particle pairs in the reactions. We use this to estimate the number of neutrons produced per pulse n_p as well as the neutron production rates. We then multiply this number by an equation which factors in the pellet mass m_{pell}, the number of grams per mole of the substance g_{pm} (i.e. 5 for D^3He), Avogadro constant A_v, the fraction of neutron producing reactions f and the burn-up fraction of the fuel ϕ to determine the number of neutrons

$$n_p = \frac{m_{cap} A_v \times f\phi}{g_{pm}}. \tag{5.52}$$

This gives the number of neutrons per pulse which is also multiplied by the number of engines used in any design model within the code. We can then find the number of neutrons per second from $n_{pell} = n_p \times f_{hz}$. In principle if the number of neutrons produced n_{pell} is known then this can also be correlated with the number of thermonuclear reactions. So that for example in the DT reaction a 14 MeV neutron is produced in conjunction with a 3.52 MeV ^4He particle. Multiplying this energy by the number of reactions gives an estimate for the total thermonuclear energy output within a pellet, and therefore an independent measure on the gain. Measuring the neutron production rate for this purpose is typically how laser facilities like NIF will estimate the energy output in a pellet.

Finally, it is necessary to estimate the mean free path of a high energy 14 MeV neutron as it transitions through a fuel. This is essential in estimating how many neutrons will be emitted at the pellet surface. The fuel will be defined with an initial density ρ_o and then compressed to a density ρ_c so that $\Gamma = \rho_c/\rho_o$. The number density of the compressed material is then given by the Advogadro number N_A and the atomic number A so that the mean free path is then given by

$$\lambda_{14MeVn} = \frac{1}{n\sigma_c} = \frac{A}{N_A \rho_c \sigma_c} = \frac{A}{N_A \rho_o \Gamma \sigma_c}. \tag{5.53}$$

For a Daedalus second stage capsule filled with D^3He the neutrons will be absorbed into the ^3He. We can calculate the range of the neutron before it is absorbed. Assuming $\rho_o = 0.089$ g/cm^3, $\Gamma = 2000$, $A = 3.01605$, $\sigma_c = 1.2 \times 10^{-24}$ cm^2 implies 3.55×10^{25} atoms cm^{-3}, and so $\lambda_n = 0.195$ mm. For the Daedalus first stage capsules with $\Gamma = 1000$ leads to $\lambda_n = 0.391$ mm.

5.3 Laser Driver Requirements

A fusion engine will release thermonuclear energy given as a power output by $P_{out} = \alpha\phi f_{hz} m_{cap} c^2$. This compares to the jet power of the engine which is given as a function of the mass flow rate of fuel into the engine reaction chamber $\dot{m}$ so that $P_j = \dot{m} v_{ex}^2/2$. Comparing the output power with the jet power these are defined by an inequality since there will be energy losses from the system such as due to high energy neutrons and Bremsstrahlung radiation. Therefore, the nozzle jet efficiency η_j defines the losses from the system. Balancing the two, and rearranging leads to a thermonuclear minimum mass requirement as a function of these propulsion parameters

$$m_{cap} \geq \frac{\dot{m} v_{ex}^2}{2 n_e \eta_j \alpha \phi f_{hz} c^2}. \tag{5.54}$$

Where it is important to note that m_{cap} only defines the thermonuclear part of the capsule. We note also that we have divided by the number of engines n_{eng} which is important in a parallel staged thrust system, or one would get the combined capsule mass of all engines. External to the fuel will be another shell layer known as an ablator which may be thick and dense and this is what provides the inertia for the fuel but also acts as rocket blow off during laser implosion to provide the necessary momentum for the compression. However, if the shell is overly massive this will add substantial mass to the vehicle and so affect the mass ratio, so care must be taken to minimise the shell mass where possible.

We assume that one of two scenarios arise for the ablator shell. In the first, by the time of thermonuclear energy release all the ablator shell has been ablated and does not contribute to the exhaust velocity. In the second, particularly in the case of a thick shelled design, the ablator shell may contain some component of a hydrogen-based expellant which does contribute to the exhaust velocity.

We note that the mass flow rate per engine is equal to $\dot{m} = m_{cap} f_{hz}$ and so the total mass flow rate for all engines will be $\dot{m}_{tot} = n_{eng}\dot{m}$. This is particularly useful since one of the problems with using ICF is that the capsules tend to be very low in mass, which means that unless they are to be detonated at a very high pulse frequency the mass flow rate will be limited. This will then also limit the vehicle thrust since $T = \dot{m} v_{ex}$. However, a way in which this can be mitigated is to couple the thermonuclear material to another material known as an expellant. This is known as thrust augmentation and the various methods by which this can be achieved are discussed further in Chap. 8.

Next we discuss energy gain algorithms, relating to the ratio of output energy to input energy delivered to the target. Various algorithms have been derived for gain models, including by Orth [23] who determined a model where the driver energy scales to some sub-linear power as $G = 877 E_{dr}^{1/3}(MJ)$. However, this model was on the assumption of $G = 1500$ for VISTA which at this current time is unrealistic although may be possible in the future. Instead, a different algorithm has been derived by this author by balancing the output energy with the driver energy so that $G = 4.385 E_{dr}^{1/3}(MJ)$. This then gives

$$E_{dr}(MJ) = \left(\frac{\alpha\phi m_{cap}c^2}{4.385 \times 10^6}\right)^{3/4}. \tag{5.55}$$

This algorithm correctly predicts a 2 MJ driver energy requirement for the current experiments on NIF [24, 25] but also is normalised to the Daedalus design [26, 27]. It could be argued that whilst the Orth model is overly optimistic, this model is pessimistic since it is likely that gains on the order of G >100 will eventually be achievable such as through methods like shock assisted ignition. For any actual designs in the future, they will be some average of the two extremes.

The model explored by Orth [23] relates the capsule gain to the driver energy E_{dr} (MJ), and the capsule compression efficiency ϵ_c as follows

$$G = c\epsilon_c(\epsilon_c E_{dr}(MJ))^n. \tag{5.56}$$

Where the coefficient $c = 6000$ and power index $n = 1/3$, so that for a driver energy of 5 MJ the gain would be $G = 1200$. Others have then presented different types of gain scaling laws assuming a typical compression efficiency of 0.2. This includes Rosen-Meyer-Ter-Vehn (MTV) [28] and others [12, 23, 29, 30]

$$G = 6000 \times \epsilon_c(\epsilon_c E_{dr}(MJ))^{0.3}, \tag{5.57}$$

$$G = 30,000 \times \epsilon_c(\epsilon_c E_{dr}(MJ))^{0.4}, \tag{5.58}$$

$$G = (1300 E_{dr}(MJ) - 1,040)^{0.53}, \tag{5.59}$$

$$G = 16 \times E_{dr}(MJ)^{2/3}, \tag{5.60}$$

$$G = 400 \times E_{dr}(MJ)^{0.4}. \tag{5.61}$$

All of the above are for hot-spot ignition models. Efforts have also been made to develop gain models for shock assisted ignition using numerical simulations such as by Perkins [15] where an implication of the research is that a lower driver energy would be required as compared to hot-spot ignition, as given by

$$G = 126 \times E_{dr}(MJ)^{1/2}. \tag{5.62}$$

This gain model went up to capsules scaled for maximum energy/power attainable on NIF, i.e. 2 MJ / 500 TW. At higher laser power the ignition may be killed by laser-plasma instabilities. Gain models have also been developed for fast ignition based on numerical simulations [31, 32]. For example, simulations predict that using a collimated monoenergetic electron beam a minimum energy of 100 kJ using a

1.05 μm laser would be required. For a 1 MJ and 2 MJ driver energy this would correspond to a gain of $G = 155.5$ and $G = 215.8$ respectively

$$G = \frac{743}{1 + 30/E_{dr}(kJ)^{1/3}}. \tag{5.63}$$

Most gain models scale as driver energy to the ~0.3–0.7 sub-linear power. Since they are fitted to both experimental data and numerical simulations relating to NIF there is a question on the applicability of these scaling laws outside of 2 MJ/500 TW laser regime. At this point it is all speculation subject to further experimental data, such as from the Halite-Centurion tests.

Finally, the main issue with the use of lasers in space propulsion is the very low efficiency which is projected to reach a few percent for Nd:glass but may be as high as ~20% for DPSSL [33]. They must also be applied within a wide range of thermal limits which presents problems for optics and instrumentation. The development of efficient laser systems as energy drivers is a critical issue for the feasibility of propulsion applications since it would allow a greater amount of energy delivered to the target and greater gain.

Problems of Chap. 5

5.1 A Daedalus-like capsule mass of $m_{cap} = 0.288$ g of DT is heated to $E = 50$ keV. (i) What is the minimum heat energy needed for a hot spot mass 1% the total fuel mass? (ii) How much energy would be required to heat the entire fuel capsule?

5.2 A fusion engine has an exhaust velocity of $v_{ex} = 10{,}000$ km/s, and the engine detonates at $f_{hz} = 100$ Hz pulse frequency for a mass flow rate of $\dot{m} = 1$ g/s. The burn-up fraction is $\phi = 0.333$ for a DT fuel. (i) What is the jet efficiency of the engine design? (ii) What is the ICF capsule mass in grams?

5.3 For the capsule mass problem calculated above (i) What will be the driver energy requirement deposited onto the target prior to any capsule losses? (ii) For a laser 15% efficient, what is the size of the energy that must be generated at the source, such as from a nuclear reactor or as a part of an induction loop cycle?

References

1. Hyde RA (1983) A laser fusion rocket for interplanetary propulsion, UCRL-88857. Presented 34th IAC, IAF-83-396, October 1983. University of California, Livermore
2. Lawson JD (1955) Some criteria for a power producing thermonuclear reactor. Technical Report for the Atomic Energy Research Establishment, Harwell, England, UK, AERE GP/R 1807. Published Proce. Phys. Soc. B 70(1)

3. Pick MA et al (1998) The technological achievements and experience at JET. JET-P(98)34, presented 20th symposium on fusion technology
4. Long KF, Lamontagne M (2025) Project icarus: pellet injection system for the pegasus spacecraft concept design. J Br Interplanetary Soc **78**(6):206–215
5. Nuckolls J, Wood L, Thiessen A, Zimmerman G (1972) Laser compression of matter to super-high densities: thermonuclear (CTR) applications. Nature 239:139–142
6. Lindl JD (1998) Inertial confinement fusion, the quest for ignition and energy gain using indirect drive. Springer/AIP Press, Berlin/Melville
7. Pfalzner SP (2006) An introduction to inertial confinement fusion. Taylor & Francis, Milton Park
8. Dewald E (2016) Status of indirect drive ICF experiments on the national ignition facility. LLNL, TR-686482
9. Zylstra AB, Hurricane OA, Callahan DA et al (2022) Burning plasma achieved in inertial fusion. Nature 601(7894):542–548
10. Fleurot N (2005) The laser megajoule (LMJ) project dedicated to inertial confinement fusion: development and construction status. Fusion Eng Design 74(1–4):147–154
11. Atzeni S (2008) Fast ignitor target studies for the HiPER project. Phys Plasmas 15:056311
12. Tabak M, Clark DS, Hatchett SP (2005) Review of progress in fast ignition. Phys Plasmas 12:057305
13. Betti R, Zhou CD, Anderson KS et al (2007) Shock ignition of thermonuclear fuel with high areal density. Phys Rev Lett 98:155001
14. Betti R, Theobald W, Zhou CD, Anderson KS, McKenty PW, Skupsky S, Shvarts D, Goncharov VN, Delettrez JA, Radha PB, Sangster TC, Stoeckl C, Meyerhofer DD (2008) Shock ignition of thermonuclear fuel with high areal densities. J Phys Conference Series 112:0220224
15. Perkins LJ (2009) Shock ignition: a new approach to high gain inertial confinement fusion on the national ignition facility. Phys Rev Lett 103:045004
16. Perkins LJ (2011) Validation of the scientific basis of shock ignition on the national ignition facility. LLNL-Prop-467878, 27 January, Lawrence Livermore National Laboratory
17. Atzeni S (2009) Laser driven inertial fusion: the physical basis of current and recently proposed ignition experiments. Plasma Phys Control Fusion 51:124029
18. Atzeni S, Davies JR, Hallo L, Honrubia JJ, Maire PH, Olazabal-Loume M, Feugeas JL, Ribeyre X, Schiavi A, Schurtz G, Breil J, Nicolar P (2009) Studies on targets for inertial fusion ignition demonstration at the HiPER facility. Nucl Fusion 49:055008
19. Gaidos G, Lewis RA, Meyer K, Schmidt T, Smith GA (1999) AIMSTAR: antimatter initiated microfusion for precursor interstellar missions. Acta Astronaut 44(2–4):183–186
20. Lindl JD (1972) Laser compression of matter to super-high densities. Nature 239
21. Spitzer L (2006) Physics of fully ionized gases. Dover Publication, Garden City. Originally published 1956
22. Bond A, Martin AR (1978) Project daedalus: the mission profile, final study report. J Br Interplanetary Soc Special Supplement:S37–S42
23. Orth CD (2003) VISTA: a vehicle for interplanetary space transport application powered by inertial confinement fusion. LLNL UCRL-TR-110500, 16 May
24. Long KF (2024) Minimum condition requirements in laser driven ICF for advanced space propulsion applications. J Br Interplanetary Soc 77(1)
25. Long KF (2023) Operational performance parameter range in ICF propulsion theory. Acta Astronaut 210:303–310
26. Martin AR, Bond A (1978) Project daedalus: the propulsion system part 1: theoretical considerations and calculations. J Br Interplanetary Soc Special Supplement: S44–S62
27. Bond A, Martin AR (1978) Project daedalus: the propulsion system part 2: engineering design considerations and calculations. J Br Interplanetary Soc Special Supplement:S63–S82
28. Rosen MD, Lindl JD, Thiessen AR (1984) Simple models of high-gain targets - comparisons and generalizations. UCRL-50021-83. University of California, Livermore

29. Lindl JD, Bangerter RO, Mark JW-K, Pan Y-L (1978) Review of target studies for heavy ion fusion. UCRL-94382. University of California, Livermore. Proceedings of the International Symposium on Heavy Ion Fusion, 24 October 1986
30. Lindl JD (1995) Development of the indirect-drive approach to inertial confinement fusion and the target physics basis for ignition and gain. Phys Plasmas 2(11):3933
31. Solodov AA, Betti R, Delettrez JA, Zhou CD (2005) Gain curves for fast-ignition inertial confinement fusion. Phys Plasmas 12:110702
32. Solodov AA, Betti R, Delettrez JA, Zhou CD (2007) Gain curves and hydrodynamic simulations of ignition and burn for direct-drive fast-ignition fusion targets. Phys Plasmas 14:062701
33. Bayramian A, Bopp R, Deri B et al (2016) High-energy diode-pumped solid-state laser (DPSSL) for high-repetition-rate petawatt laser systems. Presented at conference on high intensity lasers and high field phenomena, CA, USA

Chapter 6
Reactor Engineering

It is important to note that scientific and technological progress are usually non-linear - often startlingly so. The science and technology of the future that would be available for building interstellar space ships would indeed be quite different from those imagined from the perspectives of the early twenty-first century. Nevertheless, it is a good idea to start thinking what it will take to mount such an undertaking.... and also so that we can begin exploring various scientific and engineering possibilities now - rather than waiting endlessly for 'the right time' to come.

— *Yoji Kondo, 2003 [1]*

6.1 Description of a Fusion Engine

The performance of a fusion driven spacecraft engine will be strongly dependent on the type of fuel adopted for the mixture since this is one of the factors that will determine the exhaust velocity. For example, for a DT fusion fuel at assumed 50:50 mixture the primary thermonuclear reaction will be

$$D + T \rightarrow {}^{4}He(3.5\,\text{MeV}) + n(14.1\,\text{MeV}). \tag{6.1}$$

This will produce an expected energy release of around 3.41×10^{11} J/g. This will then determine the performance by a consideration of the atomic mass of the products and the reactants. The goal is to compress the fuel to a state of high temperature T and high areal density ρr. The efficiency of the implosion is measured by the burn-up fraction. Once a state of ignition is achieved within a fusion capsule, the amount of energy released depends on the mass of the fuel m_{cap}, the specific energy of the fuel being reacted α and the burn-up fraction.

As shown in Fig. 6.1 the fusion reactions may be initiated in an open (a) or closed (c) reaction chamber depending on how the charged particles are controlled with the use of magnetic fields, where (b) shows the internal geometry of one hemisphere.

K. F. Long, *Stellar Engines*, Springer Aerospace Technology,
https://doi.org/10.1007/978-3-032-18062-9_6

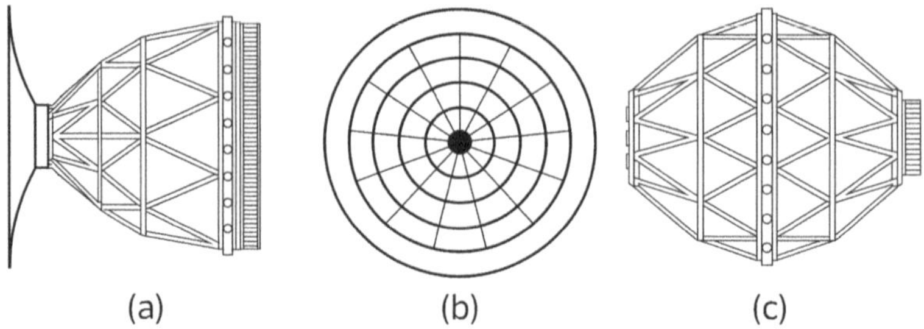

Fig. 6.1 Types of engine reaction chambers[KFL]

The earlier Fig. 3.4 illustrated a basic spacecraft conceptual layout for a fusion propelled spacecraft with emphasis on the propulsion elements. This is a modified drawing from one originally presented by Spencer [2, 3] but then significantly modified by this author and to include a laser-fusion element. The fusion material is ignited within a reaction chamber by a set of laser beam lines which serve the purpose of facilitating a spherical implosion symmetry. The plasma is controlled within a magnetic field generated by superconducting coils and magnetically directed through the nozzle for thrust. Any propellant may have to be stored in cryogenic tanks. The reaction chamber is surrounded by an insulation layer to capture some of the high energy neutrons and x-rays. Any waste heat is ejected through the radiators.

A significant consideration in fusion ignition is the several pathway energy losses along the way and in propulsion theory this tends to be expressed as a power cycle. This begins with any losses in the beamer system in turning it on and powering it up to the sufficient charge and the loss due to any system integrations. So that if a system is considered that delivered E_{in} but it experienced losses ϵ_1 then the energy delivered to the target E_{del} will be

$$E_{del} = (1 - \epsilon_1)E_{in}. \tag{6.2}$$

Then there are the losses associated with the coupling of the laser to the capsule surface which facilitates its implosion. As described above some of that energy will be reflected due to the laser-plasma interactions. The losses associated with this laser coupling are ϵ_2 so that the target energy is

$$E_{tar} = (1 - \epsilon_2)E_{del} = (1 - \epsilon_1)(1 - \epsilon_2)E_{in}. \tag{6.3}$$

Then is the losses associated with the direct compression and heating of the fuel which is represented by the burn-up fraction. If the total of these losses is ϵ_3 then the input that goes into the thermal energy of the implosion E_{th} is given by

$$E_{th} = (1 - \epsilon_3)E_{tar} = (1 - \epsilon_1)(1 - \epsilon_2)(1 - \epsilon_3)E_{in}. \tag{6.4}$$

Once the capsule is successfully ignited it will release a large amount of thermonuclear fusion energy. This results in a multiplication factor called the gain G. If $G>1$ then the system is generating more energy than it is producing in fusion energy. Although relative to the thermal energy of the compression it will be producing a positive gain, relative to the original input energy it may not be, or it may just be at breakeven. It is therefore desirable to have a state where it is a lot greater than unity.

$$G = \frac{E_{fus}}{E_{th}} >> 1. \tag{6.5}$$

Although it is possible to define this differently such as the fusion energy relative to the input energy onto the target $G_{in} = E_{fus}/E_{in}$ or as the fusion energy relative to the wall-plug electrical energy $G_{wp} = E_{fus}/E_{wp}$. Designers must be clear in what form of gain term they are using and to define the inefficiencies.

For a commercial civilian power reactor, a gain on the order of $\sim$10 may be deemed appropriate. For the application of space propulsion, a gain of at least $\sim$10s–100s would be a more desirable requirement. Instead of energy, these relationships can also be written as power expressions, since power is just energy output over some time interval. These are typically expressed as overall power flow diagrams in a space propulsion context. This might include showing the original efficiency of the input laser, laser coupling losses, neutrons, x-rays and plasma debris. This is discussed in Chap. 11.

Figure 6.2 shows an illustration of the type of reaction chamber arrangement that might be used within a fusion propulsion vehicle. This is a hemispherical reaction chamber like those used on the Daedalus design, where capsules are injected into the chamber at a high repetition rate. The laser driver beam lines are located on a

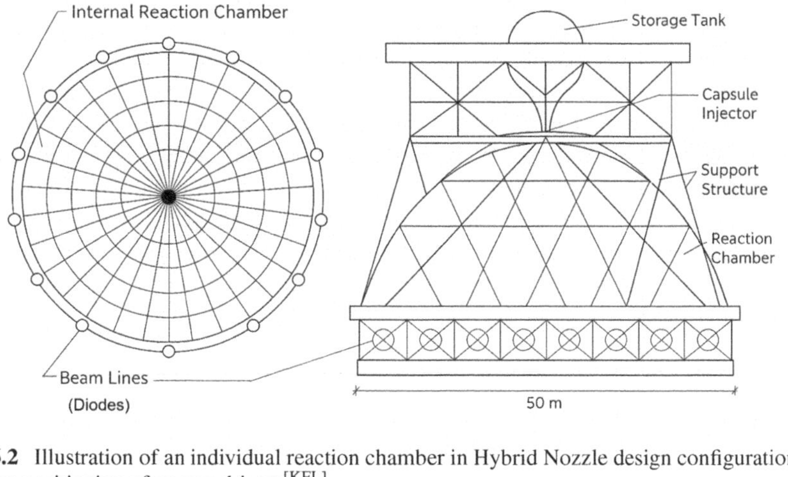

Fig. 6.2 Illustration of an individual reaction chamber in Hybrid Nozzle design configuration and circular positioning of energy drivers[KFL]

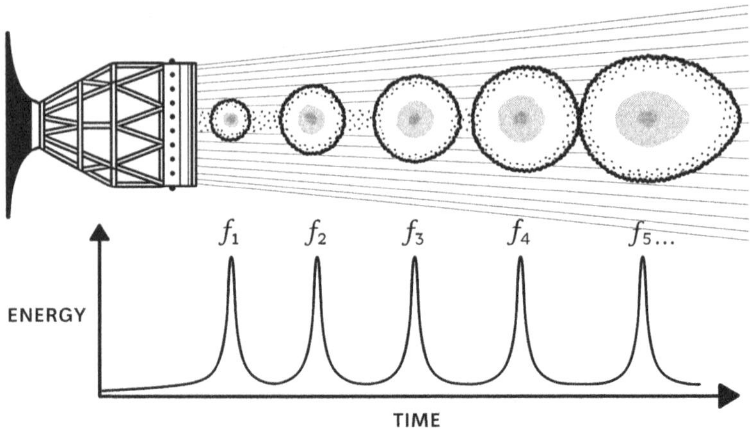

Fig. 6.3 Illustration of a fusion engine under pulse detonation of a magnetic nozzle[KFL]

circular array around the reaction chamber. The system is held together by a support structure which is connected to the spacecraft.

A large plasma envelope which forms around the reaction chamber expands as a spherical wave around the engine but is dragged out due to the motion of the spacecraft thus becoming an elongated ovoid in the spacecraft frame that gradually expands outwards. The key problem with this arrangement is the minimisation of the neutron and x-ray dosage from the capsule detonations. Layers of insulation and radiation shielding can be constructed around the individual engines to mitigate the radiation dose. In addition, smart capsule designs can be adopted so that much of the emitted neutrons and x-rays are absorbed within a higher Z layer at the outer shell.

As illustrated in Fig. 6.3, each time a fusion pellet is detonated within a reaction target chamber, the number of these detonations occurring per second is the engine pulse frequency. This may be as little as 1 Hz or as many as 1,000 Hz. At the lower limit this implies a very low mass flow rate and so a low acceleration of the spacecraft leading to an elongated mission profile. At the higher limit this puts considerable stress on the engineering structure and may lead to material fatigue from the heat loads for example. So, this goes from the extremes of practical but not very useful, to very useful but not very practical. Pulse frequencies for next generation reactors on the planet which have the aim of generating electricity for a national grid may typically be in the range $\sim$1–50 Hz. Spacecraft on an interstellar trajectory may typically be in the range $\sim$100–200 Hz.

Electrical Supply

The fusion engine on-board a spacecraft will likely require a separate power reactor first to initialise it. But once it is underway, some of the energy released from each detonation can be bootstrapped to help power the next cycle such as in a magnetic field induction loop. The power source is likely to be a nuclear fission reactor of some form. Borrowing the equations from the performance of electrical ion drive engines, the electrical power is given as a function of the efficiency ϵ as follows

$$P_e = \frac{1}{2}\frac{v_{ex}T}{\epsilon}. \tag{6.6}$$

If we had an engine with a performance of 60% for an exhaust velocity of 1,000 km/s and a thrust of 10 N, then the electrical power requirement would be 8.33 MWe. This then implies a jet power of $P_j = v_{ex}T/2 = 5$ MW, where $\epsilon = P_j/P_e$. We can then estimate how long it would take to accumulate an associated energy and store it within a nuclear reactor, where the power is given by the burst energy output E, energy accumulation period τ and processing efficiency η

$$P_e = \eta\frac{E}{\tau}. \tag{6.7}$$

Therefore, for the case of a 8.33 MWe reactor power, producing 150 MJ of energy at a processing efficiency of $\eta = 30\%$ would result in an energy accumulation time of 5.4 seconds, which could be stored in some form of capacitor. If we assume a reactor rating of 1–2 kWe/kg, then this implies a reactor mass in the range 13.8–27.5 tons. It goes without saying that the reactor power must meet the total spacecraft power demands and this is a part of closing the power cycle.

The above estimates for power are based on ion-drive like systems and are just for the auxiliary electrical loop and does not take account of the lasers or the fusion jet for example. It will not go unnoticed by the reader however, that an engine of this sort is producing an enormous amount of energy rearwards of the vehicle, much of which is lost as waste heat. It would therefore make sense to attempt to capture some of this energy and return it to the system for energy storage, so that it can be used to power the next pulse cycle. Indeed, this is how many nuclear pulse engines work, and we refer to it as an induction loop cycle.

The principle is to use magnetic pick-up coils, so that a current is passed down a set of transmission lines and electrical energy can be accumulated. In principle, this then means that the main nuclear reactor need only be used for the initial start sequence, power management of other systems and as a back-up in case of engine destabilisation. This is not an ideal process though, and it is likely that much energy will be lost during the induction sequence. Good models for these sorts of engineering mechanisms do not exist, which is an opportunity for budding research engineers and inventors.

If we take the example of problem 3 from the last chapter, where it was found that a power requirement of 431 MJ must be sourced at the laser before firing, allowing for its inefficiency, if the efficiency of the induction process was only 50%, then this means we would have to induct around 862 MJ of energy from the engine reactions. The excess heat would then be emitted using waste radiators. Therefore, an induction loop system solves the problem of providing continuous power to each pulse cycle but comes with the penalty of being inefficient and may push up the

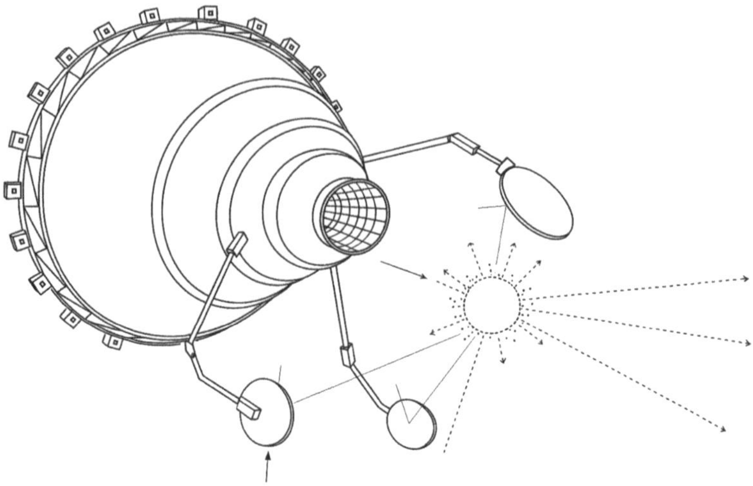

Fig. 6.4 Illustration of an engine with pellets injected into a target zone of a reaction chamber and then under laser impingement with mirror reflectors[KFL]

mass of the vehicle. The use of induction systems in power management is discussed further in Chap. 10.

Pellet Injection

Figure 6.4 illustrates fusion fuel pellets being injected just external to an engine system and then hit by multiple lasers, like the VISTA spacecraft design. This process would require extremely high precision and in a rapid time. For an advanced propulsion system, since the pulse frequency is likely high this implies the injection velocity must also be high. For example, for a 10 or 100 Hz pulse frequency this implies a pellet must be injected every 0.1 to 0.01 seconds respectively. The fundamental idea is that each pellet must have sufficient time to undergo injection into the target chamber, ignite and expand away from the detonation point, prior to the next detonation occurring.

Further, with each pellet there will be a plasma expansion wave that originates from the point of detonation, travelling at an average speed on the order of 10,000–15,000 km/s. This means that for a 10 m diameter reaction chamber (5 m radius) the expansion time would be 0.5 μs and 0.3 μs respectively. At the defined pulse frequencies this is not such an issue, but it starts to become an issue as the pulse frequency goes up in order of magnitude, or if the reaction chamber size gets smaller. Hence it should be clear that pulse frequency, capsule fuel mass, reaction chamber size is all interrelated.

Since the pellets must be injected on a timescale that corresponds to the pulse detonation frequency, this also has implications for the pellet injection velocity. This also depends on the barrel injection length, since by implication a shorter barrel length will have less time to accelerate up to the required injection velocity. If we take the case of the 0.1–0.01 seconds detonation time, for a barrel length of 10 m this implies an injection velocity in the range 0.1–1 km/s. As the pulse frequency increases into the 100s Hz this can rise to many km/s depending on the design.

The key consideration is how to inject pellets into the reaction chamber at such high velocity. A common idea is to use some form of electromagnetic rail gun, and to then coat the surface of the pellet with magnetic properties, such as with a superconducting material, and this was suggested by the Project Daedalus study. However, other ideas have also been considered such as using hydrogen gas guns and spinning gravity wings and these were discussed in a paper by this author [4]. In general, any timing jitter within the injection system must be far below the hydrodynamic timescale (~10s ns) to preserve drive symmetry.

We must also consider the vast number of ICF pellets and how these can be adequately manufactured for a particular mission. Since we are mainly referring to the thermonuclear fuel component we will use the terminology of capsules in this discussion. The number of capsules for a given mission is given by

$$n_{cap} = \frac{m_{prop} f_{hz}}{\dot{m}}. \tag{6.8}$$

In general, it would be a desirable requirement that the time to manufacture all of the capsules for the mission t_{man} is a lot less than the total mission time t_{tot} otherwise the timescales to launch will be elongated and one is also then faced with issues of other more advanced versions of the technology maturing. This can be expressed as a requirement $t_{man} << t_{miss}$. For the mission time, we can express this as the sum of the boost phase time t_b and the cruise phase time t_c which is then given which respect to the boost distance S_b and cruise distance S_c respectively

$$t_{miss} = t_b + t_c = \frac{m_{prop}}{\dot{m}} + \frac{S - S_b}{v_c}, \tag{6.9}$$

$$t_{man} << \frac{m_{prop}}{\dot{m}} + \frac{S - S_b}{v_c}. \tag{6.10}$$

Since the production rate is given by

$$R_p = \frac{n_{cap}}{t_{man}} >> \frac{n_{cap}}{t_{miss}}, \tag{6.11}$$

$$t_{man} >> \frac{m_{prop}}{m_{cap} R_p}. \tag{6.12}$$

These are for the situation where the capsules are manufactured pre-flight. What about the scenario when the capsules are manufactured on-board the mission. In this case we would require the opposite condition where the capsules are at least manufactured at the same rate as their usage, or ideally greater, such that $t_{man} \geq t_{miss}$. In which case the conditions would be:

$$t_{man} \geq \frac{m_{prop}}{\dot{m}} + \frac{S - S_b}{v_c}, \tag{6.13}$$

$$R_p = \frac{n_{cap}}{t_{man}} \leq \frac{n_{cap}}{t_{miss}}, \tag{6.14}$$

$$t_{man} \leq \frac{m_{prop}}{m_{cap} R_p}. \tag{6.15}$$

Which by implication then also places a minimum condition on the pulse frequency requirement as a function of the production rate and production timescale for a given mass flow rate and total propellant mass

$$f_{hz,min} \geq \frac{\dot{m} t_{man} R_p}{m_{prop}}. \tag{6.16}$$

It is useful to examine the Daedalus design [5, 6]. The Daedalus had a second engine stage which had a total propellant mass of 4,000 tons consisting of 0.288 g mass capsules, burned over 1.76 years at a pulse frequency of 250 Hz using a mass flow rate of 0.072 kg/s. This computes a total number of 1.3888×10^{10} capsules. If these were to be manufactured over a 10-year period, this would imply a production rate of 44 capsules/s sustained over that period. Similarly, the Project Daedalus first engine stage had a total propellant mass of 46,000 tons consisting of 2.84 g mass capsules, burned over 2.05 years at a pulse frequency of 250 Hz using a mass flow rate of 0.713 kg/s. This would compute a total number of 1.6129×10^{10} capsules and a production rate of 51 capsules/s.

Table 6.1 shows the production rates for these two engine stages over different production time scales. It is observed that to hit a production rate of less than 10 capsules/s the capsules must be manufactured over a time scale of order half a century. Similarly, to manufacture the capsules in a rapid time less than around 5 years implies a production rate of $\sim$100 capsules/s. In manufacturing the capsules therefore, it is a trade-off between short timescale for maximum production rate versus long timescale for low production rate which is then further traded with mission launch date and mission duration.

Figure 6.5 shows that an optimum production rate would appear to be of order $\sim$50 capsules/s manufactured over $\sim$5–10 years. Also shown in Table 6.1 and Fig. 6.5 are estimates for production rates of 100 tons typical of a NIF capsule with a mass of 1 milli-g [7, 8]. An optimum production rate for these sorts of capsules would be of order $\sim$100 capsules/s manufactured over $\sim$30 years and would involve a total of $\sim 10^{11}$ capsules. In general, it is observed that the production rate (capsules/s) is inversely proportional to the manufacturing timescale $R_p(cap/s) \propto 1/t_{man}$ (Years).

Table 6.1 Calculation of ICF capsule production rates (capsules/s) over different manufacturing time scales for different capsule masses

Capsule production time scale (years)	Daedalus 1st stage $m_{cap} = 2.84$ g $m_{prop} = 46{,}000$ tons (capsules/s)	Daedalus 2nd stage $m_{cap} = 0.288$ g $m_{prop} = 4{,}000$ tons (capsules/s)	NIF capsule $m_{cap} = 0.001$ g $m_{prop} = 100$ tons (capsules/s)
1	514	440	3,171
5	103	88	634
10	51	44	317
20	26	22	158
30	17	15	106
40	13	11	79
50	10	9	63
100	5	4	32

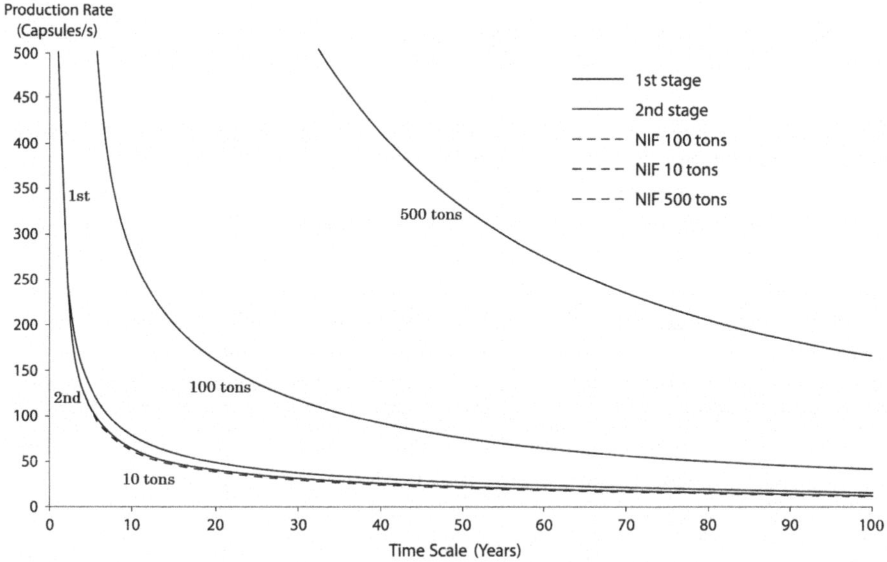

Fig. 6.5 Production rates (capsules/s) for ICF capsules over varying timescales with those for Daedalus engine stages

These results may suggest that it would be better to manufacture the individual ICF pellets on-route whilst the gas is stored within large propellant tanks. But then, this comes with two implications. The first is that this would mean that the manufacturing rate would have to be equal to or greater than the utility rate. In other words, if the engine is firing at a pulse frequency of 250 Hz like the Daedalus design, then the pellets must be produced at a similar rate or exceeding this, which doesn't seem reasonable. This suggests that a spacecraft which manufactured its pellets on-route would likely utilise a low engine pulse frequency and so would have a longer mission profile. The second implication is that manufacturing on-route suggests a need for substantial on-board production technology which would then increase the overall mass of the spacecraft. This then might suggest that spacecraft which manufacture their pellets on-route are likely to be mass heavy compared to those that do not. For any probes that stopped at destinations en-route for the function of fuel mining (stopping time) this would impact the total mission time and elongate the mission.

6.2 Magnetic Field Control

The transition of a high temperature plasma through an engine system is a complex process involving different particles, velocities and the interaction of electric and magnetic fields. A large plasma envelope which forms inside the reaction chamber expands as a spherical wave around the engine but is dragged out due to the motion of the spacecraft thus becoming an expanded ovoid. The key problem with this arrangement is the minimisation of the neutron and x-ray dosage from the pellet detonations. Layers of insulation and radiation shielding can be constructed around the individual engines to mitigate the radiation dose.

Since fusion energy relies on the release of charged particles within the reaction chamber as a constituent of the plasma that is formed, these particles must be directed magnetically for thrust generation. This is the equivalent of converting the kinetic energy of the charged particles into the kinetic energy of the thrust. This is facilitated using magnetic fields since otherwise the charged particles contact the material structure. The plasma must then detach from the magnetic field, or ions and electrons may spiral back towards the structure and reduce the thrust. This is illustrated in Fig. 6.6.

As the plasma expands adiabatically it has a Maxwellian velocity distribution. This is where the particles are described by the following equation

$$f(v) = \left(\frac{m}{2\pi k_B T}\right)^{3/2} 4\pi v^2 exp\left(-\frac{mv^2}{2k_B T}\right). \tag{6.17}$$

From which by using integration it is possible to derive the particle speeds in terms of the most probable, mean and root mean squared values where the velocity is strongly dependent on the temperature and particle mass

$$v_{mp} = \sqrt{\frac{2k_B T}{m}}, df(v)/dv = 0, \tag{6.18}$$

$$\bar{v} = \int_0^\infty vf(v)dv = \sqrt{\frac{8k_B T}{\pi m}}, \tag{6.19}$$

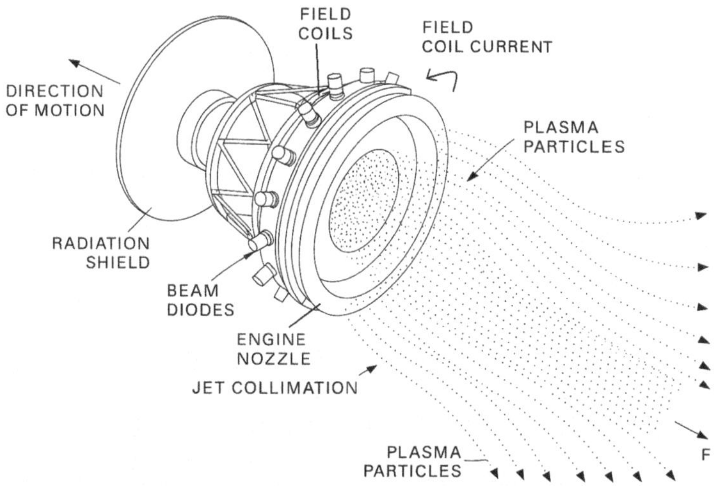

Fig. 6.6 Illustration of a magnetic nozzle.[KFL]

$$v_{rms} = \sqrt{\bar{v}^2} = \left(\int_0^\infty v^2 f(v) dv \right)^{1/2} = \sqrt{\frac{3k_B T}{m}}. \tag{6.20}$$

Since particles of different velocity will be at a different expansion radius this sets up a gradient in the plasma density. This then sets up a diamagnetic current on the plasma expansion surface in addition to an induced diamagnetic current J in the plasma and in the external magnetic field B to generate a Lorentz force for the electric and magnetic fields via $J \times B$, where the equation of motion for a particle of charge q and velocity v is given by the force density

$$F = q(E + v \times B). \tag{6.21}$$

Dividing both side by volume dV and where the current density per unit volume is $J = \rho v$ and the charge density per unit volume $\rho = dq/dV$ then

$$\frac{dF}{dV} = \rho E + J \times B. \tag{6.22}$$

Where the $J \times B$ acts perpendicular to the current direction and the magnetic field direction, thus converting the plasma flow in the thrust direction. Plasma detachment in a nozzle was demonstrated in numerical simulations by Kojima [9] for the case when the ratio of the plasma kinetic energy to the magnetic field energy is unity. Cassibry [10] has studied novel ways of winding field coils, longitudinally rather than azimuthally, to promote a higher force in the axial direction.

The principal way to achieve the required magnetic field is with the use of large field coils to provide magnetic field strengths of 10s T to provide the containment of the charged particles and then redirect them axially towards the rear of the spacecraft. The system is known as a magnetic thrust chamber and the rebounding of the magnetic field to push the plasma is analogous to the motion of a spring.

A novel method adopted for the Daedalus spacecraft design was to employ a highly optimised hemispherical structurally conducting thin-shell wall configuration around the reaction chamber which could also act as a form of shock absorber. When the detonation of an ICF pellet occurs, the plasma will expand into the target chamber rapidly. Since the expanding plasma compresses an existing weak magnetic field into the shell this causes its field strength to be raised from sub-T to 10s T within microseconds time duration. The design of the wall geometry as it goes into tension, would cause it to rebound and then push back on the plasma. This sets up a situation of steady state oscillation commensurate with the pulse detonation frequency but may lead to excessive heat loads and material fatigue.

For Daedalus the pulse detonation frequency was 250 Hz or one detonation every $1/f_{hz} = 0.004$ s. The expansion speed of the plasma within the reaction chambers was estimated to be ~15,000 km/s. This implies that for the two reaction chamber sizes of 50 m (first stage) and 20 m (second stage) that the time for the expansion of the plasma from the detonation point to the chamber walls $t = R/\bar{v}$ would be 3.3 μs and 1.3 μs for the first and second stage respectively. These times are far smaller than the detonation times and so there should be plenty of time between each detonation so that each plasma ball does not interfere with the next one.

Important for any space propulsion engine is the magnetic field strength used to confine the plasma within the engine exhaust after each detonation. The plasma pressure is defined by

$$P = nk_B T. \tag{6.23}$$

Where n is the particle species number density, and T is the temperature. This is essentially a measure of the kinetic energy density of the plasma as it expands from the detonation point. The plasma must be confined using magnetic fields. The Magnetic pressure is the pressure of the generated magnetic field is given by

$$P_B = \frac{B^2}{2\mu_o}. \tag{6.24}$$

Where B is the magnetic field strength and the permeability of free space $\mu_o = 1.256 \times 10^{-6}$ NA^2. The ratio of these two terms is then the plasma pressure over the magnetic pressure and it called the plasma beta

$$\beta_B = \frac{2nk_B\mu_o T}{B^2}. \tag{6.25}$$

Since $\beta_B \propto 1/B^2$ the stronger the field strength the lower the value of the plasma beta. However, the plasma beta is also a function of the particle density and temperature and so in principle within the engine nozzle should be very low which implies a high degree of magnetic confinement of the plasma since the magnet pressure dominates. For the corona within the Sun, it will typically have a value of around $\beta_B \sim 1\%$. For a magnetic Tokamak as proposed for a potential commercial fusion reactor it may be as high as $\beta_B \sim 5\%$. Hence, we target $\beta \ll 1$ near the nozzle throat and $\beta \rightarrow 1$ downstream.

If the magnetic confinement is weak within the engine nozzle or as the exhaust products expand away, then there is a risk that the energy of the individual charged particles will cause them to interact with the material structure therefore causing radiation and heat loads. It is desirable therefore to maintain a high degree of decoupling and so control the flow of the charged particles away from the engine to prevent them spiralling back around the magnetic field lines towards the vehicle.

Again, in the Daedalus study the authors derived a minimum condition for the magnetic field pressure as a function of the ion charge number Z, ion mass m, electron charge e, plasma expansion velocity $\bar{v}$ and distance from the plasma to the engine chamber wall R_w so that

$$B_o \geq \frac{3m\bar{v}}{ZeR_w} \geq \frac{1}{R_w}. \tag{6.26}$$

For the first engine nozzle $R_w = 50$ m, and $B_o = 0.0333$ T. For the second engine nozzle $R_w = 20$ m and $B_o = 0.0667$ T which implies an initial magnetic pressure within the reaction chamber (prior to pellet detonation) of 441 J/m^3 and 1,770 J/m^3 for the first and second stage respectively. These values are 1.6 and 1.3 times the minimum conditions for the two stages respectively. In these calculations it was estimated that $\bar{v} = 15{,}000$ km/s.

The fuel for the system was to be D^{3}He and was ignited at a temperature of 60 keV. The initial number density of the 50:50 fuel mixture was 0.089 g/cm^3 so that the nominal number density would be $n = N_A\rho/A$, where the Advogadro number $N_A = 6.022 \times 10^{23}$ mol^{-1}, and the average atomic mass $A = (2.01410 + 3.01603)/2 = 2.515$ so that $n = 2.131 \times 10^{28}$ m^{-3}. The first and second stage fuels were to be compressed by a factor of 1,000 and 2,000 respectively. This implies a compressed density for the two stages of $n_1 = 2.131 \times 10^{31}$ m^{-3} and $n = 4.262 \times 10^{31}$ m^{-3} and so a plasma pressure of 2.048×10^{17} J/m^3 and 4.097×10^{17} J/m^3.

Then in order for the plasma beta $\beta_B \rightarrow 1$, we can estimate the magnetic field strength required from the magnetic pressure. This would require an instantaneous magnetic field strength of $>>$10s T for both the first and second stage respectively and assumes significant compression of the magnetic field against the field adjacent to the reaction chamber wall. To gauge the plausibility of this would require more detailed calculations than presented here. An independent analysis of the Daedalus reaction chamber and associated plasma expansion was given by others [11] where it was found that the peak pressure at the chamber wall would be around 100 MPa and that the plasma was found to expand almost spherically up to turnaround. If we equate this to a magnetic pressure that would imply a peak magnetic field strength of 15.85 T which is consistent with the "10s T" claim by Daedalus but also far too low to be consistent with a unity value for the plasma beta.

For a space propulsion engine $\beta_B \rightarrow 1$ around the deflection point where the magnetic field pushes the charged particles. Since the plasma will then expand rearwards of the spacecraft B should decrease which means that β_B will increase. This ensures the charged particles spiral along the field lines and towards the rear of the vehicle and do not come back to the spacecraft. If the value of β_B moving away from the spacecraft does not continue to increase, then this will reduce the decoupling efficiency of the charged particles and may also decrease the jet efficiency.

The gyroradius, also known as the Larmor radius, defines the circular motion of a charged particle following the path of a magnetic field. It is defined by the charged particle mass m, the perpendicular component of the velocity $v_\perp$, the proton number Z, electric charge q and the magnetic field strength B as follows

$$r_g = \frac{m v_\perp}{ZqB}. \tag{6.27}$$

If the plasma contacts the chamber walls this leads to significant heating. Typically, when designing a propulsion engine, one should specify a magnetic field strength that is sized to the least magnetized relevant species population, which is usually the heaviest and fastest ions that still carry significant momentum.

We can estimate the gyroradius for the charged particles within the plasma. In a D^3He fuel such as used in the Daedalus design this is mostly for protons, Deuterium and Helium-3 particles. The results of calculating these for the Daedalus reaction chamber assuming a magnetic field strength of 15.85 T and $v_\perp = 1.5 \times 10^7$ m/s is shown in Table 6.2 along with the gyrofrequency of oscillation within the field. The calculated Larmor radius for the different charged particles is the minimum distance that must be maintained between the expanding plasma, and the reaction chamber walls if direct interaction is to be avoided. Calculations by Elliott [11] suggest the closest approach distance would likely be greater than 3 cm.

It is useful to illustrate with an example calculation for a problem so that we can get a feel for the various parameters. Let us assume we have a cylindrical reaction chamber with a diameter of 30 m (15 m radius). We assume a particle species for ^4He which has a mass $m = 6.6447 \times 10^{-27}$ kg. Let us further assume a plasma

Table 6.2 Larmor radius of charged particles within Daedalus reaction chamber assuming field strength $B = 15.85$ T and velocity $v_\perp = 15{,}000$ km/s

Parameter	p^+	D	T	^3He	^4He
Z	1	1	1	2	2
Z_e	1.0072	2.0135	3.0155	3.0160	4.0026
m (kg) $\times 10^{-27}$	1.673	3.344	5.007	5.008	6.646
q (C) $\times 10^{-19}$	1.602	1.602	1.602	3.204	3.204
r_g (cm)	0.988	1.975	2.958	1.479	1.963
ω_g (rad/s) $\times 10^8$	15.181	7.594	5.071	10.139	7.64
$r_g \times SM$ (cm)	1.976	3.95	5.916	2.958	3.926

expansion speed of 15,000 km/s, a representative particle density of $n = 10^{19}$ m^{-3} and a magnetic field of B = 10 T, where $kT \approx mv^2/3$. We can then calculate some of the properties of this plasma as follows

$$P_B = \frac{10^2}{2 \times 4\pi \times 10^{-7}} = 3.9789 \times 10^7 Pa = 39.8 \text{ MPa}, \tag{6.28}$$

$$P_p = \frac{1}{3} \times 10^{19} \times 6.6447 \times 10^{-27} \times (1.5 \times 10^7)^2 = 4.9835 \times 10^6 Pa = 5.0 \text{ MPa}, \tag{6.29}$$

$$\beta = \frac{5}{39.8} = 0.125(12.5\%), \tag{6.30}$$

$$r_L = \frac{6.6447 \times 10^{-27} \times 1.5 \times 10^7}{2 \times 1.602 \times 10^{-19} \times 10} = 0.0311m = 3.1 \text{ cm}. \tag{6.31}$$

Comparing the Larmor radius to the reaction chamber radius we find $r_L/R \approx 0.031/15 = 0.0021$ which is a strongly magnetized plasma. This places the flow comfortably magnetically dominated at the throat ($\beta << 1$) but then relaxing B by a factor of $\sim$2–3 would push $\beta \rightarrow 1$ for detachment further downstream. The overall goals are to keep the charged particles away from the reaction chamber surface structure whilst also directing them magnetically towards the exit for momentum conservation and thrust.

Figure 6.7 illustrates this scenario for the defined reaction chamber size. In (a) we have the chamber radius and capsule detonation wave originating from the

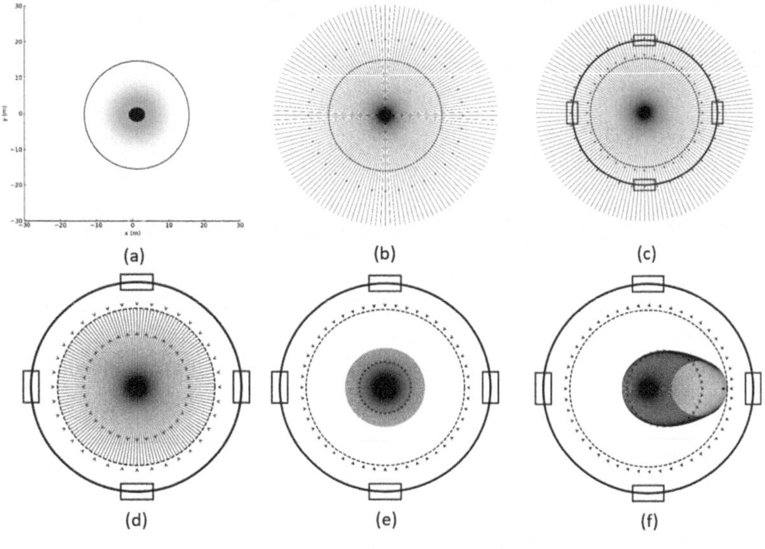

Fig. 6.7 Illustration of plasma expansion and field control within a magnetic nozzle[KFL]

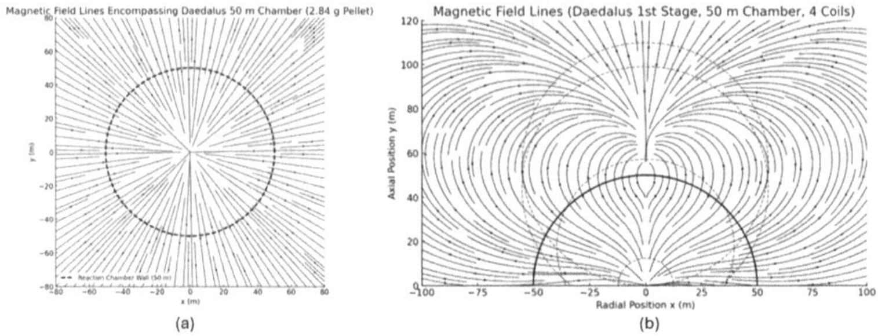

Fig. 6.8 Illustration of magnetic field arrangement around a reaction chamber[KFL]

centre. In (b) the magnetic field lines associated with the plasma stream outwards unconstrained. In (c) we have added magnetic coils to the external structure, but the associated magnetic pressure is too weak. In (d) the field strength of the field coils is increased to constrain the outward field lines. In (e) the magnetic pressure dominates, and the plasma field lines are squeezed into a converging bottle scenario. In (f) not only are the field lines constrained but they also diverge towards some exit direction, which we have illustrated here using a slightly offset x-direction.

Once the plasma has expanded into the reaction chamber, it will have its own self-generated magnetic field which the charged particles spiral along, in addition to that generated externally by the presence of field coils. Figure 6.8 shows an illustration of the magnetic fields around the Project Daedalus first engine stage 50 m reaction chamber as an example. (a) shows the field associated with the plasma that is streaming outwards from the point of detonation, and (b) shows the suppression of the internal field lines due to the externally generated fields from four field coils that direct the field lines rearwards and away from the spacecraft structure.

6.3 Jet Efficiency

The plasma debris power available P_{deb} from the fusion reactions is some fraction of the total fusion power P_{out} (minus losses) and together with the nozzle jet efficiency η_j it defines the jet power P_j of the engine. The thrust power P_{th} is that which is remaining from the plasma debris power which is then also split off into the bootstrap power P_{bs} such as is needed for an inductor power system to power the next cycle or pulse detonation. Nozzle Jet efficiency then is a measure of the difference between the power of the exhaust jet that leaves the nozzle and the power produced within the thrust chamber by the fusion reactions after extraction of the losses. In other words

$$\eta_j = \frac{P_j}{P_{deb}} = \frac{P_j}{P_{th} + P_{bs}}. \tag{6.32}$$

The plasma debris power is what remains from the power amplification of the energy release (i.e. through gain) minus laser coupling losses, direct losses to space and the emission of neutrons and x-rays so that $P_{deb} = P_{out} - P_{losses}$. A good design will optimise this power cycle for maximum jet efficiency and minimum radiation to the spacecraft structure since this comes with thermal load implications.

As shown by others [3] the amount of fuel which can be utilised in a nuclear reaction is less than the theoretical limit, as defined by the burn-up fraction, within the limit $\phi \leq 1$. Firstly, we write expressions for the kinetic energy of the jet exhaust E_{jet} and the energy output E_{out} from the fusion reactions as

$$E_{jet} = \frac{m_{cap} v_{ex}^2}{2}, \tag{6.33}$$

$$E_{out} = \alpha \phi m_{cap} c^2. \tag{6.34}$$

We then equate these to give an estimate for the exhaust velocity

$$v_{ex} = (2\alpha\phi)^{1/2} c. \tag{6.35}$$

When defining a value for the jet efficiency one must indicate where the energy has been lost. From the earlier definition of jet efficiency, we can state that $P_j = \eta_j P_{deb}$ and write the following for the debris power

$$P_{deb} = P_{out} - P_x - P_n - P_{las}. \tag{6.36}$$

Then the debris power that does not go into the exhaust jet is what will be remaining to heat up the walls of the vehicle. This is given by the absorbed power

$$P_{abs} = P_{deb} - P_{jet} = P_{deb} - \eta_j P_{deb} = (1 - \eta_j) P_{deb}. \tag{6.37}$$

Then the power absorbed into the walls of the spacecraft will be

$$P_{abs} = (1 - \eta_j)(P_{out} - P_x - P_n - P_{las}). \tag{6.38}$$

If the debris from the exhaust jet was to be absorbed into the walls then $\eta_j = 0$, which implies $P_{abs} = P_{out} - P_x - P_n - P_{las}$. Alternatively, in a perfect design with no losses and all of the exhaust debris was well collimated into the jet then $\eta_j = 1$, which implies that $P_{abs} = 0$. We have somewhat simplified the physics here, since the losses from the system are not just due to neutrons, x-rays and laser loss, but also the divergence of the jet itself in terms of collimation. There are other losses we will just mention. This includes inefficiencies due to the dispersion in velocity distribution of the different particles and how they interact with the magnetic field. The magnetic field itself may not be perfectly conducting and may diffuse into the plasma prior to the plasma decoupling from the field which may reduce the efficiency around the deflection point. The pulse detonations from the engine may vibrate the structure causing some energy absorption. These are not as large as the main ones however so in general we will neglect them.

Authors have often selected values of jet efficiency without any rigorous basis. It is a recommendation of NASA that values in the range $\eta_j \sim 0.7$–0.8 are appropriate [12, 13]. This depends on the type of design and Daedalus stands out as different from other designs due to its reaction chamber and minimal energy losses. The nozzle jet efficiency will be a measure of the energy produced from the fusion reactions and then how much of that was converted into useful work for propulsion. By implication $1-\eta_j$ will represent the losses. The momentum efficiency is defined by $\sqrt{\eta_j}$. The jet power for the spacecraft can be expressed as a function of the exhaust velocity v_{ex}, thrust T and the mass flow rate $\dot{m}$ as follows

$$P_j = \frac{\dot{m} v_{ex}^2}{2} = \frac{v_{ex} T}{2}. \tag{6.39}$$

The mass flow rate is given as a function of the ICF capsule mass m_{cap} and the pulse frequency f_{hz} via $\dot{m} = m_{cap} f_{hz}$ so making the appropriate substitution

$$P_j = \frac{m_{cap} f_{hz} v_{ex}^2}{2}. \tag{6.40}$$

To find the thermonuclear power output from the capsule detonation (with gain amplification of power) we use the fusion energy release for a given capsule as a function of the fractional energy release α and burn-up fraction ϕ via $E_{out} = \alpha \phi m_{cap} c^2$. This can be multiplied by the pulse frequency to get the energy release for all ICF capsules per second which is equivalent to the power output

$$P_{out} = \alpha \phi f_{hz} m_{cap} c^2. \tag{6.41}$$

Since all the losses are wrapped up in the jet efficiency term, as well as the burn-up fraction and mass defect, collectively this will define the total efficiency of the system performance in terms of ejection velocity. We can then bring the jet efficiency into the exhaust velocity as another factor to degrade the performance

$$v_{ex} = (2\alpha\phi\eta_j)^{1/2} c. \tag{6.42}$$

Then re-arranging for the jet efficiency we get

$$\eta_{j,min} = \frac{v_{ex}^2}{2\alpha\phi c^2}. \tag{6.43}$$

As shown in Fig. 6.9 an observation from this equation is the jet efficiency appears to be inversely proportional to the burn-up fraction. This is useful and it is the goal in an engine to successfully liberate as much of the energy from the fuel as possible which means a sufficiently high burn-up fraction. But the higher the burn-up fraction, the lower the jet efficiency, since the detonation of the pellet

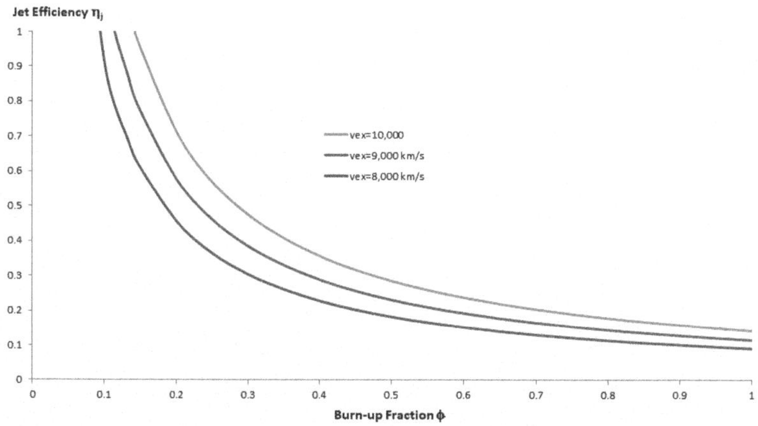

Fig. 6.9 Propulsion nozzle jet efficiency with ICF capsule burn-up fraction for lines of constant exhaust velocity assuming D^3He Fuel ($\alpha = 0.0039$)

will be more explosive and the angle of the jet less collimated which reduces the efficiency of the nozzle and therefore more power loss. It is important to understand however that $\eta_j \propto 1/\phi$ only holds true here since the exhaust velocity was assumed to be a constant. In which case a higher burn-up for the same exhaust velocity means that the engine is wasting more of the fusion energy, and so the jet efficiency must go down. However, see the later discussion in Sect. 7.2.

These equations suggest that the burn-up fraction and the nozzle jet efficiency is independent of the pulse frequency of detonation. Calculating the jet efficiency for Daedalus with $\alpha = 0.0039$ (D^3He), for the first stage assuming a burn-up fraction of $\phi = 0.175$ and exhaust velocity $v_{ex} = 10{,}600$ km/s which computes to $\eta_j = 0.915$. For the second stage assuming a burn-up fraction of $\phi = 0.133$ and exhaust velocity $v_{ex} = 9{,}210$ km/s this computes to of $\eta_j = 0.909$. The nozzle efficiency for different theoretical and experimental concepts is shown in Table 6.3.

Problems of Chap. 6

6.1 An advanced propulsion system has an exhaust velocity of 1,500 km/s and generates a thrust of 0.1 MN. The propulsion system is only 30% efficient at converting the energy to thrust. (i) What would be the minimum power required to start up the engine? (ii) Assuming a power plant reactor that had an efficiency of 1%, after one hour of operation how much energy would it generate? If we assumed this power output was produced by a reactor with a mass 50,000 tons, what rating would this correspond to in terms of specific power output?

6.2 A 25 m reaction chamber has a thrust profile that requires a pulse detonation frequency of 500 Hz. (i) What is the time between pellet detonations and assuming a plasma expansion wave of 15,000 km/s what is the time it would take for the plasma to reach the external part of the reaction chamber? (ii) What would be the minimum recommended pellet injection velocity assuming the detonation point is 5 m from

Table 6.3 Estimates for nozzle jet efficiency for different systems

Concept/nozzle	η_j
Proper DeLaval nozzle	0.98
Halyard (1997)	0.98
Classic triangular chemical nozzle	0.95
Perakis 4 ring Daedalus	0.85
Project Daedalus	0.91
Pegasus (a, c)	0.85
Pegasus (b)	0.81
Discovery-II MCF	0.80
NASA TIM	0.70
Orion flat plate	0.50
VISTA (1987)	0.34
VISTA (2003)	0.32
VISTA single ring	0.65
VISTA double ring	0.75
Hyde (1983)	0.42
SunVoyager	0.38
Discovery-III	0.10

the reaction chamber exit hole? (iii) Assuming a 10 m barrel length, what would be the acceleration associated with this velocity? State your answers in multiples of Earth gravity.

6.3 For the reaction chamber problem at the end of Sect. 6.2, (i) What magnetic field strength would we need corresponding to a Larmor radius of 1 m for the ^{4}He particles? (ii) Using the same parameters calculate the magnetic field strength for hydrogen? (iii) Comment on the difference between them.

References

1. Kondo Y, Bruhweiler FC, Moore J, Sheffield C (eds) (1971) Interstellar travel and multi-generation space ships. Apogee Books Publication, Burlington
2. Spencer DF (1963) Fusion propulsion system requirements for an interstellar probe. Technical report No.32-397, NASA CR-50964, Jet Propulsion Laboratory
3. Spencer DF, Jaffe LJ (1963) Feasibility of interstellar travel. Acta Astronautica 9:49–58
4. Long KF, Lamontagne M (2025) Project Icarus: pellet injection system for the pegasus spacecraft concept design. JBIS 78(6):206–215
5. Bond A, Martin AR (1978) Project Daedalus: the mission profile, final study report. JBIS Special Supplement:S37–S42
6. Martin AR, Bond A (1978) Project Daedalus: the propulsion system part 1: theoretical considerations and calculations. JBIS Special Supplement:S44–S62
7. Lindl JD (1998) Inertial confinement fusion: the quest for ignition and energy gain using indirect drive. Springer/AIP Press, Berlin/Melville
8. Dewald E (2016) Status of indirect drive ICF experiments on the national ignition facility, LLNL, TR-686482

9. Kojima T, Morita T, Yamamoto N (2020) Analysis of plasma detachment in the magnetic thrust chamber using full particle-in-cell simulation. High Energy Density Phys 36:100814
10. Cassibry J, Winterling B, Schillo K (2018) Pulsed magnetic nozzle for fusion propulsion. JBIS 71:119–125
11. Elliott JO, Terry WK (1985) Plasma expansion in the Daedalus first stage engine. JBIS 38(3):120–128
12. Williams CH, Dudzinski LA, Borowski SK, Juhasz AJ (2005) Realizing 2001: a space odyssey: piloted spherical torus nuclear fusion propulsion, NASA/TM-2005-213559, AIAA-2001-3805
13. Orth CD, Turchi PJ, Carpenter SA, Santarius JF, Frisbee RH, Williams CH (2004) Recommended design practices for conceptual nuclear fusion space propulsion systems, special project report, SP-108-2004, AIAA

Chapter 7
Principles of Aerospace Design

We are like 19th century individuals trying to imagine how to get to the Moon. Travel to the stars is extremely difficult and definitely expensive, but we did get to the Moon and we can get to the stars.

— Brice Cassenti, 1982 [1]

7.1 Principles of Design

The main goal of starship engineering , can be defined as:

To design a spacecraft that can carry the maximum payload mass for the mission requirement, achieving maximum science return and minimum total trip time to another planet, star or celestial object, optimised for minimum propellant mass.

All problems must start off with a requirement. Using the Project Icarus example discussed in Chap. 3, we might imagine that the requirement is to send a vehicle with payload mass 150 tons to an interstellar target within 100 years and decelerate it into full orbital velocity.

Once the requirement has been understood and the implications considered, one must immediately turn towards the generation of numbers to properly scope out what is the domain of the potential solution space but also what are the constraints on the mission. This method then leads to a determination of a required velocity increment. One then goes through the propulsion options and the associated performance to determine the exhaust velocity (or equivalent) so that the mass ratio can be calculated. Following from this we find the propellant mass requirement.

Proceeding from here one must begin to construct a mass budget definition for the various systems and subsystems. This list will evolve as a greater fidelity of analysis and calculations are conducted for each system so that overall, the concept converges towards a proper reflection of the vehicle description. As the systems are researched, a vehicle configuration then also emerges so that ideally form follows function as is standard practice in engineering design.

So now the designers have a vehicle configuration description, list of systems, mass budget breakdown and a basic mission profile and performance description.

K. F. Long, *Stellar Engines*, Springer Aerospace Technology,
https://doi.org/10.1007/978-3-032-18062-9_7

This is the first level of a concept design, and is a sufficient starting point from which to evolve more detailed calculations, trade and optimisation studies. It is necessary to allow for margins in the various systems since the integration will involve an inevitable loss and gain of mass between each. Also, it is necessary to reduce the uncertainty in each system so that the allocated mass better approximates the likely requirements. This is not easy however for starships, since we are often dealing with technologies of the future not yet invented, or at the very least extrapolated from the existing state of art and beyond what may be appropriate.

From here it is good to form a team with individuals allocated to specific specialisms so that they can focus on those specific areas. This also requires project management to ensure individuals communicate and share results, but also that they are not duplicating work, or producing work that contradicts the direction of another system. Separate to a project manager would be the adoption of a programme manager, who helps to direct the overall research and thinks about timelines for delivery, stage gates and milestones, and the technical publications output of the team. The project manager oversees the design team's workflow and integration. The programme manager, in contrast, directs the overall roadmap—which may include managing the funding, milestones and long-term technical strategy.

The above is an ideal philosophy that would work in a professional aerospace company where there is a clear chain of command and individuals are paid employees. Most starship projects are undertaken by volunteers, nobody is paid, and they also have other full-time work commitments. So instead, these sorts of starship projects tend to become more like social events, balancing the technical needs to discuss the project with the conversations between friends. Also, since the subject matter is highly interesting, the 'get-togethers' are filled with enthusiasm and are a 'hotbed' for creativity and ideas but often lacking in discipline to turn those creations into a usable product. Thus, again it falls to the project manager to create these networking opportunities for teams to collaborate, whilst also forging a sense of discipline to keep some level of focus. Starship engineering is difficult from a technical perspective, but it is also difficult at the human level, and some might argue it a major impediment towards success since one is also now dealing with social-cultural and political-philosophical challenges.

Some consideration should also be given towards the construction of the machine. The actual programme to build a starship is no small undertaking and involves many technology steps to get to the first launch. Such a programme is further complicated when one moves to crewed missions, which is ultimately the goal. One way we might prepare for such missions is to learn lessons from history and Project Apollo is arguably the biggest program achievement in the history of human space exploration.

After we had successfully landed on the Moon, the American space agency NASA asked some of the key engineers involved on the ground to define what had made Project Apollo so successful. The answers were written up in a report which is worth reading [2]. It is useful to cite some of the reasons given and then place them into the context of a starship project:

1. *Attention to detail for the spacecraft development and mission operations.* Scrutiny of each component of the mission to maximise reliability and increase success. For a starship, which is likely to be an even more complex machine, attention to detail would be needed across all systems. Yet this is not always possible since one is dealing with existing technologies but also envisaged technologies and so full systems integration is not yet possible.
2. *Simplicity in design, using established technology with hardware reliability.* Minimizing in-flight maintenance and simplifying operations, minimising functional interfaces between complex systems. Learning lessons from the previous missions and programs. This would all be true for a starship except for in-flight maintenance which would have to be undertaken by the autonomous on-board repair factory and computer. Although individual components could be designed so that their maintenance is made easier to achieve.
3. *Redundancy in design, to prevent mission failure in the event of one component being faulty.* For the starship a failure of the entire vehicle should require a sequence of failures in series rather than a single system and it should be possible to isolate those faulty components and repair or replace them.
4. *Comprehensive testing of all systems individually and as integrated units.* It is expected that for the starship the degree of testing and technology validation prior to launch will be high. In addition, precursor missions to shorter distances using the same technology would have been launched making the mission more reliable.
5. *Strict configuration control management to ensure consistency and reliability.* For the starship any changes to the hardware or software systems would have to be carefully tracked and managed to prevent conflicts occurring. Due to the distance involved this would ideally be managed by the on-board artificial intelligence computer during the mission.
6. *Extensive training and simulation, flights, use of simulators.* For the starship it would be possible to run millions of simulations to model different scenarios and how the computer responds. Since the computer is artificial intelligence, it will have its own learning algorithms so can react to each scenario based on its built experience. This implies the computer will be constructed ahead of the actual vehicle. In the same way that for Project Apollo crews were highly trained before launch, the starship's artificial intelligence must be pre-trained and validated through millions of mission simulations to ensure autonomous reliability.

It is an observation that for Project Apollo the principles of spacecraft design involved a combination of aircraft design practice and elements of missile design technology (the rockets). Similarly for the starship, the principles of spacecraft design involve a combination of a Voyager like probe with much larger mass and elements of a massive launch vehicle with powerful high thrusting engines which must operate for years at a time. In both cases, all the learning from our previous efforts and initiatives are brought together in the construction of a new machine.

In the design of advanced spacecraft, one is faced with challenges that go beyond the current design problems. This especially pertains to the environmental conditions for which the different materials of the vehicle will go through, due to the significantly high temperatures, powers, energies, accelerations and velocities. This also pertains to the fact that we are dealing with technologies that are not yet invented, although some will be based on an extrapolation from the existing state of art. To do otherwise would verge on science fiction. This means that in the design of advanced spacecraft we must straddle the two domains of rigorous analytical practicality and speculative non-mature ideas. This also justifies the sub-title of this book and the reference to *extreme aerospace engineering*.

To navigate our way through this it is good to establish some design guidelines so that mistakes can be minimised and any design solutions have some credibility in the real world. The principles below were developed by this author having worked on multiple design projects and are written for guidance.

1. When given a requirement, before proceeding towards options, one should run some back-of-envelope calculations and question the requirement, especially if it contains internal inconsistencies. It may be that you need to help develop the requirement towards practical realisation, but without closing off options.
2. When first generating a concept, remain at the top level and don't try to go too deep in any system on the first iteration. Since the concept is likely to undergo significant configuration changes which will affect the different systems. This also means keeping any margins large to allow for compromise on integration later.
3. When dealing with futuristic technologies, it should be the aim to extrapolate from known technologies as much as possible, if the accusation of doing science fiction is to be avoided. Keep in mind that the longer out in time the prediction, the greater the uncertainty in the assumptions. Extrapolating more than a few decades is the likely limit of prediction.
4. In generating ideas the goal should be to start broad and converge narrow as you head towards down-selecting of the likely options. This is a process that must be managed. Late-stage ideas, whilst creative, can add turbulence to the process. Similarly, early-stage decisions can constrain other subsystem options, so should be done rarely.
5. In the construction of design options, one should seek to build models. This can be done in a spreadsheet, on paper or using numerical programming tools. This allows for scrutiny of the model through transparency of the operation. The model should ideally be constructed in multiple tools simultaneously, since they will highlight deficiencies or errors in each other. In addition, the model should be applied to alternative designs or designs that go outside of the convergence path, since this will also highlight sensitivities and limitations in the model which helps you to understand the credibility of the design.
6. In the development of design models, properly label them and keep track of any changes so that earlier models can be revisited. You should aim to go through cycles of model development and model freezing, with the aim that over time the model will become more credible and achieve a higher level of numerical definition in terms of systems and performance.

Here we present a simple process that can be followed in the conceptual design of a new spacecraft concept. This is especially aimed at designs characterised by low TRL systems and so the technology will not yet be matured to the required level. We can outline a 12-step process as follows:

1. Define the mission requirement and interpret any implications to the design. Note that it may be that the requirement as defined is not feasible and may need revision. This is a back and forth with the designer and the client. Generate a hierarchy of requirements to include system and subsystem level.
2. Perform back-of-envelope calculations on the time of flight and expected motion to get an idea for the domain of vehicle performance being discussed. Brainstorm ideas within a team to help in the generation of concepts.
3. Construct a preliminary definition for the vehicle and likely systems needed as a form of vehicle configuration layout. From this derive a first mass estimate, then estimate the required exhaust velocity (specific impulse) required for the mission and thereby the propulsion options and types of fuels, from a preliminary literature review. This is the initial Δv close-out step for the model.
4. Generate several concept ideas, perhaps between ~3–5, for which they all meet the mission requirement but have different vehicle configurations and propulsion options. They must all close on Δv and mass at the back-of-envelope level.
5. Calculate the concept ideas to derive vehicle mass budget and mission performance. Sketch the configuration layout for each option.
6. Compare technology to existing and envisaged TRLs to meet the mission and down select options to a smaller subset, perhaps ~2–3 options. Build preliminary models such as in a spreadsheet. Construct a trade matrix for criteria like mass, performance, TRL, complexity, cost. Derive a criteria to eliminate options so that any down-select is defensible.
7. Move to refined calculations where each system is defined to a higher level of detail. As changes are made, record them so earlier models can be regenerated. Conduct an extensive literature review on the technologies for each option and assess experimental validation to support the technology performance required.
8. Down select to the preferred design concept. Conduct more extensive calculations for each system and assess margins and uncertainties in expected performance and determine failure modes. Conduct additional calculations using other numerical tools to help understand the physics and engineering constraints and expected environmental conditions for the mission. If possible generate driving design equations to aid discussions and optimisation of parameters.
9. Build a rigorous mathematical model for the design, in a spreadsheet or in a numerical programming tool. It should be possible to perturb each system within the model so as to scope out the design space for optimisation of systems. Aim to minimise mass for maximum performance. The performance should be efficient and not waste energy and also not over strain the technologies towards a point of failure through sensitivity analysis.

10. Generate 2D orthographic drawings and 3D CAD models of the vehicle concept design. This should include a list of all system and sub-system components where known. It should be accompanied by a rigorous mathematical model.
11. Produce a comprehensive description of the performance and mission architecture, which covers the design, construction, assembly, launch and mission utilisation. May also want to include operational concepts like ground segment and autonomy level, verification and validation plans, ground to flight demonstration tests. May also need to consider end-of-life plans for the spacecraft should it ever be launched, safety and planetary protection protocols.
12. Develop a strategic and technological roadmap which defines the pathway for the development of the vehicle up to in-service use for the mission requirement. This should include how each critical technology climbs the TRL and identify what are the driving technologies. Include in this a cost analysis and the definition of critical Capital Expenditure (CapEx) items in the design and determine which technologies lead the cost, and which ones must come down to what value in order for the mission to become feasible by what date.

In 1999 a US NASA / DOE workshop took place called the Technical Interchange Meeting (TIM) [3]. This was in response to the observation that there was a need to develop best practice recommendations when it comes to designing advanced space propulsion using fusion reactions. They noted inconsistencies in design assumptions, variations in technological extrapolation and risk, and differences in the fidelity of system mass properties. There was also a difference in the way designers analysed concepts using different figures of merit. There was a clear need to standardise conceptual designs of fusion propulsion systems.

One recommendation was designers compare their designs to one of three reference missions: (1) a crewed Mars round trip in $\leq$6 months, 150 tons total useful payload to Mars orbit (2) crewed Jupiter/Europa round trip in $\leq$2 years, 175 tons total useful payload to Europa orbit (3) robotic gravitational lens mission to 550 AU rendezvous only in $\leq$10 years, 10 tons useful payload.

The working group did not develop a reference mission over interstellar distances. We suggest that one of the following might be appropriate (4a) robotic mission to 4.3 ly flyby only in $\leq$50 years, 100–150 tons useful payload (4b) robotic mission to 4.3 ly rendezvous in $\leq$100 years, 100–150 tons useful payload.

The working group also defined a set of Figure of Merits (FOM) that should be adopted and agreed that these should be mission-level rather than engine/reactor-level specific. The figures of merit are:

- *Mission Level FOM*: Total trip time t_t, spacecraft total wet mass m_w, useful payload mass m_{pay}.
- *Propulsion Level FOM*: Trip distance S (or cruise velocity δv), dry mass m_d, specific impulse I_{sp} (or exhaust velocity v_{ex}), specific power P_{sp}, engine jet efficiency η_j.

The main reason behind this choice of parameters was because designs tended to be optimised around them, such as minimum trip time and maximised payload mass fraction. Any designs therefore presented in this book should strive to calculate these properties and where possible justify them. This will help to bound the specific propulsion method under study which pertains to advanced propulsion.

However, for the application of an advanced nuclear pulse engine we propose the addition of three new parameters. This includes the thermonuclear fuel capsule mass m_{cap}, pulse detonation frequency f_{hz}, capsule burn-up fraction ϕ and laser efficiency ϵ, which are suggested as mission level FOMs. Collectively these help to describe the type of mission and the operation of the engine. These parameters directly influence the overall vehicle power cycle, propellant throughput, and thrust continuity. They therefore connect the microscopic physics of fusion events to the macroscopic mission architecture. The laser efficiency in particular will be a major driving factor on the mass of the waste heat radiators and therefore determine whether such a mission is feasible.

It is not practical to review the full recommendations of the working group, but it is necessary to highlight some of the relevant recommendations for best practice in fusion propulsion design. The wording of these recommendations has been altered slightly for this book:

- Designers should analyse the performance FOM of concepts against at least one reference mission. Critical technologies used in concept should have been demonstrated at the sub-system proof-of-concept level or conservatively extrapolated by up to three decades.
- A generic mass property breakdown should be included in a system concept design. Designers should assess a 30% mass growth allowance on the total dry mass of the propulsion system and any payload (This was based on observed trends in aerospace programs for a similar 15–30% mass growth was typical).
- Flight performance reserve should be 1% of main impulse propellant and fuel. Residuals and losses should be 3% of total tankable propellant and fuel.
- The calculation background for the specific impulse (or exhaust velocity) should be explained. Designers should produce a power flow description for engine cycle with losses clearly defined. For jet power producing systems and loss mechanisms, a jet power reduction of 30% ($\eta_j = 0.7$) should be assessed unless it is already demonstrated to have a different value.
- The calculation of specific power should be performed based on the dry mass needed for transforming it into thrust, but only for that dry mass which scales with the power (This implies the propulsion mass which would be the engine/reactor and associated elements, auxiliary power systems and thermal management such as the heat radiators).

In taking account of the working group's recommendations, any design project should attempt to adopt these design practices where possible. It is also important to highlight the opinion of appropriately qualified people in the design of advanced fusion propulsion systems, and this includes Dr. Charles Orth of LLNL. In a 2000 paper [4] he stated: "*A detailed systems-analysis code is considered to be essential for the analysis of any credible concept for fusion space propulsion. Such a code allows parameter and sensitivity studies, which are critical to determine the parameters that have the most leverage upon a particular design and its performance....A systems analysis code permits more well-defined values for vehicle performance once the details for the various systems are known. Lacking such information, no credible performance estimates are possible.*"

For the Daedalus study in particular, the team was extremely constrained in how far they could model the compression and ignition of the pellets. They stated "*Simulation of the implosion of large pellets of* D^3He*... via the use of such codes was an area to which this study could not contribute, and the lack of such results proved frustrating. The remedying of this situation would be a valuable service to the cause of nuclear pulse propulsion*" [5].

The best way to approach this problem therefore is to construct a modelling tool. In particular, the model should contain two main parts. The first pertains to the spacecraft design and configuration definition and calculated mission profile. The second pertains to the energy driver and propulsion system and any specific ignition mechanisms. These are then integrated into a systems level analysis tool so that designs can be perturbed and evolved.

Ultimately, starship design lies at the intersection of physics, systems engineering and cultural interests. A disciplined design process, underpinned by transparent modelling and sharing data within the team, will be as critical to success as any breakthroughs in physics or engineering that are likely to occur.

7.2 Driving Design Equation

In general, for ICF propulsion engines it is observed that the nozzle jet efficiency is inversely correlated with the burn-up fraction since a more explosive detonation will result in a less collimated exhaust jet. The jet is defined by a half angle θ divergence term $\lambda = [1 + cos(\theta)]/2$ so that the exhaust velocity is given as $v_{ex} = \eta_j^{1/2} v_{exp}$, where the expansion velocity is given by $v_{exp} = (2\alpha\phi)^{1/2}c$ and it therefore follows that $\lambda = \eta_j^{1/2}$. In an idealised engine $\theta = 0$ would be representative of a perfectly collimated jet exhaust with no divergence or expansion away from the reaction chamber nozzle. So that the thrust $T = \dot{m} v_{ex}$ is then defined by a driving design equation for the case of no expellant and with expellant as follows

$$T = m_{cap} f_{hz} (2\alpha\phi)^{1/2} c\lambda, \tag{7.1}$$

$$T = (m_{cap} + m_{exp})^{1/2} f_{hz} (2 m_{cap} \epsilon_{mix} \alpha\phi)^{1/2} c\lambda. \tag{7.2}$$

Then simplifying

$$T = (2\epsilon_{mix}\alpha\phi[m_{cap}^2 + m_{cap} m_{exp}])^{1/2} f_{hz} c\lambda. \tag{7.3}$$

Although to be exact these equations must be corrected with a momentum weighting factor which takes account of the different mass in the reaction species and is worth laying out. The propulsion operational performance parameters are specified as a function of the ICF capsule mass and pulse frequency. Before we examine a full capsule mass of thermonuclear fuel, we firstly consider the fusion of two individual nuclei. This approach was previously taken by others in relation to the design of an interstellar flyby probe [5, 6]. The kinetic energy of an individual ion will be given as a function of the ion mass for a particular species m_i and its velocity v_i by

$$E = \frac{1}{2} m_i v_i^2. \tag{7.4}$$

The energy release from an ion due to fusion reactions is given as a function of the burn-up fraction ϕ, energy fraction α and the speed of light c as follows

$$E = \phi \alpha m_i c^2. \tag{7.5}$$

Then equating the above expressions and re-arranging

$$v_i = (2\phi\alpha)^{1/2} c. \tag{7.6}$$

For the reaction $D(^3He,H)^4He$ since the energy per ion will be the same each component of the reaction can be written as follows

$$m_{3_{He}} v_{3_{He}}^2 = m_D v_D^2 = m_H v_H^2 = m_{4_{He}} v_{4_{He}}^2 = 2\phi\alpha m_i c^2. \tag{7.7}$$

Where m_i is the mean ion mass, $m_{3_{He}}$= 3.016029, m_D= 2.014102, m_H= 1.007276, $m_{4_{He}}$= 4.002603 in atomic mass units (amu). In terms of the energy fraction for $D(T,^4He)n$, $\alpha = 0.0038$; $D(D,T)p$, $\alpha = 0.0011$; $D(D,^3He)n$, $\alpha = 0.0009$; and $D(^3He,^4He)p$, $\alpha = 0.0039$. The following was shown [5] through substitution but rewritten here in a simplified form as the effective expansion velocity

$$v_{eff} = \Big(A(1-\phi) + B\phi\Big)\sqrt{\phi}\sqrt{\left(\frac{\alpha}{2}\right)} c. \tag{7.8}$$

Where the constants A and B are the average sum of the nuclei masses for the reactants m_1 and m_2 and the products m_3 and m_4.

$$A = \frac{\sqrt{m_1} + \sqrt{m_2}}{\sqrt{m_i}} = \frac{\sqrt{m_1} + \sqrt{m_2}}{\sqrt{(m_1 + m_2)/2}}, \tag{7.9}$$

$$B = \frac{\sqrt{m_3} + \sqrt{m_4}}{\sqrt{m_i}} = \frac{\sqrt{m_3} + \sqrt{m_4}}{\sqrt{(m_3 + m_4)/2}}. \tag{7.10}$$

Where the effective expansion velocity for D^3He would be as follows

$$v_{eff} = \left(1.9899(1 - \phi) + 1.8982\phi\right)\sqrt{\phi} \times 1.3248 \times 10^7. \quad (7.11)$$

The conical nozzle design is defined by a correction factor λ_d to the nozzle exit momentum for an ideal rocket. It defines the ratio of the gasses within a finite angle nozzle and the momentum of an ideal nozzle in the axial direction as a function of the half angle θ for the divergence

$$\lambda_d = \frac{1}{2}\left(1 + \cos(\theta)\right). \quad (7.12)$$

Then the exhaust velocity is given by

$$v_{ex} = \frac{1}{2}\left(1 + \cos(\theta)\right) \times v_{eff}. \quad (7.13)$$

A nozzle that has little or no divergence will give rise to an efficient axial jet but would require a longer nozzle which increases the mass of the system. A nozzle that has a large divergence angle will have a less efficient performance and lower specific impulse but also require a shorter nozzle and so lower mass. In any nozzle design a trade between nozzle mass and nozzle efficiency therefore must be made. A more extensive discussion on this issue is given in the classic text by George Sutton [7].

Figure 7.1 illustrates how the velocity ratio changes with nozzle divergence angle. Figure 7.2 illustrates the exhaust velocity with nozzle angle as a function of burn-up fraction. The scenario of $\theta = 0°$ would represent a perfectly collimated

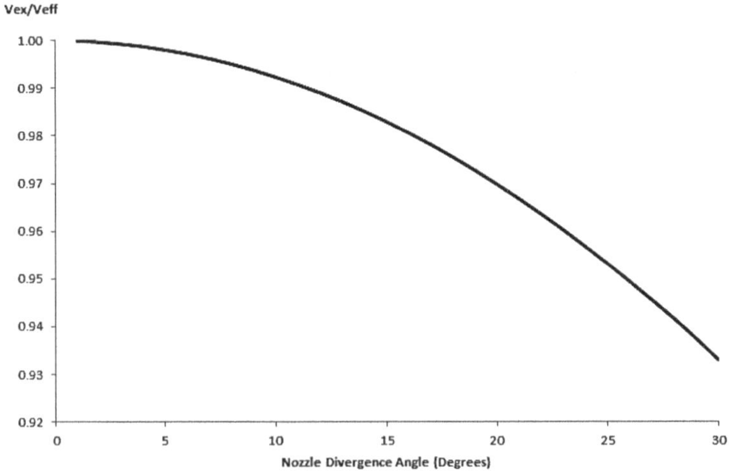

Fig. 7.1 Calculation of Ratio v_{ex}/v_{eff} with nozzle divergence angle θ

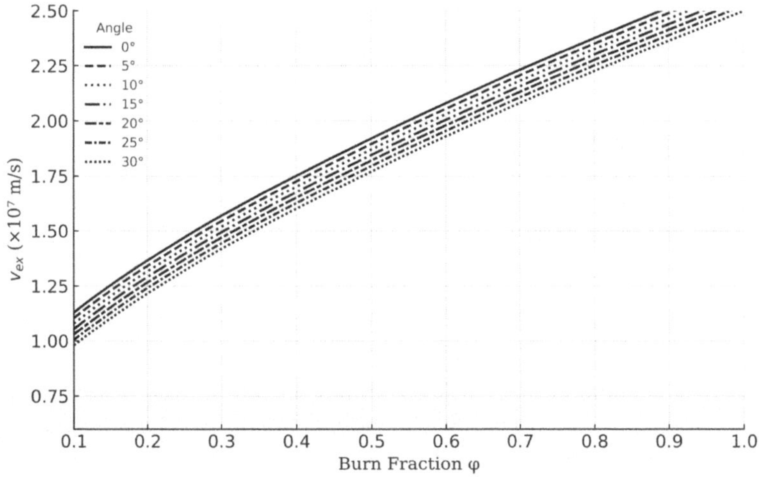

Fig. 7.2 Calculation of exhaust velocity v_{ex} with burn-up fraction ϕ for values of divergence angle θ

Table 7.1 Calculation of velocities assuming $\phi = 0.155$ and $\theta = 20°$

Reaction	α	m_i	A	B	$(\sqrt{\alpha/2})c$ (m/s)	v_{eff} (km/s)	v_{ex} (km/s)	v_{exp} (km/s)
D(^{3}He,^{4}He)p	0.0039	2.515	1.9899	1.8982	1.324	10,297	9,987	10,424
D(T,^{4}He)n	0.0038	2.515	1.9899	1.8984	1.307	10,165	9,858	10,289
D(D,^{3}He)n	0.0009	2.014	2.0000	1.9322	0.636	4,981	4,831	5,007
D(D,T)p	0.0011	2.014	2.0000	1.9321	0.703	5,507	5,341	5,536

jet with no divergence, whereas the scenario of $\theta = 30°$ would represent a widely divergent jet and therefore inefficient performance. Calculations for the different reactions are shown in Table 7.1 which are computed assuming a burn-up fraction of $\phi = 0.155$, nozzle divergence angle $\theta = 20°$.

For comparison, we might look at the Daedalus engine for the same divergence angle that utilised D^{3}He fuel. For the first stage with a burn-up fraction of $\phi = 0.175$ we compute v_{eff} = 10,931 km/s, v_{ex} = 10,602 km/s, v_{exp} = 11,076 km/s. For the second stage and $\phi = 0.133$ we compute v_{eff} = 9,548 km/s, v_{ex} = 9,260 km/s, v_{exp} = 9,656 km/s

In the context of space propulsion, it is more useful to speak in terms of energy release from an ICF fuel capsule rather than an individual ion. The energy output from a capsule is given as a function of the capsule mass m_{cap}, the burn-up fraction ϕ and the fractional energy release α as follows

$$E_{out} = \alpha\phi m_{cap}c^2. \tag{7.14}$$

The burn-up fraction is highly design dependent and involves calculations for the hydrodynamic implosion, hot-spot ignition and thermonuclear burn of the fusion fuel. The kinetic energy of a capsule will be given as a function of the capsule mass and the debris velocity $\tilde{v}$ of the individual plasma ions

$$E_{kin} = \frac{m_{cap}\tilde{v}^2}{2}. \tag{7.15}$$

Then equating the above the expansion velocity (often approximated as the exhaust velocity) is given by

$$v_{exp} = (2\alpha\phi)^{1/2}c. \tag{7.16}$$

However, if one calculates the value of Eq. (7.16) and compares it to the more accurate Eq. (7.8) there is a $\sim$0.5–1.2% difference and a slightly lower value. So the use of Eq. (7.8) correctly considers the average mass and momentum difference between the various ions in regard to masses m_a and m_b as the species mass of the two reactants and m_c and m_d as the species mass of the two products in any reaction. Whereas Eq. (7.16) does not. The difference between the two equations is calculated from a ratio which we term the momentum weighting factor M_w as follows

$$M_w = \frac{v_{eff}}{v_{exp}} = \frac{\left(A(1-\phi) + B\phi\right)\sqrt{\phi}\sqrt{\left(\frac{\alpha}{2}\right)}c}{(2\alpha\phi)^{1/2}c}, \tag{7.17}$$

$$\therefore M_w = \frac{A(1-\phi) + B\phi}{2}. \tag{7.18}$$

So that the value of the momentum weighting correction for each of the reactions would be 0.9879 (D^3He), 0.9880 (DT), 0.9948 (DD) and 0.9946 (DD) respectively. Then the average debris velocity $\tilde{v_{av}}$ is given as a function of the expansion velocity and the momentum weighting factor so that

$$\tilde{v_{av}} = M_w v_{exp} = M_w(2\alpha\phi)^{1/2}c. \tag{7.19}$$

The exhaust velocity is then given as a function of the plasma nozzle divergence half angle θ since the plasma flow through the engine nozzle will not be axial and perfectly collimated but will tend to diverge in the range $\theta = 15{-}20°$. The exhaust velocity is then

$$v_{ex} = \frac{\tilde{v_{av}}}{2}\Big[1 + \cos(\theta)\Big]. \tag{7.20}$$

Therefore the exhaust velocity can be written as $v_{ex} = M_w v_{exp}\lambda_d$ or

$$v_{ex} = \frac{M_w(2\alpha\phi)^{1/2}c}{2}\Big[1 + \cos(\theta)\Big]. \tag{7.21}$$

We also highlight the fact that the combination of the nozzle divergence term λ_d and the momentum weighting factor M_w is equal to the square root of the jet efficiency $\sqrt{\eta_j}$. For ease of clarity we can make the statement that

$$v_{ex} = \lambda_d \tilde{v_{av}} = \lambda_d v_{eff} = M_w \lambda_d v_{exp} = \sqrt{\eta_j} v_{exp}. \quad (7.22)$$

Since the vehicle thrust is $T = \dot{m} v_{ex}$, then using the fact that

$$f_{hz} = \frac{m_{prop}}{m_{cap} t_b} = \frac{\dot{m}}{m_{cap}}. \quad (7.23)$$

The mass flow rate is given by $\dot{m} = m_{cap} f_{hz}$ so that

$$T = m_{cap} f_{hz} \frac{M_w (2\alpha\phi)^{1/2} c}{2} \Big[1 + \cos(\theta)\Big]. \quad (7.24)$$

For a larger capsule mass and larger pulse frequency the thrust will be higher. Similarly, the jet power is given as a function of the exhaust velocity and thrust but also as a function of the nozzle jet efficiency η_j and power output P_{out} as follows

$$P_j = \frac{1}{2} v_{ex} T. \quad (7.25)$$

Then substituting for the jet power

$$P_j = m_{cap} f_{hz} \frac{v_{ex} M_w (2\alpha\phi)^{1/2} c}{4} \Big[1 + \cos(\theta)\Big]. \quad (7.26)$$

With the appropriate substitution for v_{ex} in Equation (7.6) we find

$$P_j = m_{cap} f_{hz} \frac{\alpha\phi M_w^2 c^2}{4} \Big[1 + \cos(\theta)\Big]^2. \quad (7.27)$$

The jet power is given as a function of the exhaust velocity and thrust but also as a function of the nozzle jet efficiency η_j and debris power P_{deb} so that

$$P_{deb} = \frac{P_j}{\eta_j}, \quad (7.28)$$

$$\therefore P_{deb} = m_{cap} f_{hz} \frac{\alpha\phi M_w^2 c^2}{4\eta_j} \Big[1 + \cos(\theta)\Big]^2. \quad (7.29)$$

Then since $P_{out} = P_{deb} + P_{losses}$ we get

$$P_{out} = m_{cap} f_{hz} \frac{\alpha\phi M_w^2 c^2}{4\eta_j} \Big[1 + \cos(\theta)\Big]^2 + P_{losses}. \quad (7.30)$$

Finally, the specific power is $P_{sp} = P_j/m_{eng}$, where propulsion mass is m_{eng} which includes the power supply, engine and thermal heat system. Therefore

$$P_{sp} = m_{cap} f_{hz} \frac{\alpha\phi M_w^2 c^2}{4m_{eng}} \Big[1 + \cos(\theta)\Big]^2. \tag{7.31}$$

For a larger capsule mass and larger pulse frequency the thrust, jet power and specific power will be higher. The above equations (original formulations by this author but building on the equations of Bond/Martin [5]) are useful and using terminology from aerospace engineering it would be appropriate to refer to them as the "*driving equations of icf propulsion*" theory. The equations highlight the separate roles of the energy driver through f_{hz}, the choice of fuel and radiation environment through α and M_w, the quantity of fuel through m_{cap}, the capsule design and laser plasma interactions through ϕ, the nozzle design through θ (and η_j) and finally the overall design of the ignition system, power supply and thermal management through m_{eng}. Collectively these describe how the propulsion performance drives the spacecraft.

Note that earlier in Sect. 6.3 we discussed the fact that according to the jet efficiency equation $\eta_j \propto 1/\phi$ provided the exhaust velocity v_{ex} = constant. This happens because when you have a higher burn-up fraction the micro-explosion of the fuel capsule lasts longer and so the plasma expands more violently and spherically. Any Rayleigh-Taylor instabilities will also grow faster. The magnetic nozzle might then struggle to confine the jet collimation during this chaotic early blow-off period. In other words a high burn-up fraction leads to a turbulent exhaust and donot produce a nice directional thrust. They are more analogous to a nuclear bomb detonation. This is the opposite to a commercial nuclear reactor which has the purpose of power generation and where the principal goal is high energy gain. In fusion propulsion we want to find a compromise between burning as much of the fusion fuel as we can through a high burn-up fraction and reducing the energy losses through a high jet efficiency. So, this is a trade. However, we can see from the above that the exhaust velocity is written as a dependence on ϕ. We can see this from the fact that $T = \dot{m} v_{ex} = m_{cap} f_{hz} v_{ex}$ so that

$$v_{ex} = \frac{(2\alpha\phi)^{1/2} c}{2}(1 + cos(\theta)). \tag{7.32}$$

Then if we substitute this into the equation for the jet efficiency we find

$$\eta_j = \frac{1}{4}(1 + cos(\theta))^2. \tag{7.33}$$

We notice from this that the dependence on both α and ϕ cancels out. This happened since we defined these as variables within the exhaust velocity equation but also because the geometric factor $(1 + cos(\theta))$ was multiplied outside the ϕ term so that the burn-up physics and the jet collimation geometry were implicitly

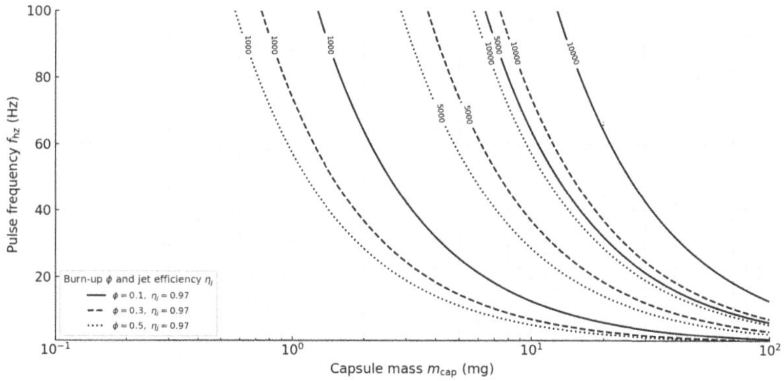

Fig. 7.3 Performance contours for driving design equation

treated differently. Yet we know that the value of ϕ will influence the value of η_j and so it is important to understand that this decoupling has occurred purely as a result of the definition for $v_{ex} = (2\alpha\phi\eta_j)^{1/2}c$. Although the real ICF physics does link the burn-up fraction and jet efficiency, treating them as decoupled parameters in the design equation is still useful since it permits engineers to independently explore how the fuel physics and exhaust collimation each influence overall thrust performance across the trade space.

Figure 7.3 shows an example of the sorts of studies that can be performed using these driving equations, where here we assume contours of constant thrust (T: 1 kN, 5 kN, 10 kN) for different values of burn-up fraction (ϕ: 0.1, 0.3, 0.5) for a given value of jet efficiency ($\eta_j = 0.97$) and from this we can see how the pulse frequency is determined from a given value of capsule mass. These are shown for a DT fuel. So, for example, using a 1 mg capsule mass assuming a burn-up fraction of $\phi = 0.5$ and for a 1 kN thrust profile this suggests a pulse frequency of around ~57 Hz which can be seen on the contour plot.

We can further examine the performance limit for ICF propulsion systems based on the equations presented above. If we adopt the simplest form, and we assume that the plasma will flow through the nozzle perfectly axially and not at an angle so that $\theta = 0$, which means $\cos(\theta) = 1$. We also assume that the momentum weighting factor $M_w = 1$ no matter the fuel. Also, for simplicity we also assume a jet efficiency $\eta_j = 1$ which means no losses are assumed for best possible performance. For the propulsion mass we assume 300 tons consistent with some of the other studies discussed below as the likely minimum limit for a fusion propulsion engine. Under these assumptions the equations reduce to an idealised ICF propulsion engine

$$T = m_{cap} f_{hz} (2\phi\alpha)^{1/2} c, \tag{7.34}$$

$$P_j = m_{cap} f_{hz} \phi\alpha c^2, \tag{7.35}$$

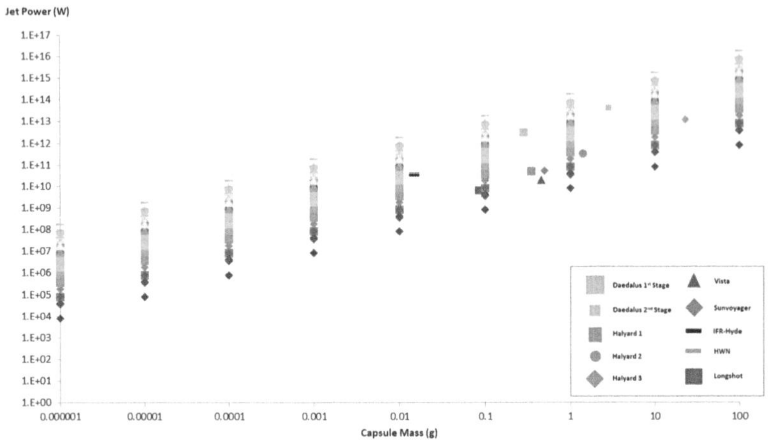

Fig. 7.4 Theoretical range calculations for spacecraft jet power

$$P_{sp} = \frac{P_j}{m_{eng}}. \tag{7.36}$$

Figure 7.4 shows the use of these equations to estimate the likely propulsion performance from a theoretical perspective. Also included is the data for actual spacecraft concepts as published in the peer reviewed literature. Where $\alpha_1 = 0.0009$ is representative of the lowest energy fraction expected from a DD reaction (for a worst-case performance estimate) and $\alpha_2 = 0.004$ is representative of the highest energy fraction expected from either a DT or D^3He reaction (for a best-case performance estimate). The indicated propellant mass is based on the mass flow rate for a 1-year burn time, where $m_{prop} = \dot{m}t_b$. These calculations are performed assuming a variable range to include the pulse detonation frequency 1–500 Hz, burn-up fraction 0.1–1, capsule mass 10^{-6}–10 g.

A scenario of low pulse frequency and high capsule mass is analogous to a nuclear pulse propulsion system [8–10] or at the least it begins to approach that domain space. For a real nuclear pulse system, the pulse units would be significantly greater in mass by orders of magnitude, each detonated perhaps at a frequency of ~1–10 Hz. However, since we are dealing with ICF they will not be this large, although we might imagine a scenario where capsules as large at 10–100 g are used, hence this is the maximum mass of our range adopted.

In contrast, a scenario of high pulse frequency but using very low capsule masses is analogous to a continuous propulsion system. This may involve 100s Hz and since the detonation rate is so high it would likely involve very small capsule masses on the order of milli-grams. Although we might imagine a scenario where capsule masses approaching those used in a laboratory environment were used and so we adopt a lower mass limit of around 1 microgram. We assume a maximum detonation rate of 500 Hz, although values higher than this may be possible.

For completeness the reverse is also computed with high capsule mass at high pulse frequency and low capsule mass at low pulse frequency, to capture any mission scenarios not yet envisaged for interplanetary, interstellar precursor or even full interstellar flight. Examining the range of parameter calculations conducted overall based upon the simplified equations, this results in a mass flow rate of between $10^{-9} - 5$ kg/s and for a propellant mass $10^{-5}-10^{5}$ tons. This range of variables was adopted as potentially representative of interplanetary and interstellar mission scenarios that might arise in the decades and centuries ahead. For the range of variables examined we have found that the range of parameters will be on the order of $\sim 10^{-3}-10^{8}$ N for thrust, $\sim 10^{3}-10^{15}$ W for jet power and $\sim 0.01-10^{9}$ W/kg for specific power. It needs to be emphasised that these are the extremes of a theoretical range hence why they are so high and it doesn't suggest it is practical to generate this performance.

In terms of the spacecraft concept designs from the literature, this contrasts with an actual range of $1.26 \times 10^{3}-7.54 \times 10^{6}$ N for thrust, $6.25 \times 10^{9}-4.0 \times 10^{13}$ W for jet power and $2.0 \times 10^{4}-4.05 \times 10^{7}$ W/kg for specific power. This reflects a much narrower range of operational parameter values that are likely. These designs had a mass flow rate range of 0.000125–1.5 kg/s for a capsule mass range of 0.015–23 g, detonated over a pulse frequency range 1.5–500 Hz.

As a caveat to the analysis no account was taken of aneutronic fuels such as $p^{11}B$, or $p^{6}Li$, $p^{7}Li$, or for the potential of using the proton-proton reaction such as proposed for the interstellar ramjet [11]. Neither were there efforts to consider the increase in performance of such designs due to vehicle staging or the use of propellant expellant for the purpose of thrust augmentation such as was used in a NASA magnetic fusion study [12]. A consideration of these issues may lead to an expansion of the ICF propulsion design space compared to that shown.

An examination of the range of estimates that are possible for the different parameters highlights the wide scatter and degree of randomness in the concept designs of the published literature. Overall, there is no general trend between them in terms of the capsule mass adopted for interplanetary, interstellar precursor or interstellar applications. This randomness speaks to the non-optimality of the various design solutions and implies a need to derive a relationship for the minimum capsule mass requirement as a function of other propulsion and mission parameters which is partly a motivation for this research.

7.3 Design Case Study

Phase I: Early Design Scoping

Now that we have established a basic process for the design of advanced vehicles, it is useful to put this into practice with an example case study. To achieve this, we will firstly describe a requirement and then take the reader through the process of leading to a concept design layout and vehicle performance description. For this case study, we define a requirement to be:

"*A flyby mission to the nearest stars carrying a 100 tons scientific payload. It is to be completed in a trip time not exceeding sixty years, and with the probe ideally arriving at the nearest stars before the year 2125. The mission is to be a test bed for multiple propulsion technologies to inform later missions*".

We title this the Voyager for Advanced Space Travel (VAST) mission and a launch date is implied by around 2060–2065. The nearest stars are 4.3 ly distance, which is 4.07×10^{16} m or 271,932 AU. As we have already seen, such a mission carrying current propulsion technology would take on the order of tens of thousands of years. To complete such a mission in around 60 years. This immediately implies a cruise speed of around 271,932 AU/ 60 years = 4,532 AU/Year = 21,500 km/s. Then if we assume an optimum mass ratio of around 4.9 [13], this implies an exhaust velocity requirement of $\Delta V_{ex} = \Delta V / Ln(4.9) = 13{,}528$ km/s.

The energy required to accelerate a 100 tons payload to this required velocity is $E = m\Delta v^2/2 = 8.32 \times 10^{18}$ J. However, this is just the payload, and we must add onto this a mass allocation for the fusion engine, propellant tanks, waste heat radiators, power supply and other propulsion technologies. The total mass that can be assembled is also dependent upon what can be constructed and lifted into orbit by the launch vehicle. Let us say that the design team has a discussion, and they decide on a total mass budget of around 10,000 tons for the entire vehicle. This then raises the energy requirement to $E = 2.31 \times 10^{21}$ J, where we have assumed non-relativistic flight here.

If this total energy was expended over an acceleration time of 10 years, then this implies a minimum power of $P = E/\tau \sim 7.33$ TW. Assuming a total propulsion mass on the order of 3,000 tons, then this would imply a specific power of 2.44 MW/kg. Note that the engine mass assumption is linked to the mass ratio since $R = m_o/m_f$, and the engine mass is to be thrown away at the end of stage burn.

If we assume that no mass is dropped other than the thermonuclear fuel mass then we can estimate the minimum fuel requirement from $m_{fuel} \approx m_{s/c}(R - 1)$, where we assume a spacecraft mass during the fusion phase on the order of 1,000 tons we can estimate the sort of fusion fuel requirement. For the mass ratios in the range $R = 4.9$, the fusion fuel would be $m_{fuel} \sim 3{,}900$ tons.

Note that the power required is for the 10,000 tons acceleration. The jet power is based on the engine performance through the exhaust velocity and mass flow rate. If we assume an efficiency of $\epsilon = 0.7$ then the electrical power required is given by $P_e = T v_{ex}/2\epsilon = 1.617$ TWe.

All the above has merely started the process to sketch out the option space for such a mission but where we have assumed a fusion stage only, although this may have been developed at different levels of fidelity and from different options as a part of the process of down select illustrated in Fig. 7.5.

Table 7.2 shows a summary of the design so far for the mission. The goal of generating this table is to illustrate the likely bounds of where the mission performance will be to inform further design discussions about what may be possible before too much time is spent on any extensive calculations. However,

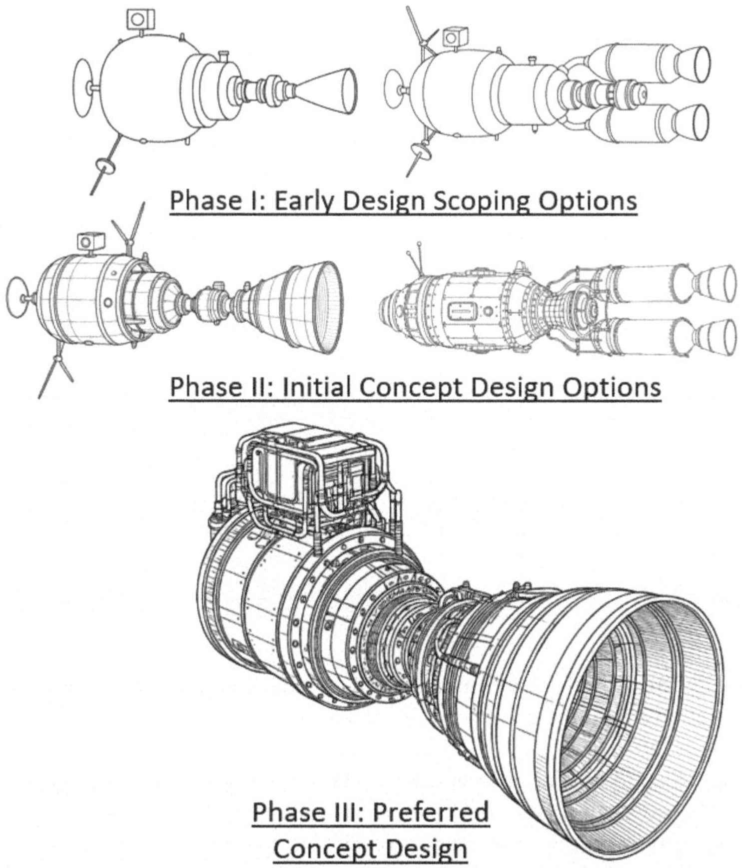

Fig. 7.5 Illustration of down select process in design[KFL]

the data derived so far is for a fusion stage only and we want to introduce other technologies, and by that also reduce the fusion requirements, so this should represent the outer boundary of the system specification.

From an examination of the table, we can reach some firm conclusions to aid the next phase of the design process. Since it is a requirement for the mission to test out other propulsion methods, then the thrust profile generated above would be more than is required for the mission. The other propulsion methods will augment the thrust profile and help to close the mission and generate the required velocity. It will be assumed therefore that the fusion engine stage will be the final stage of the mission and ahead of this we will include the two other propulsion stages.

Phase II: Initial Concept Design Options

As a part of the initial considerations several options may have been considered for each system and subsystem. For the fusion engine for example the team may have

Table 7.2 Initial performance requirements for VAST Fusion stage

Mass ratio (R)	4.9
Propulsion mass (tons)	1,000
Total spacecraft mass (tons)	10,000
Acceleration time (years)	10
Acceleration distance (AU)	4,777
Velocity (AU/Yr)	4,532
Velocity (km/s)	21,500
Cruise time (years)	50
Total mission time (years)	60
Exhaust velocity (km/s)	13,529
Fusion fuel mass (tons)	3,900
Kinetic energy (J)	2.3×10^{21}
Power required (TW)	7.33
Specific power (MW/kg)	2.44
Jet power (TW)	1.13
Mass flow rate (kg/s)	0.012
Thrust (kN)	167.3
Electrical power (TWe)	1.617
Efficiency (%)	70

been considering a single stage engine or a two-stage parallel thrust engine to reduce the acceleration time, or for some practical engineering reason.

This may lead to further development on both options to include a higher fidelity level of definition and analysis. This may also include trade studies of parameters so that the design points of compromise can be identified for maximum system efficiency. For example, by minimising the mass flow rate through the engine this may reduce the level of neutrons exposed to the structure through the acceleration phase and thus require less shielding mass and a low mass of waste heat radiators. Yet this must be traded with the fact that as the acceleration time increases, the reliability of the system drops since there is a higher chance of mechanical failure.

For this mission, and after some considerable design team discussions the two propulsion technologies also to be included are a solar sail and a laser beam propulsion system onto the same sail surface. This is where we enter the domain of system integration, since the sail must be mated to the fusion engine, but in a way so that they do not disrupt each other's performance. It is during this hypothetical scenario that the team moves towards the down select of a single engine stage to ease integration and deployment issues associated with the unfurling of the sail. This leads to an early definition of the entire spacecraft system as shown in Fig. 7.6.

Phase III: Developed Concept Design Options

This is where we will use the equations derived earlier in Chap. 1 for the general rocket equation solution which includes the external forces. In this model, whilst the fusion stage is governed by the ideal rocket term, the solar sail and laser sail are

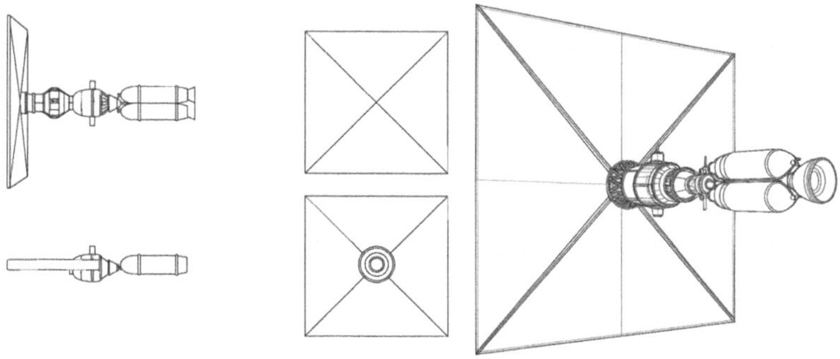

Fig. 7.6 Configuration layout of initial concept design[KFL]

seen as external forces to the propulsion. It is possible to have all three operating in unison as a form of parallel thrust, however for this example we will use each term in part as a sequence of impulse manoeuvres.

Let us imagine a situation where we have a spacecraft design that has three separate propulsion systems. The first is a fusion engine and in addition it has a solar sail, and then a laser sail for when the spacecraft gets so far from the Sun that the solar pressure drops off. We can then express our general ideal rocket equation containing the velocity contributions from the two external forces

$$\Delta v = v_{ex} Ln\left(\frac{m_o}{m_f}\right) + \delta v_{b,1} + \delta v_{b,2}. \tag{7.37}$$

Alternatively, the change in velocity is the impulse divided by the mass

$$\delta v = \frac{1}{m} \int F(t) dt. \tag{7.38}$$

Therefore, we can express it as follows

$$\Delta v = v_{ex} Ln\left(\frac{m_o}{m_f}\right) + \frac{1}{m_{ave}} \int_{t_o}^{t_{b,1}} F_{sun}(t) dt + \frac{1}{m_{ave}} \int_{t_{b,1}}^{t_f} F_{laser}(t) dt. \tag{7.39}$$

Let us now imagine that these three forces uniquely describe three separate propulsion forces on the vehicle. The first is an internally generated propulsion, based on some rocket jet principle, and this is represented by the first term in the modified ideal rocket equation of equation above. Let us next assume that the two additional external forces correspond to a solar sail and a laser sail driven system, each acting separately on the vehicle at different points in the mission. We now need to derive representative equations for these additional external forces. The equations used below are explained further in reference [14].

For a Solar Sail propulsion system, the solar pressure P_{rad} (N/m^2) is given by

$$P_{rad} = \frac{1+\eta}{c} S_r. \tag{7.40}$$

Where η is the sail reflectivity and S_r is the solar irradiance (W/m^2) given by

$$S_r = \frac{a_{1au}}{r^2} = \frac{3.04 \times 10^{25}}{r^2}. \tag{7.41}$$

The sail loading is

$$\sigma = \frac{m}{A}. \tag{7.42}$$

and the characteristic acceleration is

$$a_1 = \frac{P_{rad}}{\sigma}, \tag{7.43}$$

$$a_1 = \frac{a_{1au}(1+\eta)}{\sigma c r^2}. \tag{7.44}$$

And then the external velocity addition on the sail is described by

$$\delta v_1 = \int_{t_o}^{t_{b,1}} a_1 dt = \frac{a_{1au}(1+\eta)}{\sigma c r^2} t_{b,1}. \tag{7.45}$$

We can similarly look at Laser sail propulsion by developing a velocity increment expression. The acceleration of a vehicle of mass m and reflectivity η, driven by an incident laser power P is given by

$$a_2 = \frac{2\eta P}{mc}. \tag{7.46}$$

So that the external velocity addition is given by

$$\delta v_2 = \int_{t_o}^{t_{b,2}} a_2 dt = \frac{2\eta P}{mc} t_{b,2}. \tag{7.47}$$

We can imagine a situation where a hybrid Starship design would feature all three elements of a rocket jet exhaust, solar sail pressure, and laser sail pressure, all acting

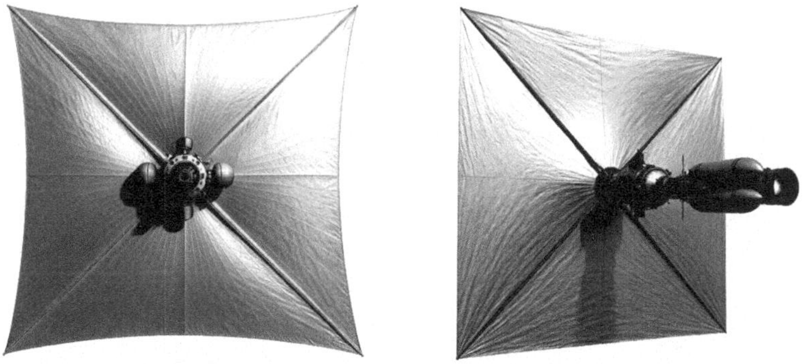

Fig. 7.7 Illustration of a hybridized advanced propulsion concept[KFL]

on the vehicle as internal and external forces in parallel. In which case

$$\Delta v = v_{ex} Ln\left(\frac{m_o}{m_f}\right) + \left(\delta v_{b,1}\right)_{sail} + \left(\delta v_{b,2}\right)_{laser-sail}. \tag{7.48}$$

Then substituting

$$\Delta v = v_{ex} Ln\left(\frac{m_o}{m_f}\right) + \frac{a_{1au}(1+\eta)t_{b,1}}{\sigma c r^2} + \frac{2\eta P t_{b,2}}{mc}. \tag{7.49}$$

This is an expanded version of the ideal rocket equation, to include both an internal force and two external forces, for a tri-propulsion problem. However, this equation implies that all three Δv terms are applied simultaneously, which is rather complex to analyse. So instead, we examine a case study design where each of the Δv is applied one after the other equivalent to three separate burn stages. Figure 7.7 illustrates such a solution. This represents a mission where the vehicle is firstly propelled from deep in the Sun's solar corona using a deployed solar sail. At some distance from the Sun when the intensity was too weak, a laser sail would then propel the vehicle further. When efficient to do so, the jet exhaust rocket system would then be deployed to complete the boost phase of the mission.

Now the intention of showing this working is merely to illustrate the point that the ideal rocket equation as we use it, is not the general solution for the space-time propulsion problem, but merely what in mathematical terms we may call a particular solution. This is because we have forced the solution to a particular subset of problems (jet thrust problems) by constraining the equation using boundary conditions. It may not be possible to find the ultimate general solution of the space-time propulsion problem, because in theory this could encompass all the laws of physics. However, it is this author's view, that by at least attempting more general solutions, and adopting hybridized design solutions which are optimal performance trade-offs between different systems, that an easier route to the stars 'may' be

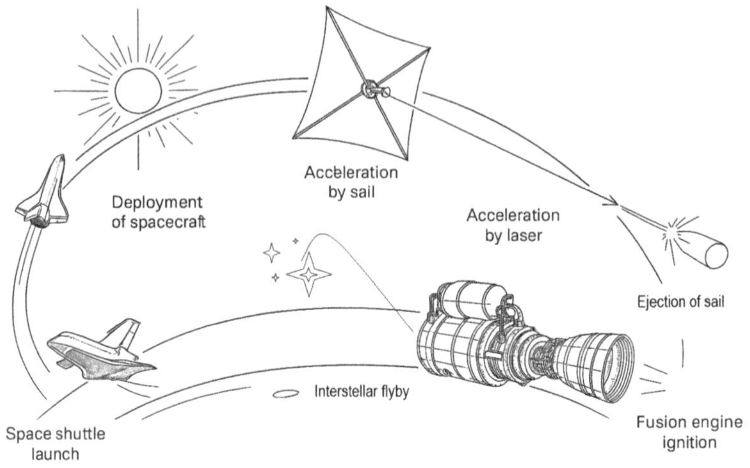

Fig. 7.8 Mission architecture for VAST spacecraft concept[KFL]

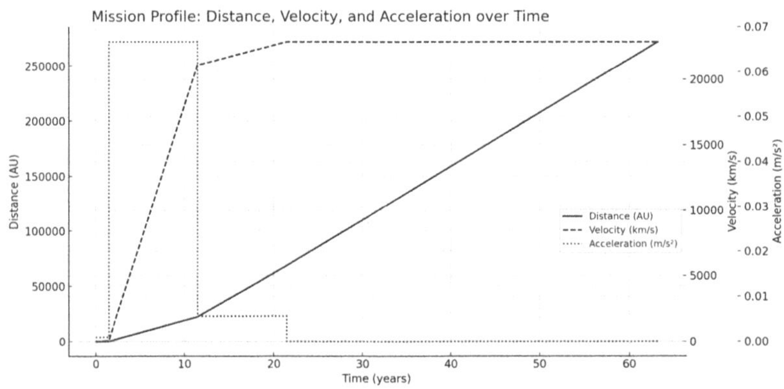

Fig. 7.9 Mission profile for hybrid propulsion system

elucidated. Until such designs are attempted, we will not know if any overall performance benefit can be obtained.

The biggest problem for many of the propulsion concepts is energy. Typical world energy consumption is in the $\sim 10^{20}$ Joules level, which is made of up of fossil fuels, nuclear and renewables. This is the total energy used by all human civilization from its industrial and technological sectors across every country. This is equivalent to a 250 tons flyby probe moving at 0.1c and so taking half a century to reach the nearest stars, if we include time for acceleration. When our propulsion technology can match these energy levels in their production capability, then our civilization will finally be able to undertake those interstellar missions.

Figure 7.8 shows the mission architecture for the VAST probe from launch at Earth orbit to insertion onto the interstellar trajectory. Figure 7.9 shows the mission

Table 7.3 Mission definition and performance for VAST concept design

Solar sail phase	
Characteristic length (km)	17.67
Sail mass (tons)	1.0
Distance of launch (AU)	0.56
Acceleration time (years)	1.47
Acceleration (m/s^2)	0.0009
Thrust (kN)	9.04
End of stage velocity (km/s)	42.1
Laser sail phase	
Specification (TW)	100
Distance of push (AU)	6.49
Acceleration time (Years)	10.0
Acceleration (m/s^2)	0.0667
Thrust (kN)	66.71
End of stage velocity (km/s)	21,038.55
Fusion engine phase	
Specification	DT fuel
Fusion fuel mass (tons)	3,000
Propulsion mass (tons)	3,000
Distance of ignition (AU)	22,264.05
Acceleration time (years)	10.0
Acceleration (m/s^2)	0.0056
Thrust (kN)	56.55
Mass flow rate (kg/s)	0.0113
Jet power (GW)	141.38
Specific power (kW/kg)	47.13
End of stage velocity (km/s) [%c]	22,864.03 [7.63]
Cruise phase	
Cruise distance (AU)	203,349.66
Cruise time (years)	42.19
Exploration phase	
Distance to target (AU) [ly]	271,931.88 [4.3]
Arrival at target (years)	63.19
Scientific payload mass (tons)	150.0
Speed through system (AU/day)	13.2
Time within 100 AU of target (days)	15.14

profile for the hybrid system. The calculations were conducted assuming $\eta = 0.1$. As shown in Table 7.3, although the probe will arrive at the target destination after 63 years it won't spend very long studying the star and its associated planets due to its high speed. Indeed, at a speed of 22,864 km/s this is 13.2 AU/day. Note that in this particular design the fusion stage did not add a substantial velocity increment on top of the earlier two stages. This suggests the configuration is not well optimised

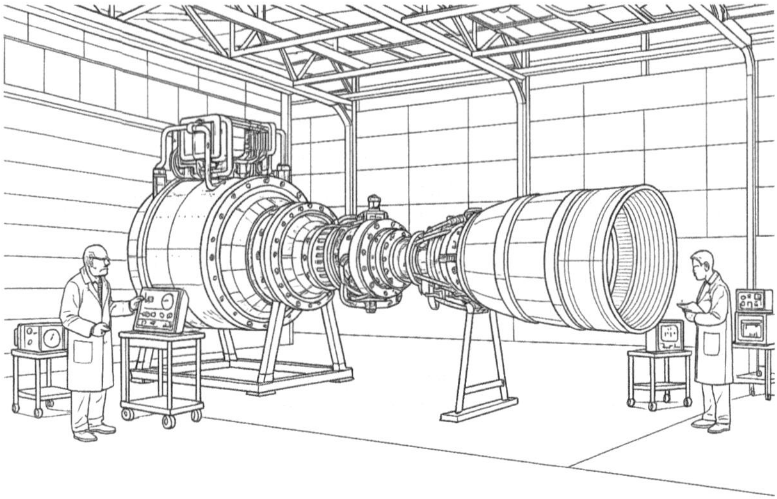

Fig. 7.10 Fusion engine under development in a laboratory[KFL]

and a redesign would be required. Yet, this is left as it is here since this is just a demonstration exercise.

If we assume that any good scientific measurements of high value begin when the probe is within 100 AU of the star, this implies a flight time to the star of around 7.57 days. It will then depart the system and arrive at 100 AU on the other side after the same length of time, which means that the science exploration phase will last a total of just over two weeks. This might suggest a redesign so that the probe can be decelerated or it could be sent at a slower cruising speed to maximize the encounter time. This is all a part of the design trades that must be made.

Prior to the spacecraft being deployed for the mission it will have to undergo significant testing to increase the technology readiness level. As shown in Figs. 7.10 and 7.11 this would involve laboratory experiments and the construction of prototypes, and this is the sort of stage many advanced propulsion technologies are at today. Once the proof of physics and engineering principles have been demonstrated in the laboratory environment, and with a predictive performance compared to numerical codes, work could then begin on the construction of the actual spacecraft, or at least a sub-scale demonstrator to further prove its utility for the mission. The development of advanced propulsion technologies is a complex undertaking and can be years or even decades in the making.

Problems of Chap. 7

7.1 A Spacecraft uses DT fusion fuel and has an ICF engine. The pellet mass is 1 gram and they are detonated at a rate of 10 per second and have an ignition performance characterised by a burn-up fraction of 25%. The 500 tons engine has nozzle divergence angle of 15°. (i) Using the equations of this chapter, calculate the

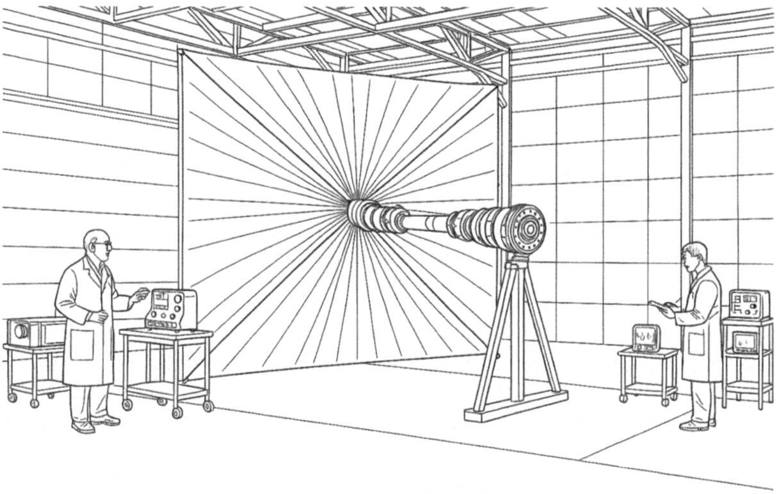

Fig. 7.11 Solar Sail under development in a laboratory[KFL]

exhaust velocity, mass flow rate of fuel, thrust, jet power and specific power for this design. Treat the entire pellet mass as reaction mass, not just the burned fraction.

7.2 For the multi-mode propulsion system examined in this chapter, imagine that the laser power available for the second phase push is only 1 TW but still over 10 years. (i) What would be the new velocity at the end of phase 2 and what distance would the spacecraft be? (ii) What would be the total velocity achieved after all three stages and how long would the mission take to reach Proxima Centauri?

7.3 If the fusion fuel allocation is now raised to 50% of the total spacecraft mass (i) What will be the velocity and distance achieved by the spacecraft at the end of the fusion phase? (ii) When would the spacecraft arrive at the destination star? Assume the 1 TW laser of the previous problem.

References

1. Cassenti B (1982) A comparison of interstellar propulsion methods. JBIS 35(3):116–124
2. Low GM, Kleinknecht KS, Simpkinson SH et al (1971) What made apollo a success? NASA SP 287. Reprinted from March 1970 issue of Astronautics & Aeronautics, a publication of the AIAA
3. Orth CD, Turchi PJ, Carpenter SA, Santarius JF, Frisbee RH, Williams CH (2004) Recommended design practices for conceptual nuclear fusion space propulsion systems, special project report, SP-108-2004, AIAA
4. Orth CD (2000) Parameter studies for the VISTA spacecraft concept, LLNL, UCRL-JC-141513, presented at NASA advanced propulsion workshop in fusion propulsion, Huntsville, Alabama, 8–10 November, 2000

5. Martin AR, Bond A (1978) Project Daedalus: the propulsion system part 1: theoretical considerations and calculations, JBIS Special Supplement:S44–S62
6. Bond A, Martin AR (1978) Project Daedalus: the propulsion system part 2: engineering design considerations and calculations. JBIS Special Supplement:S63–S82
7. Sutton GP, Biblarz O (2001) Rocket propulsion elements, 7th edn. Wiley, Hoboken
8. Everett CJ, Ulam SM (1955) On a method of propulsion of projectiles by means of external nuclear explosions, LAMS
9. Dyson G (2022) Project orion: the atomic spaceship 1957–1965. The Penguin Press, New York
10. Dyson FJ (1968) Interstellar transport. Phys Today 21(10):41–45
11. Bussard RW (1960) Galactic matter and interstellar flight. Astronautica Acta 6:170–194, Fasc.4
12. Williams CH, Dudzinski LA, Borowski SK, Juhasz AJ (2005) Realizing 2001: a space odyssey: piloted spherical torus nuclear fusion propulsion, NASA/TM-2005-213559, AIAA-2001-3805
13. Shepherd LR (1952) Interstellar flight. JBIS 11(4):149–167
14. Long KF (2012) Deep space propulsion: a roadmap to interstellar flight. Springer, Berlin

Chapter 8
Interplanetary Spacecraft Designs

> *This analysis points out the difficulty in approaching the theoretical performance for an interstellar spacecraft; however, it indicates that flight times of less than 50 years to a 5 light-year stars may be approached with a fusion-propelled vehicle, if certain engineering problems can be solved.*
>
> — *Dwain F Spencer, 1963 [1]*

8.1 VISTA to Mars Interplanetary Mission

The Vehicle for Interplanetary Transport Applications or VISTA was a study initiated out of LLNL in the United States. Although the work was first initiated in 1987 [2], work continued with the concept until at least 2003 with various revisions to the design [3, 4]. The main motivation for the work was to develop a fusion concept for missions to Mars.

Ignition via ICF was to be achieved using indirect drive methods. To ensure a symmetric implosion, the capsule could be placed within a radiation cavity known as a 'Hohlraum', and then the lasers heat a surrounding gold layer which generates a uniform x-ray bath in which to implode the capsule. This is the method proposed for the VISTA design, and it was optimistically claimed that in principle an energy gain of 1,500 may be possible. The main advantage in terms of the vehicle configuration is that only a few illumination mirrors will be required external to the axis point (VISTA has 2 or 3 depending on the version of the design), but in a direct drive scheme many more may be required to ensure uniform symmetry.

Although the VISTA papers do not give details on the capsule design hidden within the radiation cavity (hohlraum) design, they do reference an earlier paper which does contain details of that design and so we can assume this is representative of what would be inside [5]. It contains a central pocket of DT gas, surrounded by DT ice, then a CH ablator. We note that the CH ablator shell is 0.2 mm thick. Note that similar designs either tested or proposed for NIF relating to indirect drive and indirect drive capsules. As shown in Fig. 8.1, they are broadly consistent with the

K. F. Long, *Stellar Engines*, Springer Aerospace Technology,
https://doi.org/10.1007/978-3-032-18062-9_8

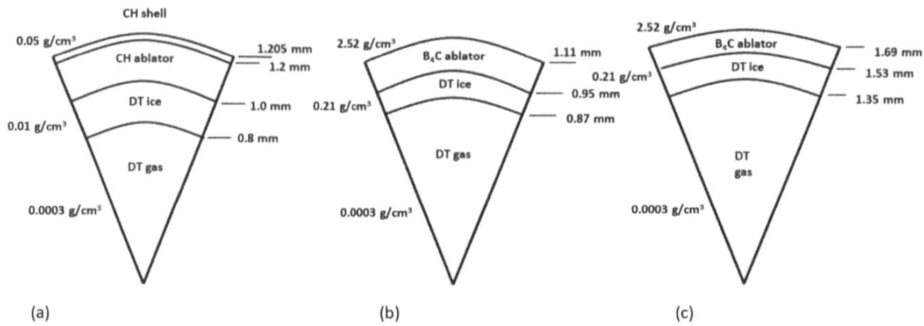

Fig. 8.1 Comparison of VISTA capsule design with NIF indirect drive and direct drive designs (relative sizes of capsules not to scale).[KFL] (**a**) ICF capsule that is the basis of Vista indirect drive design. (**b**) Typical NIF indirect drive capsule design. (**c**) Typical NIF direct drive capsule

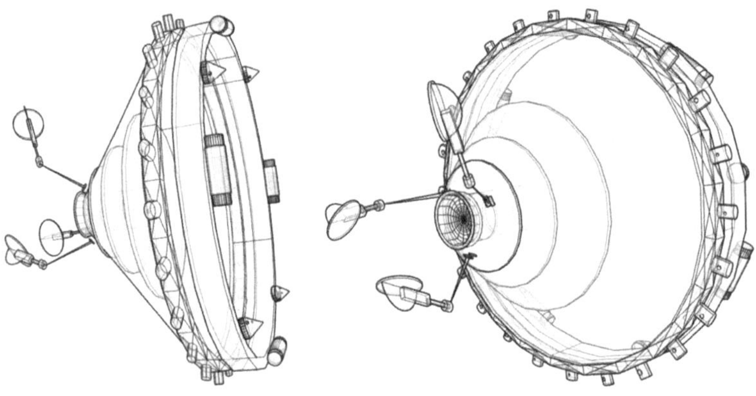

Fig. 8.2 Spacecraft VISTA concept design[KFL]

VISTA design but using a B_4C ablator shell with a thickness of 0.16 mm for both the indirect and direct drive designs.

A disadvantage of using nuclear fusion reactions is the high flux of radioactive particles from the ignition, high energy neutrons and x-rays. This acts to irradiate any material structure but also can be very harmful to any crew on board. It is necessary then to use radiation shielding to mitigate these effects. As shown in Fig. 8.2 to solve this VISTA uses an unusual conical shaped geometry for the vehicle configuration which has the effect that the ignition point is placed behind the plane of a 12 T superconducting magnet and where a neutron and x-ray/gamma-ray shadow would be cast with a 50-degree half angle. This is known as the VISTA shadow shield *shadow shield* and it has the advantage of maximising the jet efficiency of the thrust chamber.

Within the VISTA design around 3/4 of the emitted fusion energy is in the form of neutrons and x-rays which is a substantial amount of energy to mitigate through conventional shielding methods. The radiation interaction with the structure is around ~3–4% of solid angle which means it will intercept around ~3–4% of the neutron and x-ray emissions from the fusion reactions. It is worth exploring the issue of the shadow shield in VISTA just a little.

When an ICF capsule is ready to be ignited by the lasers it must be accelerated into the target detonation point on the axis. Given that VISTA has a pulse detonation frequency of 30 Hz, this means this must be conducted 30 times per second or once every 0.03 s or 30 ms. This means that the plasma debris and any associated radiation must be cleared from the axis point within this timescale. The nuclear ignition of the capsule will likely occur over ~ns and the hydrodynamic implosion and disassembly will likely occur over ~μs so this should be sufficient time. It is also complicated by the indirect drive set-up, meaning that the lasers will have to be carefully aligned to enter the bore holes within the hohlraum but this is a separate issue not examined further here. Neither are we concerned here with how the capsule is appropriately tracked for accurate pointing alignment of the lasers or on the details of beam line optics and focusing. We instead assume that the laser has successfully arrived at the capsule detonation point, ignited the capsule and we are left to deal with the consequences of that.

Instead, the interest is in assessing the effectiveness of the shadow shield and the angle at which it subtends the emitted radiation, as a function of the drift in position of the capsule detonation point. Once the capsule has detonated the plasma debris will expand away from the target firing position and interact with the magnetic fields produced by superconducting coils. This forces the plasma rearwards since it is conducting due to the presence of charged particles. This is the primary means by which the thrust is generated for the spacecraft. According to Orth [4] the effective interaction time of the plasma debris with the spacecraft structure is around ~50 μs. This means that the time for plasma debris interaction t_{pdi} with the material structure is a lot smaller than the time for an individual capsule detonation t_{det} such that

$$\frac{t_{det}}{t_{pdi}} = \frac{3 \times 10^{-2} s}{50 \times 10^{-6} s} = 600. \tag{8.1}$$

If we also approximate the motion of the x-ray radiation as moving at the speed of light, 2.9979×10^8 m/s, then within 1 μs the radiation would have transported through the vacuum around ~300 m and so encompassed the entire spacecraft structure. Within 10 μs and 50 μs the radiation would have travelled ~3 km and ~15 km through the vacuum respectively. For the fraction of x-rays that interact with the material structure they will not be moving through the vacuum but through a material with an opacity and so they will transport at a speed much slower.

The high energy 14.1 MeV neutrons are travelling at an average speed of ~1/6th the speed of light, 5.19×10^7 m/s, where the mass of the neutron is $m_n = 1.6749 \times 10^{-27}$ kg then

$$v = \left(\frac{2E}{m_n}\right)^{1/2} = 0.173c. \tag{8.2}$$

Within 1μs the neutrons would have transported $\sim$52 m through the vacuum. Within 10 μs and 50 μs the neutrons would have travelled $\sim$0.5 km and $\sim$2.6 km respectively. Once again for the fraction of neutrons that interact with the material structure, they will transport at a speed much slower. In terms of the detonation point, for the VISTA design this occurs 14.3 m distance from the toroidal magnetic thrust coil which is 26 m in diameter. If each detonation is to occur at intervals not exceeding 0.03 s then this implies that the capsules must be accelerated into the target point at a speed of 477 m/s. This would therefore require a rail gun of some sort to accelerate them at this speed.

Returning to the overall design, the VISTA spacecraft would carry a 100 tons payload and would have a payload mass fraction of around 1.5%. It had a 6,000 tons wet mass, 1,860 tons dry mass, with 3.72 tons of fusion fuel, but also with 4,100 tons of augmented expellant fuel (hydrogen) to push up the thrust performance. The engine performance would be characterised by a mass flow rate of 1.5 kg/s, thrust of 0.24 MN, jet power of 20 GW and specific power of 20 kW/kg. The spacecraft would also carry a 189 tons shield to protect the payload. The study looked at utilising other fusion reactions for a Mars mission. This resulted in roundtrip (return to Earth) times to Mars of 145 days (DT), 261 days (DD) and 241 days (D^3He) [3].

For the spacecraft designs examined in this book so far, the thrust is given as a function of the mass flow rate $\dot{m}$ and the exhaust velocity v_{ex} as $T = \dot{m} v_{ex}$. One of the issues with ICF is that the capsules tend to be very low in mass, such as $\sim$mg-g. This is a limitation on the generated thrust, although they are detonated to thermonuclear ignition and so the associated energy release may potentially be large.

To have a high mass flow rate, low mass capsules must be detonated at a high pulse frequency since $\dot{m} = m_{cap} f_{hz}$. Alternatively, to achieve a high mass flow rate larger capsule masses can be utilised at a lower pulse frequency. Although the larger the capsules the more difficult they are to drive to a state of ignition and the higher the energy driver requirements. In addition, in the limit of very large capsule mass they essentially tend towards the units used within an external nuclear pulse engine and they may not be as efficient in their energy conversion. The issue of low capsule mass can be mitigated by the inclusion of expellant propellant m_{exp} (mass/pellet) which will augment the thrust profile, and this was adopted in the VISTA design. This leads to an enhanced mass flow rate and thrust via

$$T = \dot{m} v_{ex} = (\dot{m}_{cap} + \dot{m}_{exp}) v_{ex} = (m_{cap} + m_{exp}) f_{hz} v_{ex}. \tag{8.3}$$

Figure 8.3 illustrates the different ways that expellant propellant can be utilised in different designs. The use of expellant comes at the cost of increased mass fraction to the vehicle since the pulse units are getting larger in mass when one combines the mass of the fuel with the mass of the expellant. This also assumes that the velocity associated with the fuel capsule particles and expellant propellant particles is equivalent when this may not be the case. To have the same velocity this requires a process of temperature thermalisation of the particles.

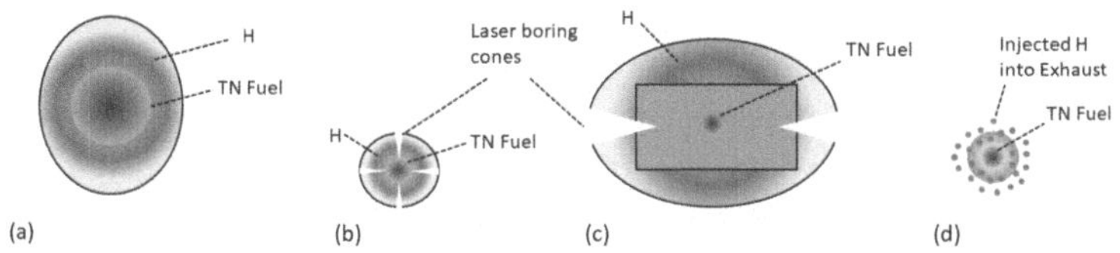

Fig. 8.3 Illustration of variations on expellant propellant concepts.[KFL]. (**a**) External nuclear pulse unit with hydrogen shaped mantle. (**b**) Direct drive ICF pellet mass with hydrogen expellant. (**c**) Indirect drive ICF Hohlram configuration with hydrogen expellant (e.g. Vista). (**d**) ICF direct drive, injected expellant (e.g. Discovery III)

When you have two separate and distinct species, the fusion particles and the expellant particles, unless they are perfectly mixed thermodynamically this will lead to a degradation in the jet efficiency. It is not easy to achieve a perfect mixing by the injection of added expellant into a hot and fast-moving plasma stream. A lot of the injected expellant mass will not be translated into available kinetic energy, and this is the equivalent of a fuel leak. Without thermalisation the individual species will have a different temperature, velocity and density.

This implies that the continued adding of expellant mass for increased mass flow rate will degrade the performance of the total fuel mixture and so lead to a reduction in the exhaust velocity. Following Orth [4], the ability to couple the expellant to the fusion reaction products can be termed a *jet coupling efficiency*. Since the jet power is the work done by the momentum change produced by the jet power in the forward direction, any particle energy not contributing to the mass exhaust stream in the forward direction is wasted energy and will not generate thrust.

Taking our example of the VISTA spacecraft design to Mars which utilises a 0.046 g DT fusion capsule augmented with a 50 g/capsule propellant wrapped around the capsule, all detonated at 30 Hz, this results in a combined mass flow rate of 1.50138 kg/s compared to what would have been a mass flow rate of 0.00138 kg/s. The capsule design is an indirect Hohlraum type and the expellant is wrapped around it as stored liquid hydrogen rather than injected. The lasers transport through bore holes to achieve the target compression and so the expellant is not involved in the compression and only reacts on detonation of the fusion capsule during its explosive disassembly.

The VISTA capsule design was described as a 'grapefruit'[4] and to assist the thermalisation with the fusion energy the LH_2 expellant material was seeded with high Z materials to maximise the neutron absorption efficiency into the expellant. It would be more difficult to add expellant for a direct drive capsule design which depended on the rocket effect of the blow-off layer from the compression. However, it may be possible to instead inject propellant into the exhaust stream at the moment of detonation.

Earlier we discussed how the exhaust velocity in a conventional (non-expellant) system can be described by

$$v_{ex} = (2\alpha\phi\eta_j)^{1/2}c. \quad (8.4)$$

The goal is for the expelled propellant to achieve the same temperature as the fusion fuel so that they are in some state of quasi-equilibrium. But the reality is that there will be some difference, and we assume here that the temperature of the expellant is doubled, and where it is treated as a thermal expansion velocity of the expellant v_{th}; and this is equated to the exhaust velocity where $v_{ex} = v_{th}$.

In the situation with expellant propellant, an equation can be derived where the kinetic energy of the combined expellant and thermonuclear gas $m_{exp} v_{ex}^2/2$ is again balanced with the Einstein energy for the thermonuclear capsule leading to

$$\alpha \epsilon_{mix} \phi m_{cap} c^2 = \frac{4(m_{cap} + m_{exp}) v_{ex}^2}{2\eta_j}. \tag{8.5}$$

Where it is observed that the exhaust velocity is reduced by the ratio of the capsule to expellant masses m_{cap}/m_{exp} when doing work to heat the gas to a thermal equilibrium state with the thermonuclear energy release. In describing the exhaust velocity, we also define a coupling efficiency factor ϵ_{mix} which relates to how well the thermonuclear fuel couples to the injected expellant material, where for a perfect mixing scenario $\epsilon_{mix} = 1$. Therefore, the modified exhaust velocity becomes

$$v_{ex} = \frac{1}{\sqrt{2}} \left[\left(\frac{m_{cap}}{m_{exp} + m_{cap}} \right) \epsilon_{mix} \alpha \phi \eta_j \right]^{1/2} c. \tag{8.6}$$

But in many cases $m_{exp} >> m_{cap}$ and so this can be reduced to

$$v_{ex} = \frac{1}{\sqrt{2}} [(m_{cap}/m_{exp}) \epsilon_{mix} \alpha \phi \eta_j]^{1/2} c. \tag{8.7}$$

So, if the mass flow rate is increased by the addition of expellant propellant for the purpose of augmenting the thrust, this will be done so at the cost of a reduction in exhaust velocity. This can also be seen from the jet power $P_{jet} = 0.5 \times \dot{m} v_{ex}^2$ where for a constant jet power if the mass flow rate is increased (as with the use of expellant) then the exhaust velocity must also go down. This also implies a modification to the jet power equation such that they are equated through an equivalence statement since the exhaust velocity must represent the sum of the two species velocities, i.e.

$$P_j = \frac{(\dot{m}_{cap} + \dot{m}_{exp})(v_{ex})^2}{2\eta_j n_{eng} \epsilon_{mix}}. \tag{8.8}$$

Substituting the mass flow rate for the individual capsule and expellant mass where $\dot{m} = (m_{cap} + m_{exp}) f_{hz}$ we find

$$m_{cap} \geq \frac{\dot{m}(2v_{th})^2}{2n_{eng} \epsilon_{mix} \eta_j \alpha \phi f_{hz} c^2} \equiv \frac{2\dot{m} v_{ex}^2}{n_{eng} \epsilon_{mix} \eta_j \alpha \phi f_{hz} c^2}. \tag{8.9}$$

We also end up with a modified jet efficiency. Then rearranging

$$\eta_{j,min} = \frac{2v_{ex}^2}{\alpha\phi\epsilon_{mix}c^2}\left(\frac{m_{cap} + m_{exp}}{m_{cap}}\right). \tag{8.10}$$

Alternatively given $m_{exp} >> m_{cap}$ and if the temperature of the expellant doubled and its molecular weight halved, we find

$$\eta_{j,min} = \frac{2v_{ex}^2}{\alpha\phi\epsilon_{mix}c^2}\left(\frac{m_{exp}}{m_{cap}}\right). \tag{8.11}$$

This is therefore a corrected version of the earlier (non-expellant) equations but in the circumstances where expellant mass is also being used and for a perfect mix. These are original derivations by this author. This is important in the design of a spacecraft since we will be using both fusion fuel and expellant propellant. Note that altering the amount of expellant injected (or included) for each capsule will also permit another method of creating thrust variability.

This can be further examined by considering the individual particle species energies. The average kinetic energy of the particle species is given as a solution of the Maxwell-Boltzmann distribution which is described as a function of temperature T and velocity v as follows which we discussed earlier in Chap. 6. For the particle distribution function, a solution is found for the root mean squared velocity which is also referred to as the thermal velocity

$$v_{rms} = \left(\frac{3k_B T}{m}\right)^{1/2}. \tag{8.12}$$

Indeed, the average thermal energy per atom is given by $E = 3k_B T/2$. As illustrated in Fig. 8.4 for particles at one temperature $k_B T$ they will have a set distri-

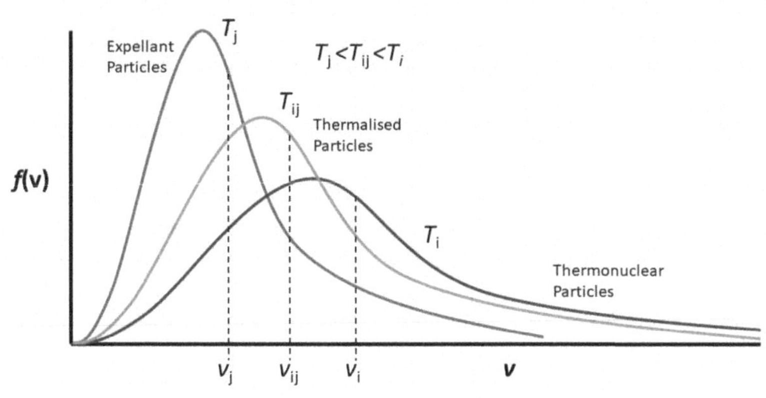

Fig. 8.4 Temperature thermalisation of particle species in exhaust plasma[KFL]

bution, but this will be shifted to a flatter distribution as the particles gain energy at higher temperature and so the velocity peak will shift to a larger value. This means that for the two gasses of interest here, the expellant particles with temperature T_j will gain energy and shift to a higher temperature, whilst the thermonuclear particles with temperature T_i will lose energy and shift to a lower temperature. At some point during this process, provided there isn't energy loss from the system, the two particle species should mutually interact and (through momentum exchange) thermalise to an equilibrium state where the average temperature T_{ij} of the combined particles is spatially uniform and temporally constant. Although the new distribution will be dominated by the expellant particles since they are so much larger in quantity.

Consider the scenario where there is fusion fuel present with species m_i but also the addition of expellant propellant as a separate species m_j. Its thermalisation is caused by the energy release from the capsule thermonuclear ignition. It takes energy to heat up the expellant propellant and that must come from the fusion energy release. As a result, the fusion temperature T_i will be reduced to a lower temperature of $T_{i'}$ and so the particles will also have a lower velocity of $v_{i'}$, and similarly for the expellant particles but in the other direction towards energy gain. Since the quantity of expellant dominates the thermonuclear fuel species ($N_{exp} >> N_{fus}$) we can neglect the fusion particles in the velocity of the exhaust stream and instead define a thermal velocity based only on the expellant

$$v_{th,j} = \left(\frac{3k_B T_j}{m_j}\right)^{1/2}. \tag{8.13}$$

If we then assume that the temperature of the expellant is raised to $T_{ij} = 2 \times T_j$. Since we are also looking at hydrogen H_2 for the expellant this will dissociate into two atoms with atomic mass $m_{ij} = m_j/2$, ($H_2 \rightarrow 2H$) leading to

$$v_{th,ij} = \left(\frac{3k_B T_{ij}}{m_{ij}}\right)^{1/2} = \left(\frac{3k_B 2T_j}{m_j/2}\right)^{1/2}. \tag{8.14}$$

The final exhaust velocity is then given as a multiplication of the thermal velocity based on how high its temperature was raised by the process

$$v_{ex} = 2\left(\frac{3k_B T_j}{m_j}\right)^{1/2} = 2v_{th,j}. \tag{8.15}$$

The difference between the thermalisation for the two species can be equated to a fuel-expellant mix coupling efficiency. If there are many species they would all have to be modelled and so the computation can become complicated. A model that assumes complete mixing and therefore thermal equilibration will simplify the computation, although this must be justified as realistic. Thermalisation may be difficult to achieve and depends upon the relaxation timescales of the mixture into an equilibrium state compared to the expansion speed of the exhaust stream.

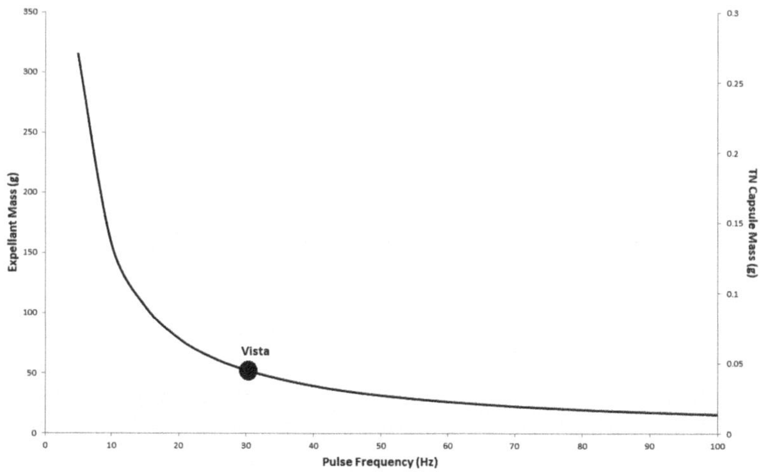

Fig. 8.5 Example calculations for ICF design parameters in VISTA

Table 8.1 Calculations for VISTA rendezvous probe to mars using different fusion fuels models carrying 100 tons payload where t_{ret} is the round-trip time

Parameter	$D(T,{}^4He)n$	$D(D,{}^3He)n$	D(D,T)p	$D({}^3He,{}^4He)p$
v_{ex} (km/s)	152.0	75.0	83.0	156.0
v_c (km/s)	80.4	39.6	43.9	82.5
T (MN)	0.231	0.114	0.126	0.237
P_j (GW)	17.58	4.28	5.24	18.52
P_{sp} (kW/kg)	13.98	3.40	4.17	14.72
t_{ren} (days)	61.91	119.92	108.88	60.46
t_{ret} (days)	123.82	239.84	217.76	120.92

Figure 8.5 shows an example calculation for the VISTA spacecraft design for how the thermonuclear capsule mass, expellant mass and pulse frequency are related. In this example VISTA has a capsule mass of 0.046 g, an expellant mass of 50 g and a pulse frequency of 30 Hz.

To explore the idea a little we can create our own model for VISTA. In the case of the two-stage burn but single structure VISTA model for a rendezvous mission to the planet Mars carrying a 100 tons payload, at 30 Hz pulse frequency using a capsule mass of 0.046 g DT fuel, we calculate a total cruise speed of 80.3 km/s completing the rendezvous mission to Mars in $\sim$62 days and so a return to Earth within $\sim$124 days. Both stages would generate a thrust of 0.23 MN, jet power 17.6 GW and specific power 13.9 kW/kg. The payload mass fraction was 1.7%. The overall results are shown in Table 8.1, where a return to Earth will be anywhere in the range $\sim$120–240 days depending on the principal fuel being used.

In terms of the final published design for VISTA, there are several differences in the time to Mars as compared to the published VISTA study due to several factors. Firstly, in the report the authors specified that the pulse frequency would run between ~1–30 Hz, where the latter of the range was considered a maximum. Therefore, the authors may have constructed an acceleration program based around a variable thrust and so variable pulse frequency throughout the flight.

A second factor is that the VISTA models evolved, and those shown here were based on the earlier models where there was a difference in the mass of the systems. We assumed 3.72 tons DT fuel, 6,000 tons wet mass, 1,860 tons dry mass and 4,104 tons non-dry mass (TN fuel + expellant). In contrast, for the final 2003 VISTA model they assumed 3.5 tons DT fuel, 6,500 tons wet mass, 1,813 tons dry mass, 4,143 tons non-dry mass. These differences would likely account for the ~15% difference in the arrival time (i.e. 124 days versus 145 days). So the VISTA models shown in Table 8.1 are most consistent with the pre-2003 VISTA models.

The analysis is complicated by the fact that there are several different models of VISTA since the project evolved towards an improved design. The design has its origins in the work by Hyde [6] and then became a published study in 1987 [2]. The design then evolved and was published in various reports such as in 2000 [3] and then eventually the final study report in 2003 [4]. Tracking all the design changes would require the generation of multiple systems models.

Although the Daedalus design was comprehensive in its consideration of all the spacecraft systems, as a propulsion study, the VISTA design may be even more reliable. This is due to the fact it was able to build on the earlier studies; it is more recent and so contains more up-to-date information on science discoveries, but it is also well informed by actual NIF experiments using lasers on ICF designs, and presumably modern numerical codes. The innovative conical spacecraft design also makes for a creative way to minimise the radiation load on the structure. Perhaps of critical importance to the VISTA studies is that we have demonstrated an ability to model fusion propulsion systems that include an expellant propellant for the purpose of thrust augmentation via an enhanced mass flow rate.

Finally on VISTA, it will be noted that the designers claimed a gain of $G = 1{,}500$ for a 0.046 g fuel mass yet driven by a 5 MJ laser through indirect drive ignition. However, this 5 MJ was the total energy delivered to the target and not the initial driver energy. For the earlier VISTA models the assumed laser efficiency was 6% but for the later VISTA models the laser efficiency was said to be 12% which would imply a total driver energy of 83.3 MJ at 30 Hz pulse frequency which would be 2.5 GW. This is on the basis that 50% of the power from the plasma debris to the inductor was lost as waste heat, resulting in 1.25 GW input power to the inductor at each cycle. These would also correspond to the high burn-up fraction of $\phi = 0.476$ and yet lead to a gain of 1,500 from the capsule mass of $m_{cap} = 46$ mg.

Achieving such a high gain may be possible in the future using alternative ignition models like shock assisted ignition. However, it is not a reasonable extrapolation of the performance of NIF capsules based on a hot-spot ignition model. Yet who is to say what may be possible in the future. In the studies that follow, we will show how the use of expellant propellant can be used in two other mission scenarios that build on from VISTA. It is useful to explore some other design concepts which employ an expellant propellant design for thrust augmentation.

8.2 Discovery-III to the Outer Planets

This work was originally based on another system, although it was not an ICF based system. This was the NASA Discovery-II spacecraft design for a mission to Saturn/Jupiter [7], which was a magnetic confinement fusion (MCF) based system where the augmented propellant expellant is injected as hydrogen slush into the reaction chamber. The inclusion of augmented propellant would certainly appear to be important for the case of interplanetary mission design, although it comes at the cost of increased mass.

In 1968 the science and science fiction writer Arthur C Clarke wrote the book 2001: A Space Odyssey [8]. This was written in collaboration with the film director Stanley Kubrick who was making the film of the same name which went on to become one of the most iconic cinematic productions ever made for the genre. In the story the astronauts travel to the outer planets on board an interplanetary spacecraft called *Discovery-I* which is shown in Fig. 8.6. The planet is Saturn in the novel but Jupiter in the film. The 'Cavradyne' propulsion system for the spacecraft was not explicitly stated in either the film or the novel but a general conclusion seems to be that it utilised some form of gas core nuclear fission reactor.

In celebration of this spacecraft (and movie), around the turn of the century the US space agency NASA decided to create their own version of this spacecraft which they called *Discovery-II*. The report on the spacecraft design and mission to the outer planets was eventually published in 2005 [7]. Overall, the NASA team gave a favourable review of the early attempts at Discovery-I although for the same mission they opted a completely different type of propulsion system which was nuclear fusion using a spherical torus magnetic confinement fusion (MCF) reactor. The designers examined a mission to Jupiter in 118 days and a mission to Saturn in 212 days. NASA has continued to have an interest in the prospect of crewed outer solar system missions in other studies where they have examined the use of magnetised target fusion [9].

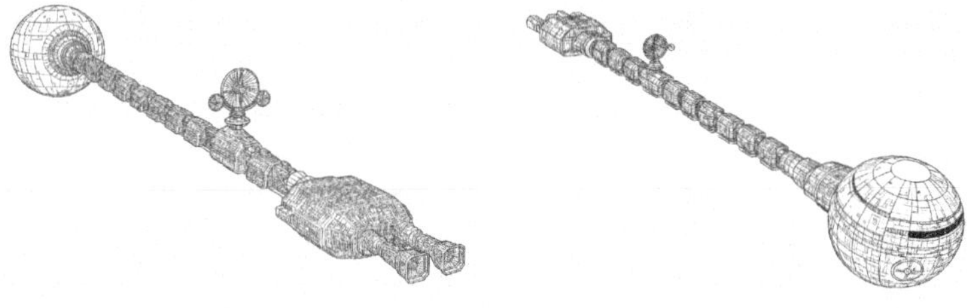

Fig. 8.6 2001 A space odyssey Discovery-I spacecraft[KFL]

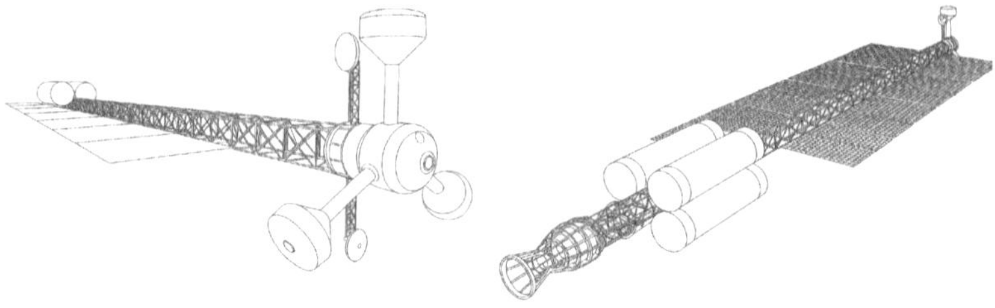

Fig. 8.7 NASA spacecraft Discovery-II concept design rendezvous interplanetary mission[KFL]

In this section it is not the intention to give a full review of either the Discovery-I or Discovery-II design but merely to demonstrate a version of this using an ICF engine. We will restrict the work to a rendezvous mission. Following the tradition of the other authors, we refer to this as Discovery-III. To be clear, the three models are then *Discovery-I*: Gas core nuclear fission Cavradyne propulsion; *Discovery-II*: Spherical torus magnetic confinement fusion propulsion; *Discovery III*: Laser driven inertial confinement fusion propulsion.

The Discovery-II spacecraft is illustrated in Fig. 8.7 and is a crewed vehicle designed to accommodate between 6–12 astronauts. The key design requirement is a rendezvous mission to Jupiter or Saturn with a design that was characterised by a payload mass fraction exceeding 5% and a trip time not greater than one year. We will not go into detail here about the trajectory or rendezvous with moons of the gas giants but instead focus on an assumed mission to the outer planets where the usual conic sections are ignored and a straight-line approximation was assumed as justified in Chap. 2.

The 240 m long spacecraft would carry a 172 tons artificial gravity payload. This would be on the end of a long structural boom, from which would also be attached the heat radiators and propellant cryo-tanks. On the rear end of the structure would be the spherical torus nuclear fusion reactor and magnetic nozzle. The total structural mass would be 646 tons.

The spacecraft would be propelled by 11 tons of D^3He fuel, which was then augmented by Hydrogen expellant propellant. From the total expellant of 861 tons, around 807 tons would be used for the main fusion engines, whilst the remainder would be used for reaction control and flight performance reserve. The use of the expellant increased the mass flow rate to a total of 0.079 kg/s for the Jupiter mission and 0.045 kg/s for the Saturn mission. The increase in mass flow rate would also increase the vehicle thrust.

In addition, the spacecraft would carry a 10 tons nuclear fission reactor for start-up of the main engine with a power output of 2 MWe. There would be 4 propellant tanks with a total mass of 6 tons. The main fusion engine would have a mass of around 310 tons and would exhibit a nozzle jet efficiency of 80%. The total wet mass at launch would be around 1,690 tons and the dry mass would be around 872 tons, with a total mass ratio of 1.9.

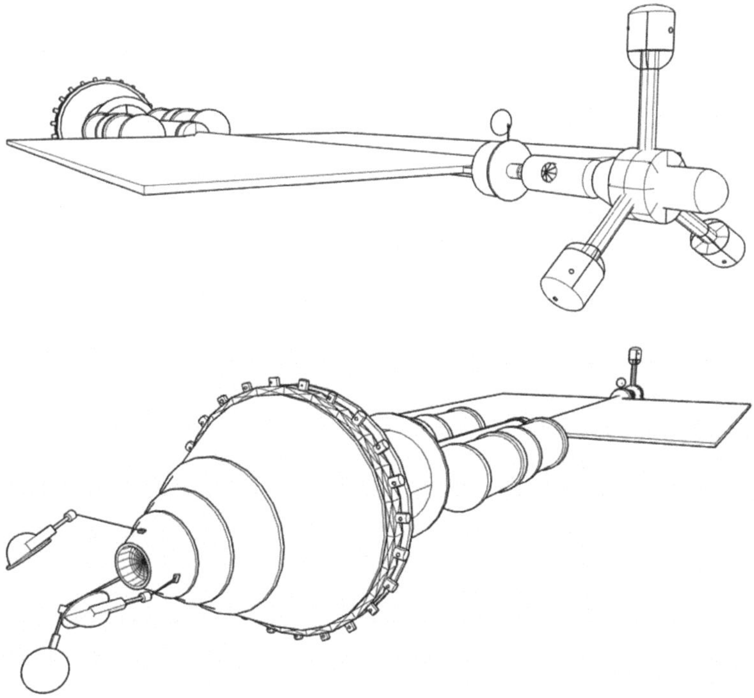

Fig. 8.8 Concept design spacecraft Discovery-III[KFL]

The 118 days mission to Jupiter at 4.7 AU would require a jet power of 4,830 MW and a specific power of 8.62 kW/kg. It would use an exhaust velocity of 347 km/s and achieve a cruise velocity after engine stage burn of 223 km/s, although it would not enter a cruise phase but immediately begin to decelerate towards its rendezvous. For the Saturn mission to 9.57 AU the Discovery-II spacecraft would take 212 days to complete the rendezvous. It would be characterised by a cruise velocity of 309 km/s.

The goal therefore is to demonstrate a version of Discovery-II but using an ICF engine. It was the aim to stay close to the initial requirements of a high payload mass fraction and a total mission time of 118 days for the Jupiter mission and 212 days for the Saturn mission. In addition, the system masses for the Discovery-II will be matched in the definition of Discovery-III where possible. All other variables such as ICF capsule mass, pulse frequency, burn-up fraction, driver mass, will then be assumptions of the modelling. The vehicle configuration for the Discovery-III is shown in Fig. 8.8 where it is a hybrid between Discovery-II and the SunVoyager probe (discussed in next section) due to the ICF engine. The main distinguishing feature being the change of payload to a crewed habitat rotating for artificial gravity designed to accommodate 6–12 astronauts for the mission. We will explore two interplanetary missions in turn.

Jupiter Interplanetary Mission
Calculations were conducted on a model for the Jupiter rendezvous mission with the aim to achieve the mission in around 118 days using the same mass assumptions. The parameters matched when compared to the Discovery-II mission included the total payload mass of 172 tons, the mass flow rate of 0.08 kg/s and the exhaust velocity of 347 km/s which guided the value of the expellant propellant set at 576 mg/pellet for every 2.88 mg fuel capsule. The other parameters assumed included $\epsilon_{mix} = 0.8$, $\alpha = 0.0039$, $\phi = 0.33$. The mission was run at a pulse frequency of f_{hz} = 138 Hz.

The mission to Jupiter required 5.27 tons, with 2.93 tons allocated to the first stage burn and 2.34 tons allocated to the second stage deceleration burn. The total augmented expellant hydrogen propellant allocated to the acceleration stage was 586 tons and similarly for the deceleration stage the expellant allocation was 468 tons, bringing the total to 1,054 tons.

For the Discovery-III mission the fuel was D^3He as an equi-mixture. For the acceleration phase around 1.017 billion capsules would be required and 0.812 billion for the deceleration phase. Although no frontal shield was needed for the mission, one was defined for Discovery-III on the order of 100 kg. The supporting power mass was defined to be 10 tons for a 2 MWe reactor, which would be needed for initiation of the fusion reactor. The neutron fluxes associated with the Discovery-III design would be 1.1447×10^{19} neutrons/pulse or 1.582×10^{21} neutrons/second. The total wet mass of the spacecraft was around 2,124 tons and the total dry mass 1,065 tons. The payload mass fraction was 8.09%.

For the mission modelled the spacecraft will accelerate for around ~85 days up to a velocity of around ~76 km/s. It would then begin the process of deceleration for the remainder of the mission arriving at the destination of Jupiter orbit at 4.7 AU after a total of ~118.3 days. The propulsion system generated a thrust of 27.76 kN, a jet power of 4.82 GW and a specific power of 8.43 kW/kg.

When compared to the Discovery-II mission the results are similar, although the Discovery-III has a substantially increased vehicle mass and so reduced velocity. However, the ICF mission required half the fusion fuel mass when compared to the MCF mission (5.27 tons versus 11 tons), although the expellant mass was greater (1,054 tons versus 807 tons).

Saturn Interplanetary Mission
We next demonstrate a model of the Discovery-III mission with an assumed ICF engine, as an equivalence to the NASA Discovery-II mission to Saturn. This is to 9.57 AU for which the NASA model completed in 212 days. Once again this was a rendezvous mission carrying a 172 tons payload.

In addition to the increased distance, the main difference between the Jupiter and Saturn missions is in the reduced mass flow rate at 0.045 kg/s. This results in changes to the propulsion performance such as the thrust and jet power. As before, the aim is to match the total mission time, mass flow rate and exhaust velocity. Back engineering the results of the NASA design also implies an exhaust velocity of 463 km/s. The main change for the Saturn mission is in the increased fusion fuel mass of

5.29 tons which we allocate across the acceleration and deceleration stage to achieve the total mission time. For the mission to Saturn the same pellet mass was assumed at 2.88 mg as well as expellant mass augmentation rate of 449.71 mg/pellet.

The NASA study indicated a minor adjustment for fuel loading but did not give further details. However, for the Discovery-III designs the total wet mass was reduced to 1,896 tons. The total fusion mass for the acceleration stage was 2.89 tons and for the deceleration stage it was 2.4 tons with a total of 5.29 tons. This then led to a total augmented expellant hydrogen propellant allocated to the acceleration stage of 451 tons and 375 tons for the second stage bringing the total to 826 tons.

For the mission to Saturn the spacecraft would accelerate for around 116 days up to a velocity of 83 km/s. It would then begin the process of deceleration for the remainder of the mission arriving at the destination of Saturn orbit at 9.7 AU after a time duration of slightly greater than 212 days. The propulsion system generated a thrust of 20.95 kN, a jet power of 4.85 GW and a specific power of 8.49 kW/kg. The calculations estimated that the likely neutron fluxes associated with the Discovery-III design to Saturn would be 1.145×10^{19} neutrons/pulse or 1.145×10^{21} neutrons/second.

Discussion of Discovery-III Concept Design

A mission to Jupiter and Saturn using an ICF engine has been generated comparable to the published NASA study. Various Discovery-III models presented in Table 8.2, as ICF equivalent concept designs to the Discovery-II are preliminary. Rather than a detailed definition, the models presented should be seen as an indication of the likely performance of such a propulsion system when applied to the same mission and payload.

In general, the results for the NASA Discovery-II MC mission, whilst the Jupiter mission appears to be highly optimised there is room for improvement in the Saturn mission. For the Discovery-III ICF mission there appears to be room for improvement in the various models in terms of optimisation. This creates uncertainty around any conclusions that ICF may appear to be competitive to MCF in terms of thermonuclear fuel usage but is potentially interesting.

Since the Discovery-III designs were run with very high pulse frequencies it is constructive to consider how these might be reduced by the change of expellant propellant. The total amount of expellant material must add up to that defined for the mission, such as the 1,054 tons for the Jupiter mission and 826 tons for the Saturn mission. For the same assumption of total mass flow rate and expellant mass, it is then possible to see how the augmented propellant mass requirement changes with engine pulse frequency since

$$M_{aug} = m_{exp} N_{pell}. \tag{8.16}$$

Where the number of pellets is related to the mass of thermonuclear fuel via

$$N_{pell} = \frac{M_{fuel}}{m_{cap}}. \tag{8.17}$$

Table 8.2 Comparison of calculations for NASA Discovery-II spacecraft concept design with Discovery-III models for Jupiter and Saturn Rendezvous missions

Parameter	D-II (Jup)	D-III (Jup)a	D-III (Jup)b	D-II (Sat)	D-III (Sat)
Fusion method	MCF	ICF	ICF	MCF	ICF
Pulse frequency (Hz)	–	138.20	100.0	–	100.0
TN capsule mass (mg)	–	2.88	2.88	–	2.88
Expellant (mg/pellet)	–	576.0	449.71	–	449.71
Total wet mass (tons)	1,690	2,124	1,809	1,699	1896
Total dry mass (tons)	872	1,065	1,065	872	1,065
Mass structure (tons)	646	693.5	693.5	646	693.5
Mass fusion fuel prop (tons)	11	5.27	4.74	11+	5.29
Mass expellant prop (tons)	807	1,054	739.68	807+	826
Acceleration (m/s^2)	0.0165	0.0103	0.0085	0.0123	0.0083
Decceleration (m/s^2)	–	0.0129	0.01	–	0.0099
Exhaust velocity (km/s)	347	347	463	463	463
Mass flow rate (kg/s)	0.08	0.080	0.0452	0.045	0.0452
Thrust (kN)	27.8	27.76	20.95	20.86	20.95
Jet power (GW)	4.83	4.82	4.85	4.83	4.85
Specific power (kW/kg)	8.62	8.43	8.49	8.62	8.49
Total mission time (days)	118.0	118.26	118.13	212.0	212.14
Total mission distance (AU)	4.7	4.7	4.7	9.57	9.57
Peak velocity (km/s)	223	76.12	75.84	309	83.17

For a given expellant mass m_{exp}, total mass flow rate $\dot{m}$ and pulse frequency f_{hz} the capsule mass required for the mission can be estimated from the following

$$\dot{m} = \dot{m}_{cap} + \dot{m}_{exp} = (m_{cap} + m_{exp}) f_{hz}, \tag{8.18}$$

$$m_{cap} = \frac{\dot{m}}{f_{hz}} - m_{exp}. \tag{8.19}$$

Then the total augmented propellant mass for the mission is given by

$$M_{aug} = M_{fuel}\left(\frac{\dot{m}}{m_{cap} f_{hz}} - 1\right). \tag{8.20}$$

And since $(m_{cap} + m_{exp})/m_{cap} = (m_{exp}/m_{cap}) + 1$ then alternatively

$$M_{aug} = M_{fuel}\left(\frac{m_{exp}}{m_{cap}}\right). \tag{8.21}$$

The relationship of the augmented propellant mass to the pulse frequency is shown in Fig. 8.9. A mission can move to a lower pulse frequency, but this would require more expellant and so push the vehicle mass up which means the mission

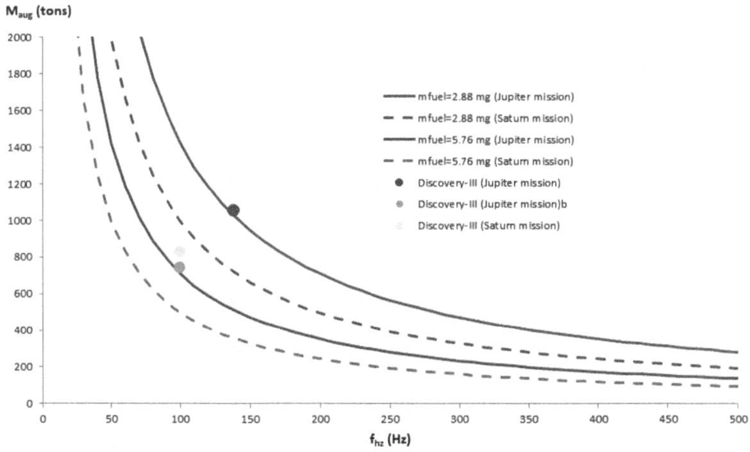

Fig. 8.9 Calculations of Discovery-III augmented propellant mass

is elongated. Conversely the augmented propellant can be maintained low, but this requires a high pulse frequency. The design point for the Discovery-III mission is also shown as an optimum between these two extremes.

In comparing ICF to MCF there are several issues that we can highlight. The first is that an MCF system would struggle to use DT fuel since the large flux of neutrons from the reaction would require significant radiation shielding to protect any superconducting magnets which form the magnetic mirror system. This is unlike an ICF engine which does not use the magnetic bottle concept and so other than surrounding structure and the engine reaction chamber there is less of a need for shielding. Any excess of neutron radiation on critical technologies that form the engine would suffer degradation such as material fatigue.

We also note that it is possible to vary the thrust of the fusion engine by simply changing the pulse frequency of detonation, since $T = \dot{m}v_{ex} = m_{cap} f_{hz} v_{ex}$. This is not a method that is possible with MCF design concepts where instead any change must be achieved by the injection (or depletion) of mass into the exhaust stream via the mass flow rate. This method available for ICF can be seen as a type of exhaust throttling. Although in fairness a continuous thrust system such as with MCF may facilitate a simpler engineering design.

The main result that seems to come out of a comparison between the MCF and the ICF design is that the latter didn't seem to require as much thermonuclear fuel material, approximately half. It is too early to reach a strong conclusion here, but this might indicate a competitive advantage of ICF based propulsion; although we also note a caveat that the total vehicle mass was also greater for the ICF engine and the difference in the level of optimisation for the designs.

We can't leave this section without at least mentioning the possibility of the USS Heinlein which is shown in Fig. 8.10, as an extension of the Discovery-III design but with an increased payload capacity and enhanced performance (higher speed). This

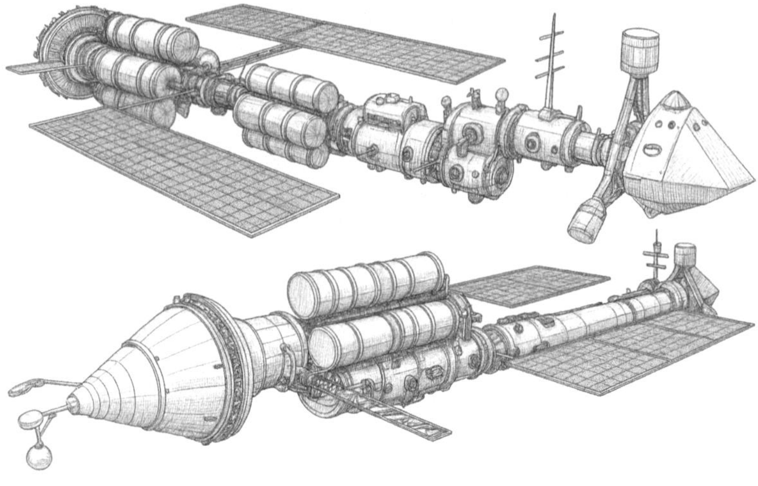

Fig. 8.10 Concept design spacecraft USS Heinlein PeaceKeeper Patrol Vessel[KFL]

is a proposed US Space Force ship that has the goal of patrolling the 'oceans' of our solar system to ensure peaceful co-existence and the safety of different colonies. It is a vessel that effectively represents a military police force as a peacekeeping vessel.

As humanity moves off world and begins the process of colonisation of the Moon, Mars and beyond, if we are not to be held together through constructive agreements like the 1967 Outer Space Treaty, then the nation state competition we see on Earth today may also spill over into the domain of space. In which case, these sorts of vessels would be a vital part of ensuring peace, but also in acting as rescue missions for deep space colonies should they get into trouble. An argument was previously made for such vessels in a separate publication by this author [10].

8.3 SunVoyager Interstellar Precursor

In the 1980s-1990s a proposal was made for a Thousand Astronomical Unit (TAU) mission driven by a 1 MWe nuclear-electric engine carrying a 1.2 tons science payload to 1,000 AU in around 50 years [11, 12]. This would require an average cruise velocity of around 20 AU/Yr or $\sim$100 km/s. One of the major scientific drivers for TAU was measurements of stellar parallax and of the local heliospheric plasma environment. Several such proposals for interstellar precursor missions have been made including for a 200 AU mission [13–16]. Indeed, the suggestions of a 1,000 AU probe has also been recently resurrected by Johns Hopkins University Applied Physics Laboratory (JHUAPL) [17].

A proposal has also been made for a Voyager 3 mission [18] that would take a space telescope to 550 AU for the goal of imaging exoplanets using solar

Table 8.3 Description of science objectives for SunVoyager mission

SO.No	Primary	SO.No	Secondary
1.1	Focal lens exoplanets	2.1	Charged Particles & Dust
1.2	Dwarf planet flyby observations	2.2	Astrometry measurements
1.3	Interstellar asteroid detection	2.3	SETI emission signatures

gravitational lensing around what is known as the Focal point. The Voyager 3 probe would have a total mass of 14 tons and carry a 0.2 tons science payload.

Others have written about this location, which relates to the dynamic solar atmosphere, and it has a range between 550–1,000 AU [19, 20]. Indeed, Maccone argues that sending a probe to this distance is a necessary precursor mission before the first interstellar flight is ever attempted [21]. The main benefit of placing an optical telescope at this position is to help identify and characterise interstellar exoplanets in terms of their size, orbital parameters, dynamical origin and composition. The practicality of an astronomical telescope observing from the solar gravitational lens has been discussed by Turyshev [22] and Landis [23].

An interesting question arises as to the nature of such a probe if it was to be driven by highly advanced propulsion systems that are not yet technologically mature. We are interested in describing what such a mission would look like in terms of its mission profile and propulsion requirements. The proposed flyby spacecraft mission is called the SunVoyager. As shown in Table 8.3 for this mission the primary science objectives are defined, which are stated primarily as a planetary science mission consistent with a weather eye towards considerations for the recent Decadal strategy for planetary science and astrobiology [24].

Three primary science objectives are defined for this mission. The first is the optical observation of the nearest stars and associated exoplanets with the use of gravitational lensing and the transit method. The second is to conduct a flyby of a deep space dwarf planet which has not been previously visited. The third is the possible imaging of deep space interstellar objects such as 1I/'Oumuamua [25], 2I/Borisov [26] and 3I/ATLAS [27], since observations of such objects could provide a direct analysis of materials formed in the disks of other stars, but also constraints on their mineralogy and elemental abundances could provide insights into their parent disks. The spacecraft is to deploy smaller sub-probes to conduct flypasts of either a Dwarf planet, interstellar asteroids or other targets of interest for a more detailed in-situ investigation. Ten will be included in the mission each with a maximum mass of 0.5 tons. In addition, three secondary science objectives are defined for this mission. The objectives include measurements of the charged particle and dust composition of the interstellar medium [28], measurements of astrometry [29]. The third is related to the Search for Extraterrestrial Intelligence (SETI) and is along the lines of the detection of power and propulsion emission signatures in deep space [30–32].

For the SunVoyager mission, we assume a launch in the latter part of the current century as a flyby probe only from Earth orbit and this timescale has

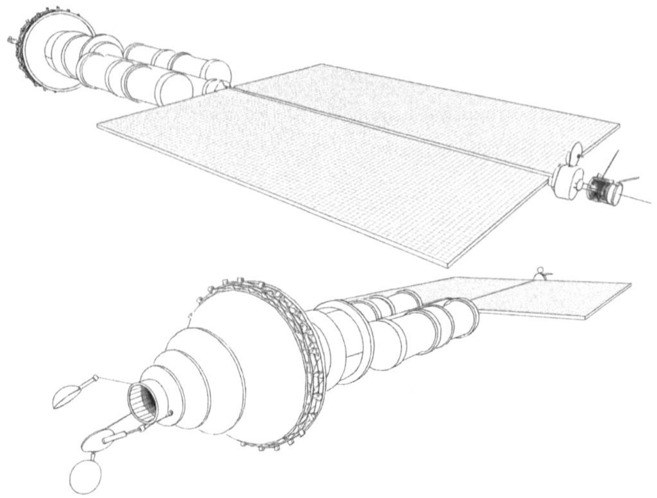

Fig. 8.11 Concept design spacecraft SunVoyager for interstellar precursor mission[KFL]

been determined in recent technological and strategic roadmap work [33, 34]. We consider a straight-line trajectory and so neglect the usual conic sections, where the gravity fields from the planets within our solar system can be neglected due to the high velocity of the spacecraft. Although it is possible to conceive of missions that utilise a gravity assist from the large gas giants, the velocity addition would have a minor influence on the mission given the large velocity.

As shown in Fig. 8.11 a similar conical propulsion system is adopted for the SunVoyager design as for the VISTA spacecraft. In earlier publications a preliminary model for the SunVoyager probe was constructed as an attempt to scope out the design space [35, 36]. An ignition model was constructed for a capsule mass of 0.00576 g with a fuel radius of 2.494 mm and was filled with an equal mixture of D^3He. It would be driven by a ~42 MJ laser corresponding to an effective intensity of $I_L = 4.184 \times 10^{17}$ W/cm^2. Each pellet would generate an energy of ~653 MJ and so represent a gain of $G = 15.55$.

The engine structure would also include a superconducting magnetic coil for the function of generating the magnetic field to direct the charged particles for thrust. The lasers are also shown in position around the engine structure, but their correct positioning requires detailed analysis, and they are only shown for illustration. The use of a 'shadow shield' as adopted in the VISTA design is required to protect the superconducting coil and the lasers from the radiation. Following on from the engine geometry, an annular radiation shield would also be contained within the engine structure to mitigate excess radiation on the axis and protect the spacecraft.

The spacecraft payload has a power to mass ratio of 36 W/kg. The design concept has a mass ratio of 1.11 for a total wet mass of 1,151 tons. The payload mass fraction is 7.82%. The spacecraft is travelling at a sufficiently slow speed ($v/c = 0.002$) not to be overly concerned about charged particle and dust impacts [37]. However, a minor particle bombardment shield is used on the front of the spacecraft to protect the payload in the event of a minor collision with larger space matter. This is constructed of Beryllium and will be on the order of $\sim$2–3 mm thick and will have a total mass of around 1 ton. This shield also acts as a detector for charged particle and dust collisions in terms of their mass measurement.

The spacecraft will have a nuclear fission reactor for initiating the fusion engine and running other systems during the cruise phase of the mission. It will be a 10 tons reactor producing a power output of 3.6 MWe, characterised by 0.36 kWe/kg. Recent requirements for NASA designs of a compact nuclear reactor are for a 10 kWe reactor to be used on a crewed Mars mission [38]. More recent discussions refer to a requirement of 40 kWe, and future requirements are likely to approach 1 MWe for small human colony missions. So, this seems to be the first goal in developing power systems for space. The reactor itself will produce residual heat generation, and this can be re-radiated into space using panels with a total mass of 161 tons and utilising a coolant of 80 tons and Brayton cycle radiators.

The spacecraft is an interstellar precursor probe. It will be assembled in Earth orbit over multiple launches, and each propellant tank may require a separate launch vehicle. With 4 propellant tanks and the remaining structural mass of the spacecraft the assembly of the entire spacecraft is likely to be comparable to something like the ISS as an international effort and may require on the order of $\sim$10 separate launch vehicles to place all parts into the necessary assembly orbit.

The spacecraft will be launched outside the vicinity of Earth orbit and will move the vehicle away for one entire day at an initial velocity of 17 km/s so that it fully escapes from Earth. This may be achieved by using booster rockets such as via nuclear thermal propulsion system or by running the fusion engine on a low pulse detonation mode on the order of $\sim$0.1 Hz. This implies an acceleration up to a distance of $\sim$0.005 AU which is around twice the mean Earth-Moon distance. This is a safe place to then initiate the fusion engine away from the Earth orbiting satellites and human space station crews.

The fusion reactor is initiated by the on-board fission reactor. A pellet is accelerated into the target chamber where the bank of laser beam lines can target it within the open reaction chamber to the point of thermonuclear ignition. A set of externally placed laser focusing mirrors are required to ensure a symmetric implosion. The plasma from the detonation will expand into the hemispherical target chamber, with the charged particles then directed by large magnetic fields internal to the chamber. These are then ejected for thrust generation whilst the next pellet is loaded onto the target ignition point.

Table 8.4 shows calculations for varying the amount of propellant augmentation where it can be observed that with increasing expellant mass there is a loss of exhaust velocity. With increasing propellant augmentation, the mass of the spacecraft becomes much greater which then begins to reduce the cruise velocity achieved. The parameter t_{1kau} is the time to make science observations through the gravitational lens zone between 500–1,000 AU, although observations would also be made leading up to this. Figure 8.12 shows similar data plotted for different

Table 8.4 SunVoyager spacecraft mission performance assuming a thermonuclear capsule mass of $m_{cap} = 5.76$ mg (D^3He Fuel at $\phi = 33\%$) driven by a 42 MJ Laser with $G = 15.55$ and $\epsilon_{mix} = 0.8$, $f_{hz} = 10$ Hz

Parameter	(a)	(b)	(c)	(d)	(e)
v_{ex} (km/s)	5,840	2,612	826	261	83
m_{exp} (g)	0.01	0.05	0.5	5.0	50.0
m_{aug} (tons)	17.36	86.80	868.05	8,680	86,805
$\dot{m}$ (kg/s)	0.00016	0.00056	0.0051	0.05	0.5
T (kN)	0.92	1.456	4.177	13.06	41.50
P_j (GW)	2.687	1.902	1.725	1.705	1.722
P_{sp} (kW/kg)	3.719	2.631	2.386	2.358	2.382
S_b (AU)	409.66	280.20	311.99	230.23	91.84
S_c (AU)	90.30	219.76	187.97	269.73	408.12
t_b (Yrs)	5.50	5.50	5.50	5.50	5.50
t_c (Yrs)	0.597	2.09	1.46	2.047	5.07
$t_{500\,au}$ (Yrs)	6.098	7.59	6.960	7.548	10.570
v_c (km/s)	720.84	498.76	612.75	625.69	382.17
t_{1kau} (Yrs)	3.29	4.75	3.87	3.79	6.21

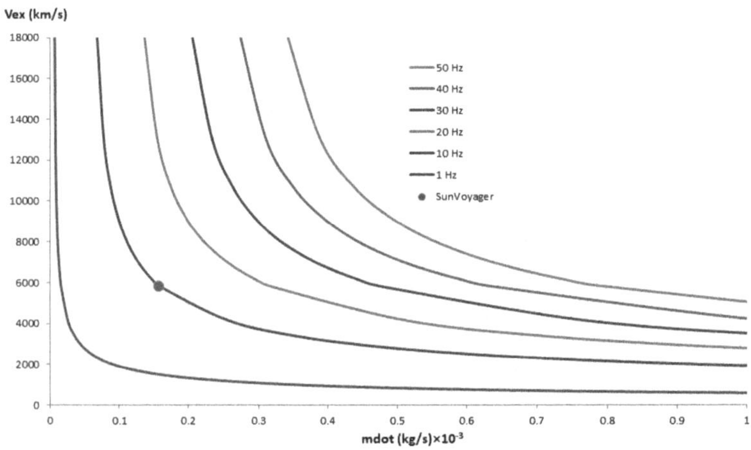

Fig. 8.12 Calculations of SunVoyager exhaust velocity as a function of increasing expellantmass for different values of pulse frequency

pulse frequencies. It can be observed that there is a trade-off between high exhaust velocity and high mass flow rate such that an optimum value may be preferred depending on the mission requirement. The 10 Hz design point chosen for the SunVoyager probe is also shown Fig. 8.12.

Once the fusion engine is ignited the spacecraft will accelerate at 0.0035 m/s^2 with a mass flow rate of 0.00015 kg/s for 5.5 years. By the end of the burn phase the spacecraft would be at a 5.258×10^{13} m or 351.47 AU. Once the burn phase is complete all propellant tanks will be ejected. During the burn phase a radiation shield protects the propellant tanks from the pellet detonations and associated neutron and x-ray radiation environment. It is expected that each detonation will produce around ~2.289×10^{19} neutrons/pulse or equivalent to ~2.289×10^{20} neutrons/s. During the use of the fusion engine, it will generate a thrust of 0.92 kN, a jet power of 2.68 GW and a specific power of 3.72 kW/kg.

The spacecraft will reach the cruise velocity of 721 km/s or ~152 AU/Yr by the end of the boost phase. The spacecraft will then cruise for a further 417 days or 1.14 years until it enters the start of the Focal lens gravitational point around 500 AU. It will then cruise for a further 1,404 days or 3.85 years until it reaches the 1,000 AU outer boundary point. The focal lens science phase of the mission will therefore take place over around ~5 years from which observations of exoplanets around nearby stars will be made.

The charged particle and dust measurements will take place throughout the burn and cruise phase of the mission since these rely on impacts with the forward particle shield which will itself act as a data recording instrument in terms of the particle size, populations and associated collision energies. The detection of interstellar objects will take place through the cruise phase up to 1,000 AU. Observations of a dwarf planet will be made on-route as a fly past. The object will depend on the trajectory taken and the time of launch but may be 90,377 Sedna (84 AU), 136,199 Eris (96 AU) or other trans-Neptunian object. The goal is to achieve high resolution images of the object and enhance our understanding of its physical properties and evolution. SETI emission signatures will also take place over the same period. The probe will use its on-board computer to filter any detected signals from known astrophysical sources.

The number of ICF pellets required for the mission is 1.736×10^9 in total. It is no small undertaking to manufacture this many pellets for use in a propulsion engine although with robotic automation and additive manufacturing this may become easier. In the future manufacturing of the pellets may even be possible in space provided the fuel can be sourced. This might be Deuterium from the world's oceans then lifted into orbit or through gas giant mining of Helium-3 [39]. For example, if the production timescale for the pellets was over 20-years, then the production rate would be ~9,900 pellets/hour or ~237,800 pellets/day. Alternatively, the fuel can be stored within the large propellant tanks but then the pellets loaded with the appropriate gas on-board and in-situ with its use. This implies that the production rate would have to at least be equivalent to the pulse frequency detonation rate or ideally exceed it [40]. Given that the SunVoyager is only operating at 10 Hz such a possibility seems reasonable.

Finally, it is worth doing a quick comparison of the SunVoyager mission compared to another fusion engine for the same type of mission. For this we compare against the Princeton Plasma Physics Laboratory's Direct Fusion Drive (DFD) which has also been designed for a 550 AU mission [41]. The engine for this mission also uses D^3He fuel but it operates on very different principles as a reversed field configuration engine. We also compare against the University of Michigan Gas Dynamic Mirror (GDM) Fusion concept which uses DT fuel for this mission [42]. For the GDM not all of the information was given and instead it was

Table 8.5 Comparison of SunVoyager with DFD, GDM and TAU to 500–550 au

Parameter	SunV	GDM	DFD	TAU
Wet mass (tons)	1,151	488	5.28	32.0
Fuel mass (tons)	10.0	126.52*	3.22	11.2
Payload mass (tons)	100	$<< m_{wet}$	0.3	5.0
Payload mass fraction (%)	8.7	–	5.7	15.6
Exhaust velocity (km/s)	5,840	1,785*	510	140
Mass flow rate (kg/s)	1.6×10^{-4}	9.41×10^{-4}	7.84×10^{-3}	7.17×10^{-5}
Thrust (kN)	0.92	1.68	4.0	10.0
Jet power (MW)	2,687	1,499	1,020	0.7
Jet efficiency (%)	37.8	–	30.0	–
Specific power (kW/kg)	3.72	6.0	2.0	0.058
Boost distance (AU)	409.66	247.07*	555.6	410
Boost time (years)	5.5	4.26*	13.0	20
$t_{1kau} - t_b$ (Years)	3.88	4.31*	4.50	16
Cruise speed (km/s)	720.84	550*	479.1	174.3
Mission time (years)	9.38	8.57*	17.5	46

reverse engineered by inference (marked in the table as *). We do not discuss the operation of these engines further here.

Table 8.5 shows a comparison between the published DFD vehicle configuration and performance in flyby mode for a similar 500–550 AU distance. Whilst the ICF engine is more massive it would appear to have a performance advantage over DFD in terms of time to destination. A comparison to GDM shows they are similar. Although as a caveat the two systems are in no way placed on a consistent basis of comparison to draw any strong conclusions. We note the significant difference in the mass flow rates, exhaust velocities and therefore thrust. It gives some idea for how the ICF driven SunVoyager compares to other engine designs. Table 8.5 also shows data for the TAU mission proposal [13] which was to be propelled by a nuclear electric engine.

Problems of Chap. 8

8.1 A conical shaped spacecraft employs a shadow shield to mitigate the radiation dosage to the structure. The exit diameter is 3 m and the angle subtended by the shadow shield is 10° (i) What is the distance of the detonation point? (ii) What is the diameter of the exit hole from the centre up to the edge of the shadow shield assuming the total angle subtended from the centre up to the shield is 15°.

8.2 A single engine spacecraft has a D^3He thermonuclear capsule mass of 0.1 g and detonates it along with injected expellant at 100 g/capsule, at a rate of 100 Hz. (i) Assuming a burn-up fraction of 1/3rd, a mix coefficient of unity, and a jet efficiency of 60%, what is the exhaust velocity of this design. (ii) Confirm the jet efficiency by calculating it using the derived equations.

8.3 Imagine you have designed a flyby spacecraft for an interplanetary mission that requires 100 tons of fusion fuel pellets. (i) Assuming each pellet is 0.01 grams in mass how many pellets will this involve during the mission? (ii) Assuming the acceleration phase of the mission lasted for 5 years, what is the minimum pulse frequency to be adopted as an average if all pellets are to be utilised by the time the spacecraft reaches the mission distance? (iii) What does this imply about pellet utilisation in-situ or being manufactured prior to the mission?

References

1. Spencer DF (1963) Fusion propulsion system requirements for an interstellar probe. Technical report No.32-397, NASA CR-50964, Jet Propulsion Laboratory
2. Orth CD, Klein G, Sercel J, Hoffman N, Murray M, Chang-Diaz F (1987) The VISTA spacecraft - advantages of icf for interplanetary fusion propulsion applications, UCRL-96676. In: Presented IEEE 12th symposium on fusion engineering, Monterey, California, 12–16 October 1987
3. Orth CD (2000) Parameter studies for the VISTA spacecraft concept, LLNL, UCRL-JC-141513. In: Presented at NASA advanced propulsion workshop in fusion propulsion, Huntsville, Alabama, 8–10 November, 2000
4. Orth CD (2003) VISTA: a vehicle for interplanetary space transport application powered by inertial confinement fusion, LLNL, UCRL-TR-110500, 16 May 2003
5. Sacks RA, Darling DH (1987) Direct drive cryogenic ICF capsules employing C-T wetted foam. Nuclear Fusion 27(3):447–452
6. Hyde R, Wood L, Nuckolls J (1972) Prospects for rocket propulsion with laser-induced fusion microexplosions, AIAA, 72-1063. In: Presented at the 8th propulsion joint specialist conference, 29 November–01 December 1972
7. Williams CH, Dudzinski LA, Borowski SK, Juhasz AJ (2005) Realizing 2001: a space odyssey: piloted spherical torus nuclear fusion propulsion, NASA/TM-2005-213559, AIAA-2001-3805
8. Clarke AC (1968) 2001: a space odyssey. Hutchinson, London
9. Adams RB, Alexander RA, Chapman JM, Fincher SS, Hopkins RC, Philips AD, Polsgrove TT, Litchford RJ, Patton BW (2003) Conceptual design of in-space vehicles for human exploration of the outer planets, NASA/TP-2003-212691, November 2003
10. Long KF (2025) The emergence of a fusion propulsion capability this century and implications for exploration and settlement strategies. In: Space education and strategic applications journal (SESA), conference proceedings, Spring
11. Etchegaray MI (1987) Preliminary scientific rationale for a voyage to a thousand astronomical units, JPL publication 87-17, NASA
12. Nock KT (1987) TAU - a mission to a thousand astronomical units, AIAA-87-1049. In: Presented at the AIAA/DGLR/JSASS international electric propulsion conference, Colorado Springs
13. Jaffe LD, Ivie C, Lewis C, Lipes R, Norton HN, Stearns JW, Stimpson LD, Weissman P (1980) An interstellar precursor mission. JBIS 33(1):3–26
14. Fearn DG (2008) Technologies to enable near-term interstellar precursor missions; is 400 *au* accessible? JBIS 61(8):279–283
15. Zurbuchen TH, Patel P, Fisk LA, Zank G, Malhotra R, Funsten HO, Mewaldt RA (2008) Leaving the heliosphere: a nuclear-powered interstellar probe, Chapter 5, AIAA
16. Bruno C, Matloff G, Simone D, Genta G, Vulpetti G, Johnson J, Long KF (2013) Technologies to enable near-term interstellar precursor missions. In: International academy of astronautics study, final report of study group III.10 to commission III, IAA-978-2-917761-23-6

17. McNutt Jr RL, Wimmer-Schweingruber RF, Gruntman M, Krimigis SM, Roelof EC, Brandt PC, Vernon SR, Paul MV, Stough RW, Kinnison JD (2022) Interstellar probe - destination: universe! Acta Astronautica 196, 13–28
18. Abdolrahimi S, Yale B, Tzounis CC, Fofrich J, Rohan P, Cabrera-Guzman J, Welsher JC, Nakhjiri N, Scott D, Johnson A (2022) Voyager 3: a concept mission to interstellar medium. J Spacecraft Rockets 59(3):975–987
19. Maccone C (2009) Deep space flight communications: exploiting the sun as a gravitational lens. Springer, Berlin
20. Turyshev SG, Anderson BG (2003) The 550-AU mission: a critical discussion. Mon Not R Astron Soc 341(2):577–582
21. Maccone C (2010) Realistic targets at 1000 AU for interstellar precursor missions. Acta Astronautica 67(5–6):526–538
22. Turyshev SG, Shao M, Alkalai L, Friedman L, Arora N, Weinstein-Weiss S, Toth VT (2018) Direct multipixel imaging of an exo-earth with a solar gravitational lens telescope. JBIS 71(10):361–368
23. Landis GA (2018) A telescope at the solar gravitational lens: problems and solutions. JBIS 71(10):369–374
24. Committee on the Planetary Science and Astrobiology Decadal Survey Space Studies Board (2022) Origins, worlds, and life: a decadal strategy for planetary science and astrobiology 2023–2032. National Academies Press, Washington
25. Fitzsimmons A., Snodrass C, Rozitis B, Yang B, Hyland M, Seccull T, Bannister MT, Fraser WC, Jedicke R, Lacerda P (2018) Spectroscopy and thermal modelling of the first interstellar object 1I/2017 U1 'Oumuamua. Nat Astron 2:133–137
26. Jewitt D, Luu J (2019) Initial characterization of interstellar comet 2I/2019 Q4 (Borisov). Astrophys J Lett 886(2):L29
27. Jewitt D, Hui M, Mutchler M, Kim Y, Agarwal J (2025) Hubble space telescope observations of the interstellar interploper 3I/ATLAS. Astrophys J 990(1):L2
28. Draine BT (2011) Physics of the interstellar and intergalactic medium. Princeton series in astrophysics. Princeton University Press, Princeton
29. Kovalevsky J, Sidelmann PK (2004) Fundamentals of astrometry. Cambridge University Press, Cambridge
30. Zubrin R (1995) Detection of extraterrestrial civilizations via the spectral signature of advanced interstellar spacecraft. In: Progress in the search for extraterrestrial life. ASP conference series Seth Shostak G (ed), vol 74
31. Freitas RA (1983) The search for extraterrestrial artifacts (SETA). JBIS 36(11):501–506
32. Viewing DRJ, Horswell C, Palmer EW (1977) Detection of starships. JBIS 30:99–104
33. Long KF (2022) On the application of pulse propulsion frequency in inertial fusion space mission design. JBIS 75(4):127–141
34. Long KF (2020) A rapid study on the development of an interstellar roadmap and planning ahead for technology maturation. JBIS 73(1):6–14
35. Long KF (2023) Sunvoyager: interstellar precursor probe mission concept driven by inertial confinement fusion propulsion. J. Spacecraft Rockets 60(3):797–811
36. Long KF (2025) Development of the SunVoyager interstellar precursor probe driven by inertial confinement fusion propulsion. J Spacecraft Rockets 62(1):55–62
37. Martin AR (1978) Project Daedalus: bombardment by interstellar material and its effects on the vehicle. JBIS Special Suppl 31:S116–S121
38. McClure PR, Poston DI, Gibson MA et al (2020) Kilopower project: the krusty fission power experiment and potential missions. Nuclear Technol 206(1):S1–S12
39. Palaszewski B (2012) Atmospheric mining in the outer solar system: resource capturing, storage and utilization, AIAA 2012-3742. In: Presented 48th AIAA/ASME/SAE/ASEE joint propulsion conference & exhibit, Atlanta, Georgia
40. Long KF (2022) Galactic crossing times for robotic probes driven by inertial confinement fusion propulsion. JBIS 75(9):118–126

41. Cohen SA, Swanson C, Mcgreivy N, Raja A, Evans E, Jandovitz P, Khodak M, Pajer G, Rognlien TD, Thomas S, Paluszek M (2019) Direct fusion drive for interstellar exploration. JBIS 72(2):37–50
42. Kammash T, Lee MJ (1996) A fusion propulsion system for near-term space exploration. JBIS 49(9):351–356

Chapter 9
Interstellar Spacecraft Designs

Travel to the stars will be difficult and expensive. It will take decades of time, GW of power, kg of mass-energy and trillions of dollars...interstellar travel will always be difficult and expensive, but it can no longer be considered impossible.

—Robert Forward, 1996 [1]

9.1 Daedalus Series Stage Spacecraft

The Daedalus spacecraft was a theoretical design study from the 1970s [2–4] and is illustrated in Fig. 9.1. The study is reliable in terms of mathematical rigour. We only discuss the concept from a system level definition and do not go into sub-system level of details.

The Daedalus spacecraft had a vehicle configuration that is made up of two structural stages. Each stage also has a single engine and a set of propellant tanks. The propellant tanks are dropped en-route soon after use. The first stage burns for 2.05 years and then is ejected. This is followed by the second stage which then burns for 1.76 years, prior to cruising to its target destination of Barnard's star at 5.9 light years distance carrying a 450 tons payload. As shown in Fig. 9.2 and in previous calculations by this author [5] it is possible to stage the vehicle structure even further, but this comes with the penalty of a shrinking payload mass if we are to hold other parameters constant.

Both stages use an engine pulse frequency of $f_{hz} = 250$ Hz, which is significantly high, and likely an order of magnitude above what is being discussed for the application of fusion technology to commercial energy generation power plants ($\sim$10–50 Hz). This high pulse frequency also places the engine structure under the risk of material fatigue from the constant pulsed cycling. As shown in Fig. 9.3 it is possible to substantially reduce the pulse frequency without over elongating the mission duration.

The ICF pellets are stored cryogenically and then accelerated along an electromagnetic rail gun into a target chamber, where dozens of electron beams will target them within the target chamber, and implode them to high density and high

K. F. Long, *Stellar Engines*, Springer Aerospace Technology,
https://doi.org/10.1007/978-3-032-18062-9_9

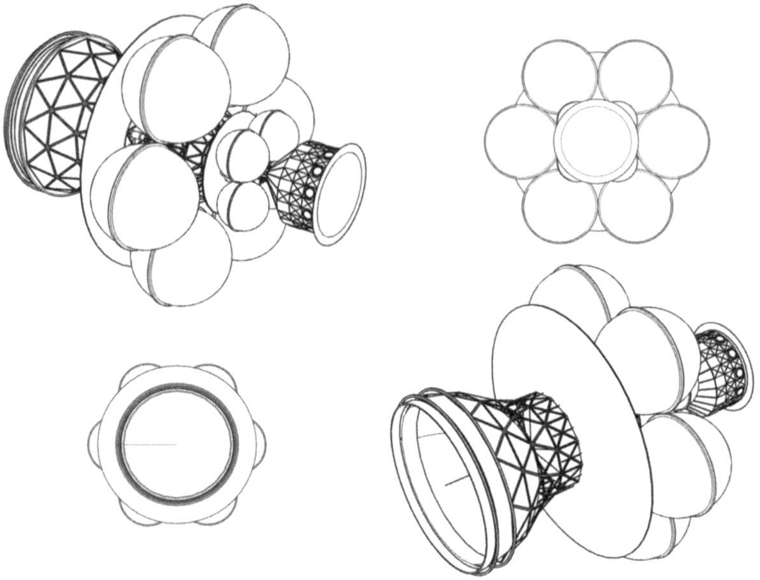

Fig. 9.1 Illustration of project daedalus interstellar spacecraft concept[ML]

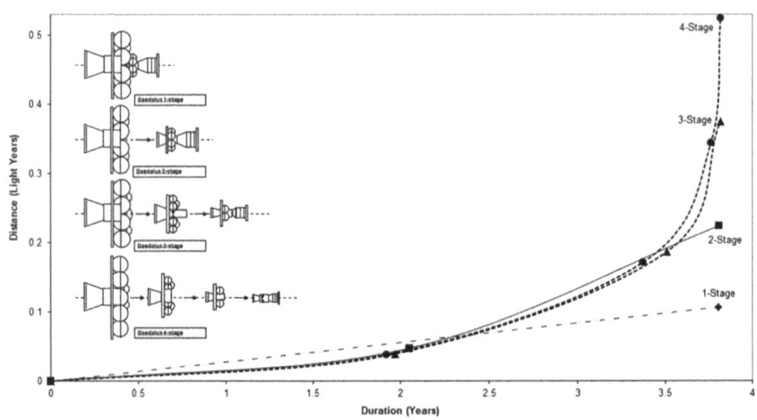

Fig. 9.2 Distance attained during the acceleration phase of the mission[KFL]

temperature so that a thermonuclear state of ignition can be reached, upon which they will detonate within the chamber, and the plasma outflow will rebound from a superconducting loop charged engine bell wall. The charged particles within this plasma will then follow the magnetic field lines of a magnetic nozzle arrangement and be exhausted for thrust generation. All contained within a hybrid nozzle design.

The mission was to be completed in just under half a century, but with only days of encounter time at the local system since the probe was flyby only. The Daedalus design had a first stage propellant mass of 46,000 tons and a second stage mass of

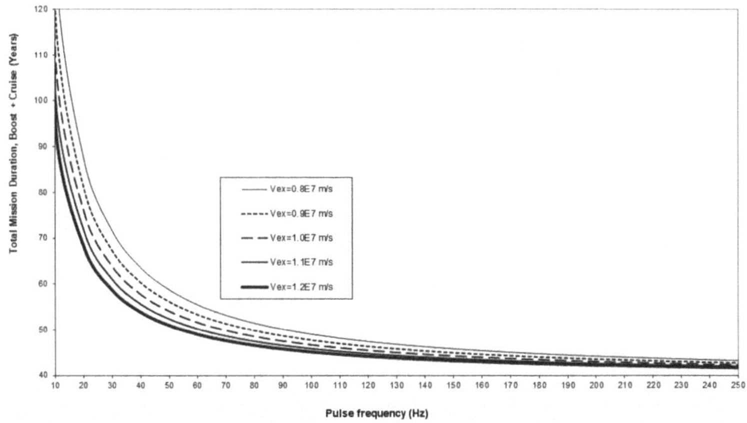

Fig. 9.3 Calculations of mission duration as a function of pulse frequency for a daedalus-like mission over varying exhaust velocity[KFL]

4000 tons. There were 6 propellant tanks on the first stage and 4 tanks on the second stage, so that the propellant per tank was 7666 tons and 1000 tons respectively. The assumed exhaust velocities were 10,600 km/s (I_{sp} =1.08 million s) and 9210 km/s (I_{sp} = 0.94 million s) for the first and second stage respectively.

The Daedalus design utilised two different ICF pellets sizes at 19.7 mm radius for the first stage and 9.16 mm for the second stage, leading to capsule masses of 2.84 and 0.288 g respectively. They would both have a cryogenic solid fuel density of 0.089 g/cm^3 and utilise a 50:50 D^{3}He fuel mixture ratio. There was to be around 16.197 billion pellets in the first stage and 13.888 billion pellets in the second stage. The burn-up fraction from each capsule was estimated to be $\phi = 0.175$ and $\phi = 0.133$ for the first and second stage respectively. Several models have been developed that are worthy of examination and which are reproductions of the Daedalus design concept. The Daedalus model (i) is what we refer to as the 1978 design as published in the Daedalus reports [2–4].

An attempt was made to construct an ignition model for the Daedalus second stage pellet mass. The driver energy was 411 MJ with an (ultra-high) laser efficiency of 81.22%. The hot-spot ρdr = 7.64 g/cm^2 and the burn-up fraction was $\phi = 0.133$. The total energy gain was $G = 33$. These are broadly consistent with the Daedalus model although the laser requirements are not practical. This was then integrated into the mission profile to generate a Daedalus-like mission.

The D^{3}He fuel was constructed of a Deuterium honeycomb with liquid Helium-3 filling. The ICF capsule designs adopted for the Daedalus study are illustrated in Fig. 9.4.

At the centre of the capsules was a small pocket of DT gas which acts as an ignition trigger based on the hot-spot ignition method. This was needed since the reactivity of DT is much higher at a lower temperature as compared to D^{3}He. However, a major issue with the use of a DT trigger is firstly it must be filled in

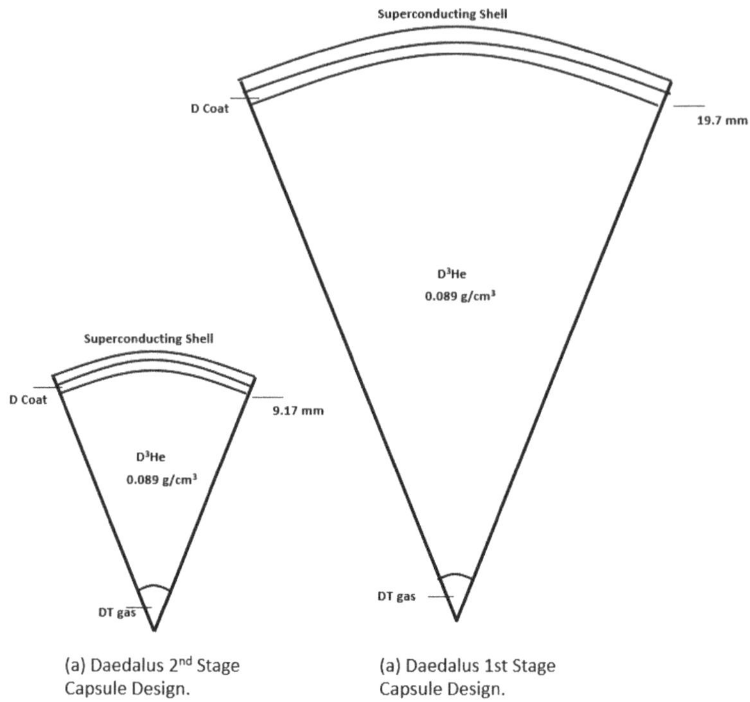

Fig. 9.4 Project daedalus ICF capsule designs[KFL]

situ prior to use. Second, the decay of the Tritium results in substantial amounts of heat which will act to expand the fuel volume when the goal is to compress an otherwise cold fuel through spherical convergence.

Since the first stage capsule is $\sim$10 times more massive and $\sim$2 times the radius of the second stage, this leads to a driver energy requirement which is $\sim$6.5 times higher. The first stage will produce a higher fusion yield per shot but would also involve longer confinement times. The main reason for using D^3He fuel is to reduce the neutron flux (compared to DT) and the use of protons and ^4He charged particles for thrust generation which leads to a high specific energy. The use of large capsule designs allows for a lower convergence ratio which improves the implosion stability and reduces hydrodynamic issues like Rayleigh-Taylor instabilities.

However, the challenge with these large designs is the need for a uniform illumination of the surface area which requires precise beam control and stable target alignment. Without the use of reflecting mirrors, or an improved coverage of the beams around the capsule sphere, it is difficult to see how a symmetric implosion could be achieved with these designs. Although it is likely that the capsules themselves are technically plausible, it is just difficult to integrate into the engineering machine. To calculate the gain G of a capsule this is given by the ratio of fusion energy output to drive energy input, in terms of a physics definition

$G = E_{out}/E_{dr}$. The total energy output from an individual capsule is given as a function of the fractional energy release or mass defect α, the burn-up fraction ϕ, the capsule mass m_{cap} and the vacuum speed of light c=299,792,458 m/s, so that $E_{out} = \alpha\phi m_{cap}c^2$.

For the Daedalus first stage capsules we have $\alpha = 0.0039$ (for D^3He), $\phi = 0.175$, m_{cap} = 2840 mg, so that E_{out} = 174,205.7 MJ/capsule which is a Gain of $G \sim 64.5$. For the Daedalus second stage capsules we have $\phi = 0.133$, m_{cap} = 288 mg, so that E_{out} =13,426.1 MJ/capsule which is a Gain of $G \sim 32.66$. This is way above the gain currently achievable at facilities like NIF, but not overly ambitious and likely achievable in the future with advanced power plant concepts. The gains reported for Daedalus are realistic, provided the issue of implosion symmetry and stability can be addressed to avoid losses, and provided the implosion engine operates successfully in the first place.

It is plausible that the high driver energies associated with these designs would allow adequate compression and heating to achieve significant burn-up fractions as indicated. Yet any losses from Bremsstrahlung radiation or x-ray preheat or an incomplete burn will reduce the potential yield. In terms of the laser (flash pump light source) paradigm however, a major issue would be finding lasers that have sufficient efficiency to deliver the required energy to the target. Also, the less efficient those lasers, the more waste heat generated implying a need for thermal heat management, as well as large power reactors for any energy storage.

We may summarise the Daedalus ICF capsules as being technically feasible from physics principles, but very challenging in terms of engineering. To minimise the instabilities the manufacturing and fabrication of each capsule must be done with high precision to avoid surface features that could lead to the onset of hydrodynamic instabilities. Then the capsules must be injected at the high repetition rate of 250 Hz and maintain this over several years.

The spacecraft starts with a wet mass of 53,204 tons and a dry mass of 3204 tons. This gives it a payload mass fraction of 0.85%. It also has a structural mass to dry mass of 44.24% and a propellant mass fraction to wet mass of 97.37%. The total dry mass was 6.02% of the total wet mass.

The spacecraft undergoes the first burn stage for 2.05 years at 0.33 m/s^2 acceleration where the probe reaches 3166 AU or 0.05 ly and a velocity of 21,341 km/s or 7.11%c. After dropping the first stage and all of the propellant tanks it then ignites the second stage engine which burns for 1.76 years at 0.27 m/s^2 where the probe reaches a distance of 9983 AU or 0.158 ly and a velocity addition of 14,973 km/s or 0.0499c, bringing the total velocity to 36,314 km/s or 0.1211c.

The spacecraft cruises for 46.99 years over a 359,959 AU or 5.692 ly, then arriving at the interstellar target destination 5.9 ly away after a total trip time of 50.8 years. At the end of the two burn periods the time dilation as measured by the spacecraft's clocks will be equal to

$$\Delta t - \tau \approx \left(1 - \frac{1}{\gamma}\right)t \approx \ 125\,\text{days} \tag{9.1}$$

Where $\tau = t/\gamma$, $\Delta t = \gamma\tau$, $\gamma = 1/\sqrt{1-(v/c)^2} \approx 1.00736$. This equation form nicely isolates the difference between the coordinate and proper time, since at low velocities where $\gamma \approx 1$, $1 - 1/\gamma \approx 0 \rightarrow$ proper time $\approx$ coordinate time, but at high velocities where $1 - 1/\gamma$ becomes appreciable and so this implies large time dilation. From the perspective of an observer on Earth approximately 17,163 days would have passed.

It is worth making some comments on the Beryllium particle shielding due to mass erosion by interstellar dust grains since the ablation thickness scales linearly with the assumed dust density which translates to shielding mass requirements. The energy flux into the spacecraft was calculated by this author to be around 40.73 J/sm^2 and the temperature on the front of the shield will be around 163.71 K. Assuming the Daedalus frontal shield area of 3217 m^2 and an equivalent emission surface area such that $A/A_o = 1$, so that it is then expected for all of the energy to be re-radiated. That said, the calculated mass ablation rate will be around 3.4×10^{-5} kg/s or $\sim$1.07 tons/Yr, and so for the approximately 50-year mission this will require a particle shield thickness of 9.16 mm which would correspond to a mass of around $\sim$54 tons. In this model the assumed mass density of the dust in the interstellar medium has not been updated, which in 1978 was assumed to be around $\sim 10^{-23}$ kg/m^3 [6], when today it is known to be an order of magnitude lower at 2.57×10^{-24} kg/m^3 [7, 8]).

In general, for the case of the original two-stage Daedalus model for a flyby mission to Barnard's star, we have been able to reproduce the original design report to good accuracy; although there are differences in the acceleration and distance attained under stage burn within the mission profile. This is due to minor differences in the mass budget definitions but also in the process for how the propellant tank drops are modelled. For example, in our model we only performed tank drops at the end of stage burn, the actual design study specifies that the tanks are dropped on-route soon after they are used, which will alter the mass ratio slightly. Hence for the model discussed here the performance would be slightly underestimated.

Most of the differences are at an error of $\sim$5% or less, giving a good match to the original calculations. The aim of the modelling was to match the published end of burn staged mass of 1690 tons (1st stage) and 980 tons (2nd stage) which has been achieved. There is also a $\sim$10% difference in the calculated second stage specific power due to the fact it is not clear what the original Daedalus team used for their total propulsion mass value, and this should include the engine, power system and any thermal management.

In another model of Daedalus, we repurposed it, and the spacecraft is bespoked towards a different astronomical target, that of Centauri A and B, at 4.365 ly, which is also consistent with a modern Project Icarus requirement [5]. Although in the 1970s evidence was beginning to support the idea of a planetary disc around Barnard's star [9] this is not the case today and instead there is more evidence in support of planetary systems present in the nearby stars and in particular around the Centauri A/B/Proxima stars [10, 11] and so this the preferred target. It is also a closer target.

For the interest of mission designers, when applied to an Alpha Centauri mission the total wet mass was 54,610 tons and the total dry mass was 4610 tons. The total payload mass fraction was 0.82%. The structural mass fraction to dry mass was 39.29%. The propellant mass fraction to wet mass was 97.59%.

The calculations for the auxiliary power supply and the thermal radiator management was the same as for the previous model and this includes the calculated radiator masses. In terms of the particle shield for this model due to dust grains the energy flux into the spacecraft will be around 43.66 J/sm^2 and that the temperature on the front of the shield will be around 166.58 K. Assuming the Daedalus frontal shield area of 3217 m^2 and an equivalent emission surface area. The calculated mass ablation rate will be around 1.5×10^{-6} kg/s or $\sim$0.048 tons/Yr, and so for the approximately 37-year mission this will require a particle shield thickness of 0.3 mm which would correspond to a mass of around $\sim$1.79 tons. If doubled to take account of charged particle impacts this leads to a total shield mass of $\sim$3 tons.

In terms of firstly running the numbers on a basic mission requirement assuming a 50-year total flight time and an exhaust velocity of 9210 km/s, the minimum required cruise velocity for a flyby probe would be 26,172 km/s or 0.0873c and would require a mass ratio of 17.14. For the same probe but as a rendezvous mission this would require a minimum cruise velocity of 52,344 km/s or 0.1746c, which would correspond to a cruise duration of at least 25 years. For the model the Daedalus exhaust velocities of 10,600 km/s and 9210 km/s were assumed and the two staged mass ratios for the mission were calculated to be 7.55 for the first stage and 5.5 for the second stage, with a total mass ratio of 41.55.

In the case of the revised two-stage Daedalus model for a flyby mission to an Alpha Centauri target carrying the same 450 tons payload, at 250 Hz pulse frequency this leads to a total cruise speed of 0.1239c. This is after a first stage burn for 2.05 years and a second stage burn for 1.76 years completing the mission in 37.35 years. The first stage had a mass ratio of 7.55 and the second stage a mass ratio of 5.5. The engine performance was not much different from the other Daedalus models but for completeness the first stage thrust, jet power and specific power as 7.52 MN, 39.88 GW and 38.06 MW/kg respectively. Similarly, the second stage thrust, jet power and specific power were 0.66 MN, 3.05 GW and 8.58 MW/kg respectively.

Although the destination has changed the same quantity of propellant (50,000 tons) has been assumed hence the main change is the reduced total mission time due to the decreased distance ($\sim$1.6 ly). No account has been made for the difference of the position of each target destination outside of the ecliptic plane which would affect the trajectory. For interest the inclination of the Alpha Centauri system is 79.2° whereas the inclination of the Barnard's star is 60.19°.

Project Daedalus remains one of the most detailed system integrated design studies ever completed for an interstellar spacecraft. Yet, it was also completed in the 1970s and so the progress in technology will inevitably make elements of the design solution redundant. The following issues pertaining to re-design considerations derive from the author's own perspective but also a post-project assessment [12] and others who reviewed the engineering and physics issues [13].

It would be desirable to reduce the cruise velocity to $\sim 0.05c$ for ease of engineering environments whilst increasing mission time. Deceleration into orbit for increased encounter time. Then expanding the total mission duration to 100 years for ease of engineering requirements.

It would also be preferred to develop alternatives to the electron beams as the target driver, where the electrons are likely to diverge in the presence of a magnetic field making focussed collimation onto a target difficult. Indeed, this is why in this book we have largely focussed on the use of optical lasers instead.

The Tritium trigger in the ICF capsules could be removed since they generate substantial self-heating through radioactive decay, but this decision would come at the cost of increased ignition requirements. Alternatively, Tritium can be manufactured on-route using a Lithium blanket via the ^{6}Li(n,α)T reaction. Another alternative is the use of a DD trigger which lowers the temperature requirement for ignition, but still within a largely D^{3}He fuel.

A consideration for the use of alternative ignition methods (i.e. shock ignition, fast ignition, indirect drive) but also smart pellet designs which also help to capture the high energy neutrons through scattering and absorption, although some neutrons will still escape within the outer layers since tritium will be created through (n,p) reactions with ^{3}He. This can be mitigated however by using depleted deuterium region towards the outer shell so there is no deuterium for the tritium to react with.

There should be an effort to reduce the 450 tons payload mass but sufficient to allow investigation of the system by sub-probes, atmospheric penetrators and landers [14]. In addition, a consideration for a change in the number of vehicle stages (i.e. see [5]) under series or parallel thrust modes. A reduction in the total propellant mass of 50,000 tons for Daedalus, which would ease fuel mining acquisition architecture requirements.

Further efforts should be made to protect the structure and mitigate against fast (high energy) neutrons and x-ray radiation which will heat the vehicle and cause mission failure. This may require insulation layers and radiation shields. Neutrons can be mitigated with only 10s cm Lithium blanket via the ^{6}Li(n,α)T reaction resulting in neutron thermalisation. Although this may add substantial mass to the vehicle structure.

The stability of the engine bell structure under enormous heat loads from the expanding plasma exhaust may lead to the build-up of fatigue on the material structure. It has been found by others using finite element modelling that the thickness of the reaction chamber may be too small and as a result a buckling load may be reached if a fuel capsule fails to ignite [15]. This might suggest a more careful design of the reaction chamber and associated wall structure.

The bootstrap of the energy from each detonation would occur through an induction loop where the energy is stored within the magnetic field and then moved down electrical transmission lines to fire up the next cycle. For Daedalus, at 250 Hz pulse frequency this suggests an induction recovery time on the order of $\sim$ms, although taking account the time to expand and eject the plasma and its associated exhaust products the induction time would more likely be on the order of $\sim \mu$s.

This must also be maintained through the acceleration phase of the mission with $\sim 3 \times 10^{10}$ induction cycles.

The accurate targeting of the ICF capsules within the reaction chamber must be achieved every 4 ms. This requires a high degree of control and alignment but also diagnostic analysis. Significant effort is also required to properly model the expulsion of the exhaust plasma and associated charged particles through a magnetic nozzle. It is appropriate to minimise the divergence of the beam for maximum thrust benefit. Different nozzle geometries should be explored in this regard.

In principle charged particles from the ignited plasma expansion should not interact with the reaction chamber wall materials. This is because the particle gyroradius is sufficiently small so that no flux of particles should breach the wall-field separation distance. This needs to be modelled in detail beyond Elliott [16].

The above are just a few comments on a complex machine. We have demonstrated some ability to calculate a representation of the Daedalus spacecraft design. Now we move on to what we term a set of Daedalus derivative designs. This includes significant perturbations or design changes to the original Daedalus configuration. The main purpose of this exercise is to help to broaden the design space.

9.2 Resolution Single Stage Spacecraft

The spacecraft called concept Resolution was the first attempt to re-design the Daedalus spacecraft. The vehicle configuration for Resolution was based largely around the Daedalus 2nd engine stage only and the goal was to demonstrate that full orbital insertion could be achieved with these assumptions, but the distinguishing factor was that the mission was to be achieved in less than 100 years and carrying a 150 tons payload [17].

In deriving the new concept designs a set of trade studies were performed that focussed on perturbations of a scaled Daedalus second stage vehicle design and considered variations in the propellant and structural mass assuming 150 tons payload mass. The initial studies suggested a propellant mass range of 21,000–24,000 tons for an approximately 1000 tons dry mass, enabling a cruise speed of 0.045–0.047c in a total mission duration of 95–100 years.

A change made for the subsequent design was to move away from the use of an electron beam as used in Daedalus, since it was realised that the charge associated with the beam would diverge electromagnetically in the presence of charged plasma ions thus preventing proper target alignment. In the assumed calculations the main driver is a set of optical laser beams.

Figure 9.5 illustrates how Resolution was imagined. Ultimately the Resolution design concept became a candidate replacement for the historical Daedalus design, yet it would have much less propellant mass and fire at a much-reduced pulse frequency. In addition, the use of only a single stage would simplify the engineering and mission architecture. Because it was also intended for a 4.3 ly mission target and it could take 100 years or less to complete the mission this presented a much more

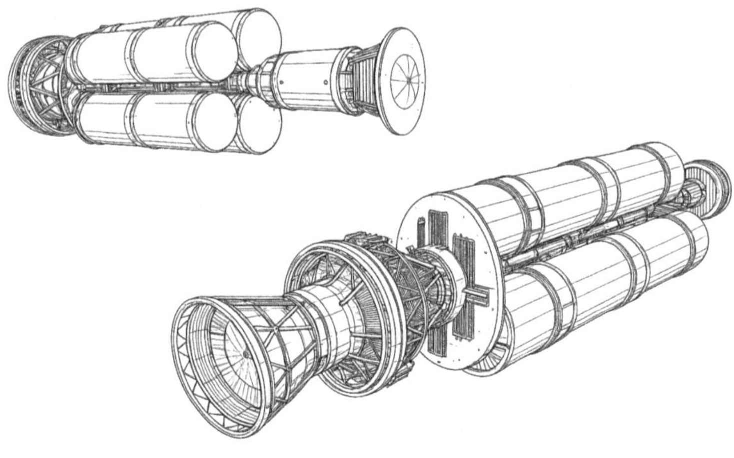

Fig. 9.5 Spacecraft resolution concept design flyby mission[KFL]

credible design solution to move forward with. The concept was being developed as a rendezvous probe for full orbital insertional into the stellar system, but it would also be useful to examine its performance in a flyby mode and this gave rise to the mission architecture which is discussed next.

9.2.1 Resolution Flyby Mission

It was thought constructive to develop a minimum flyby mission to get some feel for what may be possible if the requirement for deceleration was removed. It would also utilise D^3He mixture but would only need 2000 tons. The total spacecraft wet mass was around 2700 tons, and a propulsion mass was assumed to be 360 tons, of which around 300 tons was the engine. It is not well known how small an ICF engine could really be shrunk down to and this remains a major uncertainty for designers, yet it is vital to mass budget sizing. The $\sim$300 tons or so often quoted would appear to take its origin from the original work by Hyde, Wood, Nuckolls [18].

The payload would be fully instrumented to include high gain radio communications antennas but also optical telescopes for long-range imaging. The payload would be protected from the dust and ions of the interstellar medium by a $\sim$2 tons particle bombardment shield which was calculated via [6, 19–21] but using modern data for the interstellar medium properties [7, 8].

The propellant tanks would be stacked along the central column, also protected by radiation shields, and tanks would be dropped once used to improve the performance of the mission. The loaded ICF capsules would be injected onto a rail gun and then accelerated into the target chamber, where a bank of lasers arranged symmetrically around the target axis point would be positioned. To initiate the

Table 9.1 Calculations for resolution flyby probe to 4.3 ly using different fusion fuels model options carrying 150 tons payload

Parameter	$D(T,{}^4He)n$	$D(D,{}^3He)n$	$D(D,T)p$	$D({}^3He,{}^4He)p$
Ex. Vel. (km/s)	9057.1	4395.3	4881.7	9251.6
Dry mass (tons)	699.57	650.089	653.03	702.28
Cruise Vel. (km/s)	13,263.0	6355.1	7063.7	13,557.0
v/c	0.0442	0.0121	0.0236	0.04522
Cruise time (Years)	97.82	205.06	184.41	95.68
Trip time (Years)	100.02	207.27	186.61	97.88

fusion engine a nuclear fission reactor would be required on-board, but each cycle for the engine detonation would be powered by the energy release from the previous one in an induction bootstrap process like the Daedalus. Any excess heat would be dumped from the spacecraft through using radiator panels.

The mission would be characterised with a mass flow rate 0.0288 kg/s, a thrust 0.27 MN, a jet power 1.23 TW and a specific power of 3.42 MW/kg. The engine would use capsule masses 0.288 g detonated at a pulse frequency of 100 Hz. The spacecraft would accelerate for a boost period of 2.2 years, reaching its final cruise velocity of 13,557 km/s or 0.0452c, reaching the target destination in around 98 years from launch. Clearly, the Resolution flyby mission is a considerable improvement on the Daedalus concept, although one must caveat this statement with the observation that this mission involves a reduced mission distance, less structure mass, lower payload mass. The payload mass fraction would be 5.5% which is a considerable improvement on the 0.85% associated with the Daedalus mission.

Table 9.1 shows the results of calculating the Resolution model but for different assumptions of ICF capsule fuels, neglecting daughter reactions, but restricted to 2000 tons propellant mass. These calculations use the correct reaction cross section that is appropriate for each ignition temperature of the different fuels. The final column in the table is a model that is similar to the one described in the above paragraphs.

It is worth noting the elongated mission times for the DD models which far exceed 100 years. Separate calculations were also conducted and to get these models to within $\sim$100 years trip time the propellant mass would have to be increased to $\sim$10,000–20,000 tons with elongated burn times on the order of $\sim$1–2 decades. This is on the assumption of no increase in structure mass, no changes to the capsule mass or pulse frequency. An increase in propellant mass would also correspond to an increase in structure mass since more tanks would be required but also structural support to contain them.

9.2.2 *Resolution Rendezvous Mission*

This is like the flyby Resolution but instead the spacecraft will decelerate at the interstellar target for full orbital insertion around the star. In this model there is a single structure and a single engine. Yet, the flight involves both an acceleration and a deceleration phase of the mission profile. This is achieved by firstly burning the engine for the first phase of the mission and ejecting any used propellant tanks. The spacecraft then enters its cruise phase of the mission. As the spacecraft nears its stellar target, it will then turn itself around using on-board manoeuvring thrusters to perform a reverse thrust of the engine and therefore decelerate the probe. This comes with the advantage that we do not require two separate vehicle stages and so there is a significant structural mass saving, and this especially includes the omission of a separate engine.

In the case of the single structure but two engine burn Resolution model for a rendezvous mission to an Proxima Centauri target carrying a 150 tons payload, at 100 Hz pulse frequency using a capsule mass 0.288 g D^3He fuel for both engine burn stages, we calculate a total cruise speed of 4.91%c completing the mission in 97.27 years. The engine would generate a thrust of 0.26 MN, jet power 1.22 TW and specific power 2.86 MW/kg. The payload mass fraction was 0.87%. The structural mass to dry mass was 57.45%. The propellant mass fraction to wet mass was 98.43%.

The total wet mass for the mission was around 17,278 tons and the total dry mass was around 778 tons. As with the earlier model the spacecraft was also designed with the equivalent of the Daedalus second stage auxiliary power supply with four 650 kWe output reactors with a total output power of 2.6 MWe, with the total power supply mass coming to 6.4 tons with the additional 7.9 tons additional supporting mass. The design also included calculations for radiators associated with the fusion engine and these had a total mass of 70.31 tons. The particle bombardment shield required was estimated to be at a minimum of 2.2 tons.

The mission profile of this concept was to accelerate at 0.031 m/s^2 for 15.09 years using a mass flow rate of 0.029 kg/s to reach its cruise velocity of 14,714 km/s. The boost phase would have ended at 17,430 AU (0.276 ly). For this phase it would carry six propellant tanks each of around 2.2 tons in mass and a total dry propellant mass of 13.29 tons. Each tank would carry 7.93×10^9 pellets and there would be a total of 4.76×10^{10} pellets for the acceleration phase.

However, the spacecraft would carry an additional 4 propellant tanks for the deceleration phase of the mission utilising a total of 9.68×10^9 pellets for this phase. Therefore, the total number of ICF pellets used for the mission would be 5.7291×10^{10} pellets. For this model the propellant was split with 13,712 tons for the acceleration stage burn and 2788 tons for the deceleration stage burn.

The rendezvous mission comes with the benefit of increased encounter time and so increased science data return. For example, for the flyby mission the time for an encounter diameter of 100 AU or 10 AU will be limited to around $\sim$13 days or just over $\sim$1 day respectively. The benefit of deceleration is it would allow the possibility of deploying planetary orbiters, atmospheric penetrators and even ground landers to maximise the in-situ exploration opportunities for a science mission phase that may be years in duration.

The model for Resolution looks nice, but we have not conducted a comprehensive analysis and the likely need for large waste heat radiators so as to manage the large radiation flux to the system originating from the engines and laser drivers. Resolution was then a very simple model but not likely realistic and so it was decided to attempt a machine with more emphasis on defining the engineering requirements.

9.3 Pegasus Parallel Stage Spacecraft

A re-think over the Resolution design led to an entirely new vehicle concept known as the Pegasus which is shown in Fig. 9.6 and this was one of the main design solutions to come out of the Project Icarus study as a potential successor for the Daedalus [17]. It was designed using a code written in the language of Fortran 95 called HeliosX [22]. It had been based on an earlier concept called Endeavour which had 5 engines for the purpose of parallel thrust [23]. However, it was realised that there was a major failure mode that had been neglected, where any engines near to each other so that when the ICF pellets were detonated they would release significant amounts of Bremsstrahlung x-ray and high MeV neutron radiation. This radiation would be absorbed into the engine bell material and the surrounding structure

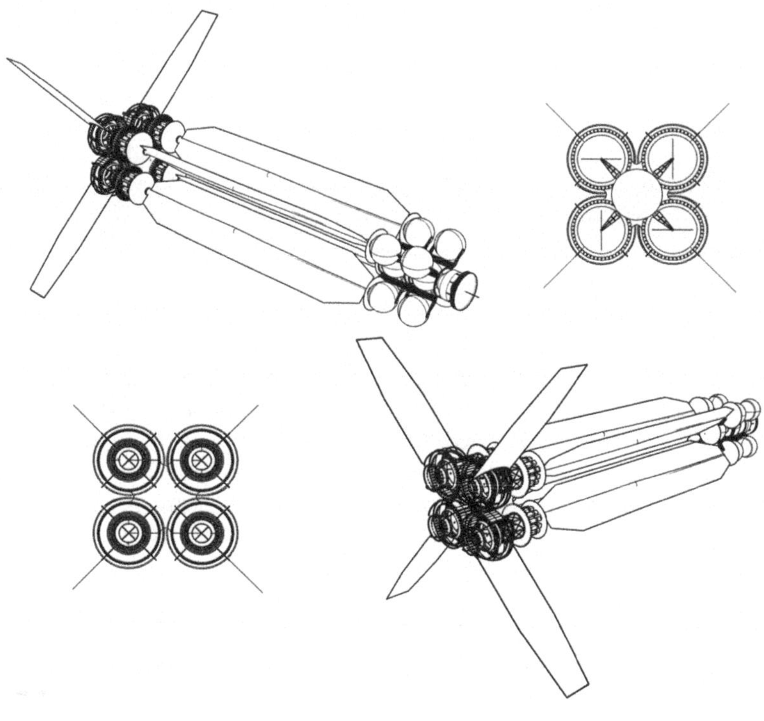

Fig. 9.6 Illustration of pegasus interstellar spacecraft concept[ML]

leading to a weakening of its material strength. It was realised that radiation shields need to be constructed around the engine bells to separate them from each other.

However, fitting the radiation shields was not easy unless one removed the central engine. For this reason, the design evolved to a four-staged parallel thrust engine design, with the radiation shields intersecting towards the centre of the engine arrangement and this is called Pegasus. In addition, it was decided that the spacecraft would also benefit from a high layer of insulation internal to the chambers to further protect it from the radiation, as a form of redundancy and risk mitigation.

Adding more engines for parallel thrust mode decreases the transit time. Calculations were conducted assuming a pulse frequency f_{hz}, capsule mass m_{cap}, total propellant mass m_{prop} and the number of engines n_{eng} where the burn time is given by the non-expellant and expellant case respectively

$$t_b = \frac{1}{n_{eng}}\left(\frac{m_{prop}}{m_{cap} f_{hz}}\right), \tag{9.2}$$

$$t_b = \frac{1}{n_{eng}}\left(\frac{m_{prop}}{(m_{cap} + m_{exp}) f_{hz}}\right). \tag{9.3}$$

As a comparison, and with f_{hz} = 250 Hz and a first stage capsule mass of m_{cap} = 2.84 g, for a single engine and a total propellant mass 46,000 tons (first stage Daedalus v_{ex} = 10,600 km/s) the burn time would be 2.05 years, but with a 4-engine configuration it would be reduced to 0.514 years. Similarly for a total propellant mass of 4000 tons (second stage Daedalus, v_{ex} = 9210 km/s, m_{cap} = 0.288 g) the burn time would be 1.76 years with a single engine and 0.44 years with 4 engines. If 10 engines were used for both stages, then the burn time would be reduced to 75 and 64 days for the first and second stage respectively. Yet, then one is faced with the problem of handling the enhanced radiation environment but also the increase in mass of the engines which affects the mass ratio.

In principle then adding more engines for increased parallelisation of thrust will result in a larger thrust and decreased mission time. However, there would be a limit to this since this involves the addition of more mass and so the performance is likely to reach a limit of diminishing returns. Some results of calculating the effect on burn time as a function of the number of engines for the Pegasus design is shown in Fig. 9.7. The results also implied that the two-stage series configuration for the Daedalus design may not be fully optimal and that instead a parallel thrust system with ~3–4 engines may have been preferable—although this needs to be traded with the additional mass increase due to the extra engines as well as the enhanced neutron radiation environment. Hence Pegasus adopted 4-engines in a parallel thrust arrangement as shown in Fig. 9.8. The propulsion system is described in more detail elsewhere [24, 25].

Another major change to appear in the Pegasus design would be the design of the ICF pellet. Although Resolution, Endeavour and Pegasus were all to adopt a similar design to the Daedalus second stage engine (minus the tritium trigger), it was thought prudent to tailor for a larger pellet shell size ablator and a larger density

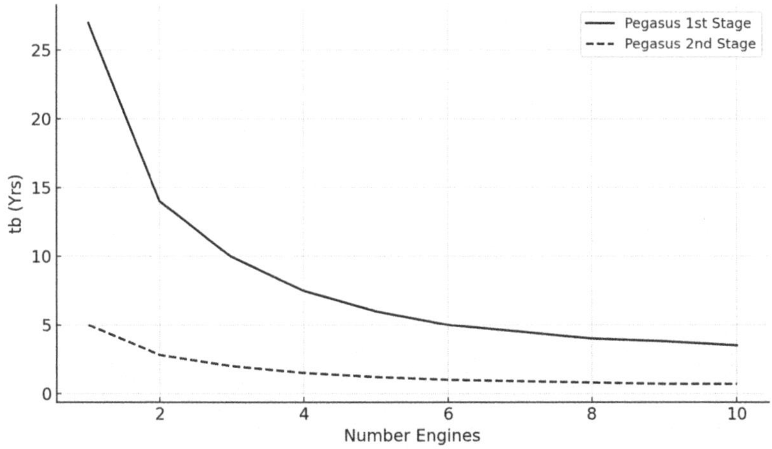

Fig. 9.7 Calculations for pegasus burn time with number of engines

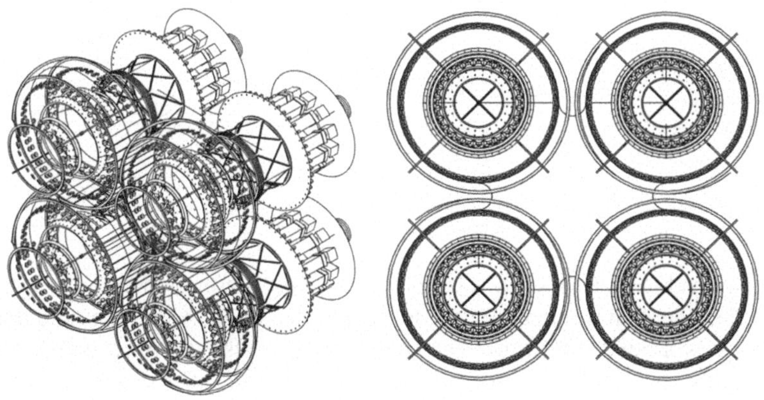

Fig. 9.8 Engineering representation of pegasus parallel thrust engines[ML]

of Helium-3 towards the outer edge of the fuel, to help mitigate the radiation release. Pegasus would mitigate the radiation release to protect the spacecraft and reduce the probability of mission failure.

Neutrons released from DD self-burn are mostly captured into ^{3}He fuel [26]. Some of the high 14 MeV neutrons sourced near the outer rim of the fuel will travel further, hence a high-density layer of Helium-3 is included to increase the capture reaction probability. Any emitted x-rays that escape from the thermonuclear burn within the fuel are absorbed into the thick ablator shell made up of a strong absorber material. Any further x-rays or neutrons that escape are then absorbed into an insulation layer internal to the thrust chamber wall. External radiator shields are included around the thrust chamber to protect them from each other and to protect the surrounding spacecraft structure.

Table 9.2 Pegasus spacecraft dry propulsion mass (which should by convention also include the 10 tons power supply mass which is in the payload)

System	Mass (tons)	Quantity	Final mass (tons)
Nuclear fission start reactor	10	1	10
Reaction chambers/Ancillaries	10	4	40
Magnetic field coils	100	4	400
Laser driver radiators	280	4	1120
Engine radiators	40.5	4	162
Radiator coolant (drivers)	87.36	4	349.4
Radiator coolant (engines)	12.64	4	50.5
Radiator pumps	40	1	40
Laser driver system	50	4	200
Beam lines	2.5	4	10
Diagnostics/Beam control	0.25	4	1
Induction/Transmission	10	4	40
Pick up coil	10	4	40
Power conversion	10	1	10
Energy storage	50	1	50
Structural bars	5	4	20
Insulation layers	1	4	4
Pumps & Valves	4	1	4
Pellet injectors	5	4	20
Total			2561

The payload mass fraction was 0.32%. The propellant mass fraction to wet mass was 93.3%. The total wet mass for the mission was around 46,383 tons and the total dry mass was around 3083 tons. The spacecraft was also designed with a nuclear power supply for the start up initiation and operation of other systems at 15 MWe rated at 2 kWe/kg. The spacecraft would carry a waste heat management system with a total mass of 1732 tons to include the radiators, coolant and pumps for the laser driver and engines. Table 9.2 shows a breakdown of the propulsion mass.

The particle bombardment shield required was estimated to be at around 1 ton, but a shield was adopted on the front and back (for reverse thrusting) and so a total mass of 4 tons was adopted as a margin. The total mass of the engines was 400 tons (100 tons per engine) and including also the reaction chambers, engine components, lasers, power supply and thermal radiators the propulsion mass was 2541 tons. The total propellant mass was 43,300 tons, with 34,445 tons allocated to the first stage burn and 8855 tons allocated to the second stage burn.

The mission is assumed to be launched from around the vicinity of the moons of Jupiter and this distance is accounted for in the mission profile. The spacecraft also moves away from its current location for 1 day at 17 km/s prior to switching on the main fusion engine. This is for the safety of any nearby human crews that may be on-board an orbiting space station. Pegasus would carry a 150 tons payload. The pulse frequency would be 1000 Hz using a capsule mass 0.288 g D^3He fuel. The total cruise speed was $0.0447c$. Both stages would generate a thrust of 2.56 MN, jet power 11.39 TW and specific power 7.12 MW/kg. After the initial boost phase, the spacecraft would cruise for around 94.87 years, followed by a deceleration period of 0.86 years. So the total mission duration would be 99.64 years over a total distance of 4.36 ly to its interstellar target, bringing the spacecraft into orbital velocity around its target star.

The mission profile of this concept was to accelerate at 0.11 m^2 for 3.9 years using a total mass flow rate of 0.288 kg/s to reach its cruise velocity of 13,400 km/s. The boost phase would have ended at 4183 AU (0.0661 ly). For this phase it would carry six propellant tanks each 2.2 tons in mass and so a total dry propellant mass of 13.29 tons. Each tank would carry 8.2×10^{10} pellets and so there would be a total of 4.92×10^{11} pellets for the acceleration phase of the mission. However, the spacecraft would carry an additional 4 propellant tanks for the deceleration phase of the mission utilising a total of 1.09×10^{11} pellets for this phase. The total number used for the mission would be 6.014×10^{11} pellets.

We can estimate the performance of a system but based on what the assumptions of the vehicle that may be achievable in the future. In particular this pertains to use of an energy gain scaling algorithm which can be adopted as an assumption on technology trends. This enables an assessment for the required driver energy for a given ICF pellet mass.

Based on the basis of this we adopted pellet models where the total energy released is kept constant for the same mission profile so that the pellet mass and pulse frequency are treated as variables. Where the pulse frequency for a given design $f_{hz,n}$ is given as a function of the Daedalus pulse frequency $f_{hz,o}$ as a ratio of the original pellet mass m_o to the new pellet mass m_1, so that $f_{hz,n} = (m_o/m_n) f_{hz,o}$.

Figures 9.9 and 9.10 show a result of calculating designs with various pellet masses corresponding to different pulse frequencies and with different burn-up

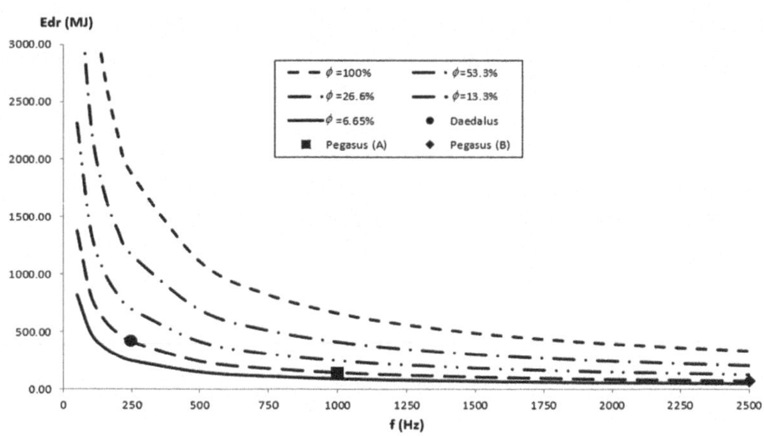

Fig. 9.9 Calculations for energy driver scaling as a function of pulse frequency for pegasus pellet designs under the assumption of different burn-up fractions

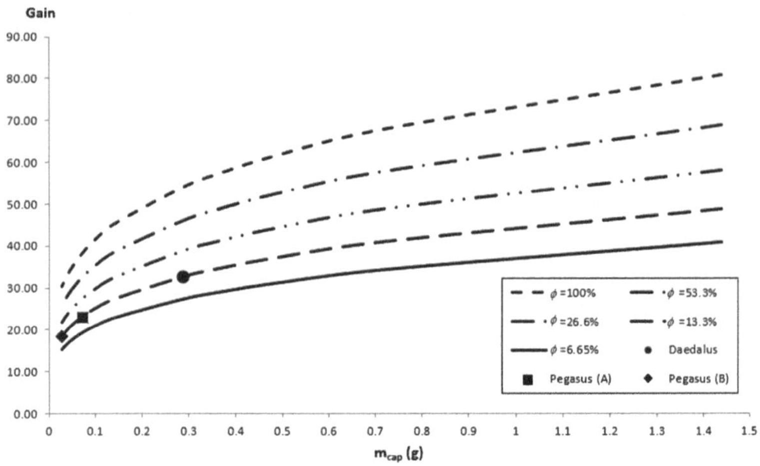

Fig. 9.10 Calculations for energy gain scaling as a function of fusion fuel mass for pegasus pellet designs under the assumption of different burn-up fractions

fractions. Each of these also correspond to a specific value of driver energy and energy gain. The results are also shown at different burn-up fractions. We adopted designs with a burn-up fraction of 13.3% so that the total energy output can be maintained at around 13,426 MJ per pellet. We adopt illustrative models at 1/4th mass scale (0.072 g at 1000 Hz for Pegasus A model) and at 1/10th mass scale (0.0288 g at 2500 Hz for Pegasus B model), as compared to the nominal Daedalus (0.288 g at 250 Hz). For both pellet designs it was assumed that a tiny amount of DD is loaded at the centre so that the D^3He fuel can be ignited at a peak temperature of around 50 KeV instead of at 60 keV.

For the 1/4th scale mass pellet model this required a driver energy of 145 MJ and peak power of 903,427 TW. In direct drive mode with a laser pulse time of around 0.1605 ns this corresponds to a laser intensity of 6.829×10^{16} W/cm^2. For the pellet design the fuel radius was 5.783 mm. In order to get the calculation to close on a credible ignition model the laser efficiency required was ~36% (which is far beyond current or projected levels). The energy output was 3344 MJ per pellet leading to a gain of $G = 23.06$. For the 1/10th scale mass pellet model this required a driver energy of 73 MJ and peak power 784,946 TW, with a laser pulse time of around 0.093 ns corresponding to a laser intensity of 8.21×10^{16} W/cm^2. The fuel radius was 4.263 mm. The laser efficiency required for the ignition model was ~32%. The energy output was 1370 MJ per pellet leading to a gain of $G = 18.82$.

In the ignition models the use of very thick ablator shells points towards a non-hot spot mode of operation for the implosion engine and may suggest designs characteristic of shock assisted ignition [??] may be more appropriate for such large systems. Only large shells would result in a substantial mass increase to the vehicle that would effectively make any missions impossible. The reason they appear to be required in the existing models is because an increasing blow-off of the ablator shell

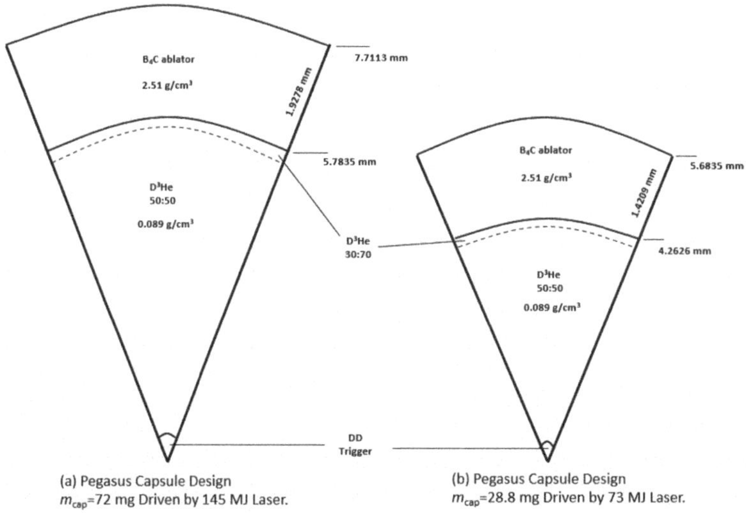

Fig. 9.11 Pegasus ICF capsule designs[KFL]

results in an increased compression force on the fuel and so a higher compression ratio. A more sophisticated ignition physics model is likely required to derive pellet designs that are thin shelled yet give the same energy output as those discussed above and then the mission would become possible and practical.

In contrast to Daedalus, for our revised capsule design where $m_{cap} = 72$ mg and 28.8 mg this results in an energy release of $E_{out} = 3.3565 \times 10^9$ J/capsule and $E_{out} = 1.3426 \times 10^9$ J/capsule, which is lower in energy by a factor of 4 and 10 respectively. This was deliberately intended from the scaling. But an implication of this is that since the energy per capsule is lower, the pulse frequency had to be increased to compensate if the total mission thrust was not to be reduced. This is why for the two Pegasus models the pulse frequency was set at 1000 and 2500 Hz, compared to the 250 Hz of the Daedalus.

Figure 9.11 shows the ICF capsule design configurations adopted for the Pegasus spacecraft. We note that these are much smaller than the Daedalus capsules, but also much larger than the NIF-like capsules. The main fuel is a region of D^3He at 50:50 equimolar mixture, with a narrow region of higher ^3He (with depleted Deuterium to prevent reactions) at the outer edge for the purpose of absorbing any high energy neutrons generated towards the outer edges, where the mean free path of a 14 MeV neutron must be less than the thickness of this narrow shell under compression. Currently, for the Pegasus designs the mean free path is on the order of $\sim$0.5 mm which means that some of the neutrons will escape.

However, we note that in a pure ^3He region the 14 MeV neutron mean free path will increase beyond $\sim$1–1.5 mm. This is because the number density is significantly lower than the full D^3He mixture and so it is less effective at stopping fast neutrons. So, the goal is to create an outer fuel buffer zone where the material is depleted of Deuterium to prevent neutron reactions via DD self-burn, which can initiate the generation of Tritium and unwanted neutrons, but also allowing it to be sufficiently thick that the increased neutron mean free path does not become overly large. For the Pegasus designs we have not specified the exact thickness of this buffer shell, since it requires more sophisticated modelling, and we merely stipulated it was likely required to minimise external neutron flux to the vehicle.

External to this is a higher density B_4C region for the purpose of mass ablation so that the pressure can be generated at the surface from which hydrodynamic shock waves compress up the fuel. This thick shell will also capture some of the escaping energy. At the centre of the capsules, a DT trigger is replaced by DD so that ignition can occur at a lower temperature, whilst not giving off excess heat.

Although we have determined the potential fuel mass to use in the Pegasus capsule design, a critical question is what sort of ablator size to use, and this depends on the In Flight Aspect Ratio IFAR which is given as the ratio of the outer fuel radius r_f to ablator thickness Δr so that $AR = r_f/\Delta r$. Since Rayleigh-Taylor (RT) instabilities can occur at the interface between the ablator and the fuel during the acceleration phase, the instability will increase with the AR. The higher the AR the thinner the ablator for a given fuel radius and so the higher the susceptibility to instabilities. Similarly, the lower the AR, the thicker the ablator and the more stable but less efficient the implosion. A thinner ablator will achieve a higher implosion velocity therefore adding in the compression and ignition of the fuel. Therefore, the thicker the ablator the lower the implosion velocity and compression.

Typically, experimental design studies will use a value in the range $AR \approx 2.5$–4 as a good compromise between implosion stability and velocity. Such designs will suppress the RT growth to manageable levels whilst achieving compression for ignition or high burn-up. They also allow for reasonable pulse shaping to control the shock timing. Thus, in the design of Fig. 9.11 we have used an ablator geometry that corresponds to $AR = 3$, which means $\delta r = r_f/3$. This is to keep the ablator thin enough to reach the high velocities required for compression but thick enough to stabilize against instabilities that would otherwise mix the ablator into the fuel and therefore quench the burn. This is because the high burn-up fractions computed for Pegasus in the representative physics model (24–28%) require high compression and stable implosions at the high driver energies of 73–145 MJ. This sort of design is also consistent with high-gain direct-drive ICF studies for advanced propulsion.

In considering ablator thickness, at the lower end (e.g. towards $AR = 2$) we have thicker ablators for greater stability and lower implosion velocity for the same driver energy, which is useful for maximizing symmetry and suppressing instabilities. At the height end (e.g. towards $AR = 4$) we have thinner ablators for higher implosion velocities but with the risk of increased RT instability. These are useful for maximizing compression and velocity for high burn-up fractions with careful drive shaping.

A possible consideration for a vehicle like Pegasus is the use of expellant propellant to augment the thrust profile. This might be needed since the quantity of fusion fuel is so large and so it might be useful to explore such ideas. As an example, Table 9.3 shows the data points for the two different Pegasus models where

Table 9.3 Augmented propellant case studies for a pegasus mission

Scale	1/4th	1/10th	1/4th	1/10th	1/4th	1/10th
m_{cap} (mg)	72	28.8	72	28.8	72	28.8
m_{exp} (mg)	0.0	43.2	72	115.2	216.0	259.2
m_{exp}/m_{cap}	–	1.5	1.0	4.0	3.0	9.0
m_{TN} (tons)	43,200	17,293	21,616	8646	10,808	32,424
m_{aug} (tons)	0	25,939	21,615	34,586	4323	38,909
f_{hz} (Hz)	1000	1000	500	500	250	250
v_{ex} (km/s)	8897	2814	3146	1989	2225	1407
$\dot{m}$ (kg/s)	0.288	0.288	0.288	0.288	0.288	0.288
T (MN)	2.562	0.810	0.906	0.573	0.641	0.405
Δv (km/s)	12,063	3815	4265	2697	3016	1907
(%c)	4.02	1.27	1.43	0.89	1.0	0.64
t_{tot} (Yrs)	110	335	300	472	423	666

we assume $m_{exp}/m_{cap} = 3$. So that for model (a) with $m_{cap} = 72$ mg the expellant mass is $m_{exp} = 216$ g. Similarly for model (b) with $m_{cap} = 28.8$ mg the expellant mass is $m_{exp} = 86.4$ g. We also assumed $\phi = 0.133$, $\epsilon_{mix} = 1.0$ (perfect mixing), $\eta_j = 0.909$, $\alpha = 0.0039$. This leads to a significantly reduced exhaust velocity.

Table 9.3 shows some example calculations of what may be possible with the use of expellant propellant for thrust augmentation, where the expellant is largely made up of hydrogen slush. The vehicle structure and mass configuration are like the one used for the Pegasus 110-year nominal mission. The calculations assume a constant total propellant mass of 43,200 tons, a 4-engine mass flow rate of $\dot{m}$ = 0.288 kg/s, a total acceleration and deceleration burn time of t_b = 4.76 years. For all these calculations we also assume a jet efficiency of η_j = 0.85. For each of the results they are shown with different ICF capsule mass scaling compared to the Daedalus second stage pellet of m_{cap} = 288 mg.

We see how the exhaust velocity changes with different m_{exp}/m_{cap} ratios but also how the mass of thermonuclear fuel to expellant mass changes and how this influences the mass flow rate, thrust and ultimately time to the destination which is now over several centuries duration due to the reduced velocity increment. Yet this is an option for optimising an interstellar mission for minimum thermonuclear fuel mass, provided we are prepared to accept elongated trip times. Figure 9.12 illustrates an engineering block diagram for how a spacecraft propulsion system like those proposed for this chapter would be arranged.

Finally, it is worth mentioning that efforts were made to also cost a programme to develop the Pegasus and this was discussed in other publications [27, 28]. The biggest cost to the Capital expenditure model (Capex) was the acquisition of helium-3 from the gas giants and the programme costs depended on whether the helium-3 mining cost was as low as \$1.4/g or even \$0.14/g. This led to an estimate for the total programme to be in the range \$165 billion up to \$310 billion, which included the development of the vehicles, technologies and missions that would lead up to

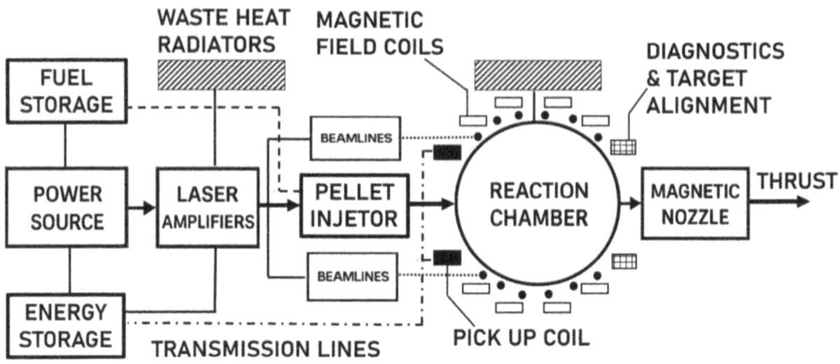

Fig. 9.12 Engineering schematic of an advanced fusion engine[KFL]

the first interstellar rendezvous flight. This was on the assumption of a circa ~2100 launch date, although it is more likely that such a mission may take place in the period ~2150–2250.

Problems of Chap. 9

9.1 A single engine spacecraft carries 35,000 tons of fuel made up of 0.5 g capsules that are detonated at a rate of 100 Hz (i) How long would this engine burn assuming there were no pauses in the pulse engine? (ii) How long would the propulsion system burn for if there were 7 engines instead? (iii) What are consequences to the mission profile with increasing the number of engines?

9.2 Imagine that we redesigned the Pegasus fuel capsules to be mostly DT fuel with a nominal density of 0.25 g/cm^3 and a molar mass of 2.0141. The fuel is compressed up to a density of 500 times the nominal to a capsule radius of 0.15 mm (i) For an assumed reaction cross section of 1.2×10^{-24} cm^2 estimate the mean free path of a 14 MeV neutron and determine if they will escape the fuel (ii) For an assumed cross section of 2.5×10^{-24} cm^2 estimate the mean free path for a 2.45 MeV neutron and whether they will escape the fuel (iii) What is the source of the two different populations of neutrons?

9.3 Although the Pegasus was a neat design solution, it came with many assumptions and a set of engineering requirements that can be considered an extreme extrapolation of the current state of art. One of the main issues was that so much power was being generated by the lasers and the engine that this would require massive waste heat radiators and for them to be maintained low they would have to be rated at around 0.5–1 kg/m^2. In addition, lasers are not very efficient, to bring down the power requirements (which influences the radiator mass) one must assume highly efficient lasers that approach ~90–99%. (i) Conduct some research on alternative types of lasers that could be used for such a mission, where the primary requirement should be high output power for high conversion efficiency and with minimum laser-plasma instabilities. Identify a promising candidate that could be explored.

References

1. Forward RL (1976) A programme for interstellar exploration. J Br Interplanetary Soc 29(10):611–632
2. Bond A, Martin AR (1978) Project daedalus: the mission profile, final study report. J Br Interplanetary Soc Special Supplement:S37–S42
3. Martin AR, Bond A (1978) Project daedalus: the propulsion system part 1: theoretical considerations and calculations. J Br Interplanetary Soc Special Supplement:S44–S62
4. Bond A, Martin AR (1978) Project daedalus: the propulsion system part 2: engineering design considerations and calculations. J Br Interplanetary Soc Special Supplement:S63–S82
5. Long KF, Obousy RK, Hein A (2011) Project icarus: optimisation of nuclear fusion propulsion for interstellar missions. Acta Astronaut 68(11–12):1820–1829
6. Martin AR (1978) Project daedalus: bombardment by interstellar material and its effects on the vehicle. J Br Interplanetary Soc Special Supplement:S116–S121
7. Draine BT (2011) Physics of the interstellar and intergalactic medium. Princeton series in astrophysics. Princeton University Press, Princeton
8. Crawford IA (2011) Project icarus: a review of local interstellar medium properties of relevance for space missions to the nearest stars. Acta Astronaut 68(7/8):691–699
9. Martin AR (1978) Project daedalus: the evidence for planetary companions of Barnard's stars. J Br Interplanetary Soc Special Supplement:S19–S23
10. Hand E (2012) The exoplanet next door. Nature 490:323
11. Witze A (2016) Earth-sized planet around nearby star is astronomy dream come true. Nature 536:381–382
12. Bond A, Martin AR (1986) Project daedalus reviewed. J Br Interplanetary Soc 39(9):385–390
13. French JR (2013) Project icarus: a review of the daedalus main propulsion system. J Br Interplanetary Soc 66(7/8):248–251
14. Crawford IA (2016) Project icarus: preliminary thoughts on the selection of probes and instruments for an icarus-style interstellar mission. J Br Interplanetary Soc 69(1):4–10
15. Reddy SK, Benaroya H (2015) Study of daedalus interstellar spacecraft reaction chamber and thrust structure. J Br Interplanetary Soc 68, 33–43
16. Elliott JO, Terry WK (1985) Plasma expansion in the daedalus first stage engine. J Br Interplanetary Soc 38(3):120–128
17. Long KF, Fogg M, Obousy R, Tziolas A, Mann A, Osborne R, Presby A (2009) Project icarus: son of daedalus - flying closer to another star. J Br Interplanetary Soc 62(11/12):403–414
18. Hyde R, Wood L, Nuckolls J (1972) Prospects for rocket propulsion with laser-induced fusion microexplosions. AIAA, 72-1063, presented at the 8th propulsion joint specialist conference, 29 November–01 December
19. Benedikt ET (1961) Disintegration barriers to extremely high speed space travel. Adv Astronaut Sci 6:571–588
20. Langton NH (1973) The erosion of interstellar drag screens. J Br Interplanetary Soc 26(8):481–484
21. Long KF (2025) Project icarus: an assessment of the particle bombardment problem on interstellar spacecraft. J Br Interplanetary Soc 78(6):216–226
22. Long KF (2023) Development of the HeliosX mission analysis code for advanced ICF space propulsion. Acta Astronaut 202:157–173
23. Long KF (2023) Application of the HeliosX ICF advanced propulsion mission analysis code to perturbed interstellar design models. J Br Interplanetary Soc 76(3):94–111
24. Long KF (2025) Project icarus: the pegasus spacecraft concept design ICF propulsion for an interstellar rendezvous mission - part A. J Br Interplanetary Soc 78(12):402–421
25. Long KF (2025) Project icarus: the pegasus spacecraft concept design ICF propulsion for an interstellar rendezvous mission - part B. J Br Interplanetary Soc 78(12):422–435

26. Halyard RJ (1997) Comparison of proposed inertial confinement fusion propulsion systems for interstellar missions. J Br Interplanetary Soc 50(4):129–136
27. Long KF (2025) Project icarus: the pegasus spacecraft concept design engineering systems description. J Br Interplanetary Soc 78(6)
28. Long KF (2025) Project icarus: strategic roadmap conclusions of the starship design study. J Br Interplanetary Soc 78(5)

Chapter 10
Propulsion Power Management

I predict that 200 years from now, barring a catastrophe, the first interstellar voyagers will begin.

—Freeman Dyson, 1968 [1]

10.1 Engine Power Cycles

The fusion engine on-board a spacecraft will likely require a separate nuclear fission power reactor first to initialise it. But once it is underway, some of the energy released from each detonation can be bootstrapped to help power the next cycle such as in a magnetic field induction loop. The fusion reactions may be initiated in a closed or open reaction chamber depending on how the charged particles are controlled using magnetic fields.

A consideration in ICF ignition is the pathway energy losses along the way and in propulsion theory this tends to be expressed as a power cycle as shown in Fig. 10.1. This begins with any losses in the beamer system in turning it on and powering it up to the sufficient charge and the loss due to any system integrations. Then there are the losses associated with the coupling of the laser to the pellet surface which facilitates its implosion. Some of that energy will be reflected due to the laser-plasma interactions. Then are the losses associated with the direct compression and heating of the fuel which is represented by the burn-up fraction. Once the capsule is successfully ignited it will release a large amount of thermonuclear fusion energy.

This results in a multiplication factor called the energy gain G. If $G < 1$ then the system is generating insufficient energy through fusion energy than the thermal energy being deposited into the system. Although if $G = 0.1$ the plasma will be dominated by the ^{4}He production and this is tending towards a state known as *burning plasma*, where alpha self-heating exceeds external heating. When $G = 1$ this can be considered as tending towards a state of *ignition*, where alpha-heating is sustained as a propagating burn. If $G > 1$ then although relative to the thermal energy of the compression it will be producing a positive target gain, relative to the original input energy it may not be or it may just be at breakeven. When $G \sim 10$ this is high gain and a propagating burn wave within the plasma will be occurring.

K. F. Long, *Stellar Engines*, Springer Aerospace Technology,
https://doi.org/10.1007/978-3-032-18062-9_10

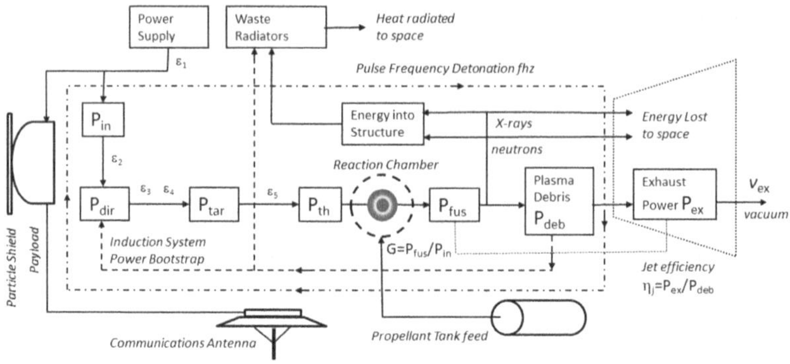

Fig. 10.1 Illustration of an ICF spacecraft and its engine power cycle with bootstrap induction loop to include other spacecraft systems[KFL]

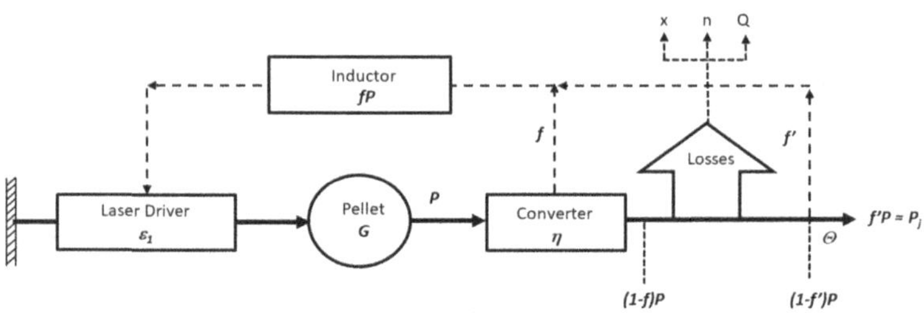

Fig. 10.2 Schematic of engine power cycle from an initial laser of efficiency ϵ_1 through an ICF pellet of gain G and the output power P to a converter (nozzle) of efficiency η, bootstrapped through an induction cycle and with associated losses[KFL]

It is therefore desirable to have a state where $G = E_{fus}/E_{th} >> 1$. System-level breakeven further requires sufficient driver efficiency and energy conversion.

For a commercial civilian power reactor, a gain on the order of $\sim$10 may be deemed appropriate. However, for the application of space propulsion a gain of at least $\sim$10s–100s would be a more desirable requirement. Instead of energy, these relationships can also be written as power expressions, since power is just energy output over some time interval. These are typically expressed as overall power flow diagrams in a space propulsion context. This might include showing the original efficiency of the input laser, laser coupling losses, neutrons, x-rays and plasma debris.

When designing advanced propulsion engines, it is important to attempt engine power flow diagrams like this to get some handle on the efficiency of the system performance. In a review of nuclear fusion power physics Gordinier [2] discusses power flow cycles in a nuclear fusion reactor and we adopt a similar model as shown in Fig. 10.2. Addressing this, but with a change of notation from the original text, the recirculating power fraction f is defined as a function of the energy gain G, the laser driver efficiency ϵ_1 to deliver energy onto the target, and the efficiency η of the

Table 10.1 Analysis of power cycle efficiency for engine designs

Parameter	VISTA	SunV	Disc-III	Daedalus	Pegasus
η_j (%)	34.1	37.8	10.3	91.7	84.9
ϵ_1	0.12	0.243	0.0769	0.99	0.99
G	1,500	20	36	33.3	24.65
η	0.5	0.5	0.8	0.99	0.9
P (GW)	225	8.7	93.02	3,330	3,587
f	0.01	0.4115	0.4515	0.031	0.045
fP (GW)	2.5	3.58	42.0	103.0	163.3
$(1-f)P$ (GW)	222.5	5.119	51.021	3,228	3,423.7
f'	0.0838	0.3069	0.0527	0.9171	0.7945
$f'P$ (GW) $[P_j]$	18.86	2.67	4.9	3,054	2,850
$(1-f')P$ (GW)	206.14	6.03	88.12	276.0	737.0
Θ	0.0915	0.4428	0.055	11.06	3.87

conversion from thermonuclear energy released by a pellet to electrical power such as used in a nuclear fusion reactor

$$f = \frac{1}{\eta G \epsilon_1}. \tag{10.1}$$

For small values of f the product of $\eta\epsilon_1 G \geq 10$. The fraction of power bootstrapped to the driver for the next pulse cycle will be defined by fP whilst the fraction that is produced from the engine for useful work will be $(1-f)P$. We can also define a fraction of jet power to total output power $f' = P_j/P$, where the notation $P \equiv P_{out}$ defined in earlier chapters, as the gross engine output power. The overall performance of the engine can also be defined by the ratio of useful work into the jet power by the total power losses to the system (due to neutrons, x-rays, induction bootstrap, waste heat as

$$\Theta = \frac{f'P}{(1-f')P} = \frac{f'}{1-f'}. \tag{10.2}$$

As shown in Table 10.1 for VISTA we find $\Theta = 0.09$. Values for the SunVoyager, Discovery-II, Project Daedalus and Pegasus spacecraft are also shown. We acknowledge the incredibly high efficiencies which are beyond the current technological state-of-art and instead should be seen as required values to make such missions feasible. This allows a useful way to compare the power cycle of an engine design to others. We note that the Daelulus and Pegasus have values greater than unity, which is manifest of the large jet power associated with these highly energetic systems.

To give an example, we can relate f to the engineering by considering the recirculating power $P_{recirc} = fP = P/\eta G\epsilon_1$ which implies an overall energy multiplication $\eta G\epsilon_1 = P/P_{recirc}$. To illustrate with VISTA, $\eta G\epsilon_1 = 0.5 \times 1{,}500 \times$

$0.12 = 90$ so that $f \approx 1/90$ and $fP \approx 2.5\,\text{GW}$, $P_j/P = 0.0838$ and $\Theta = 0.0915$. This method assumes that $(1 - f')P$ includes all non-jet channels such as neutrons, x-rays, thermal heat to radiators, induction recirculating power.

10.2 Power Cycle Line Maps

As indicated, a useful aspect of the design of any spacecraft that must be quantified is the power cycle. How much power goes into the propulsion system, how much is utilised as useful energy for thrust and how much is lost as particles or radiation? But also, how much is removed via waste heat. There is also the inclusion of an induction loop cycle to power the next pulse detonation for the pellet. A spacecraft design like VISTA will have a power cycle and this is typically illustrated with a complex diagram as shown in Fig. 10.3. This is non-trivial to analyse and so there is a need to present this information in a visual and readable way. To examine this, presented here as an original construct is the concept of a Power Cycle Line Map (PCLM) which is a mathematical representation of the power cycle but shown on a single line relative to the induction power $P(I)$—defined as the electrical power delivered to the driver after induction losses.

To demonstrate this, we will use a representation of the power cycle for the VISTA spacecraft [3] since this is considered a reliable design study discussed in Chap. 8. This assumes the thermodynamic cycle as being of induction type bootstrap

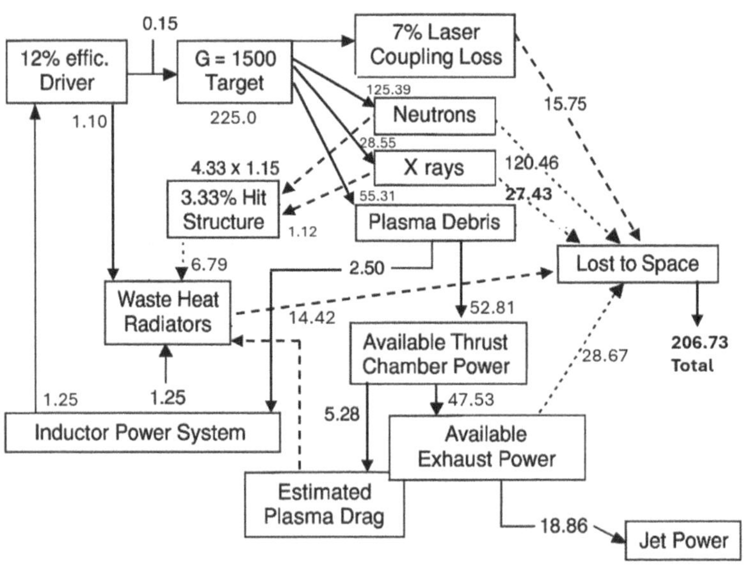

Fig. 10.3 Complex power flow cycle (GW) for VISTA spacecraft. The dotted lines are secondary effects of the cycle such as power loss to space

(i) which relies on magnetic field expansion and contraction within the nozzle chamber (rather than a Rankine type). This initiating power has an efficiency of 12% which means it suffers an effective driver loss of 88%, and then the remaining power is multiplied by the pulse frequency (Hz) at a set gain (G) around a capsule which is illustrated with the symbol $\odot$. This then leads to the fusion power output (P_f) which is divided into the fractions for laser coupling loss due to laser plasma interactions (P_l), neutron emission (P_n), x-ray emission (P_x) and finally the power due to the plasma debris (P_d); which then leads to the jet power (P_j) at a determined jet efficiency (%).

The jet power is given as a function of the jet efficiency η_j, the fraction of the plasma debris converted into thrust P_{thr} and the fraction used in the transmission lines for the purpose of bootstrap induction P_{ind} as

$$P_j = \eta_j(P_{thr} + P_{ind}). \tag{10.3}$$

The various emission mechanisms can also be separated further into direct loss into space or into space via waste heat and thermal radiators (by using subscript and superscript) but for simplicity these won't be shown in this example.

VISTA Power Cycle

For the VISTA mission to Mars the PCLM is written as the following single line equation where much of the operational performance characteristics are represented by the individual terms of the expression:

$$\begin{aligned} P(I)^{12\%}_{1.25i} \times \odot^{30\,\text{Hz}}_{1500G} &\rightarrow P_{f,225} - 0.07P_l - 0.55P_n - 0.13P_x \\ &- 0.25P_d \rightarrow P^{34.1\%}_{j,18.86}. \end{aligned} \tag{10.4}$$

The power cycle for VISTA is then illustrated in Fig. 10.4 and Eq. (10.4) where values without units are power in GW. The single squiggly lines are power loss to space and the double squiggly lines are power loss from waste heat radiators.

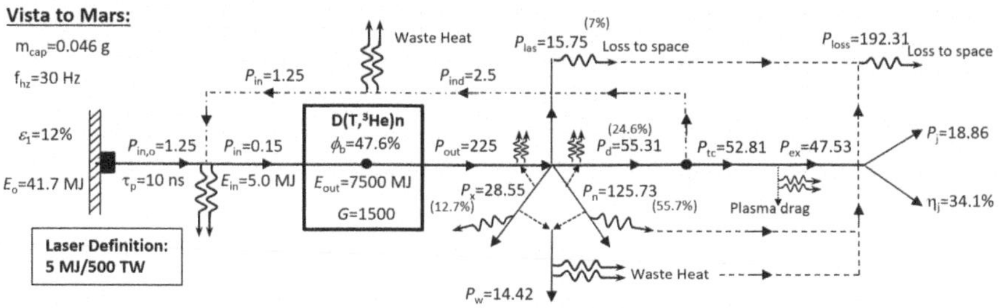

Fig. 10.4 Power flow map for engine operation of VISTA spacecraft from initial electrical energy generation through to exhaust power with bootstrap via induction for next cycle. Shown for a $m_{cap} = 0.046$ g DT fuel capsule with energy gain $G = 1{,}500$, jet efficiency $\eta_j = 0.341$, laser efficiency $\epsilon_1 = 12\%$[KFL]

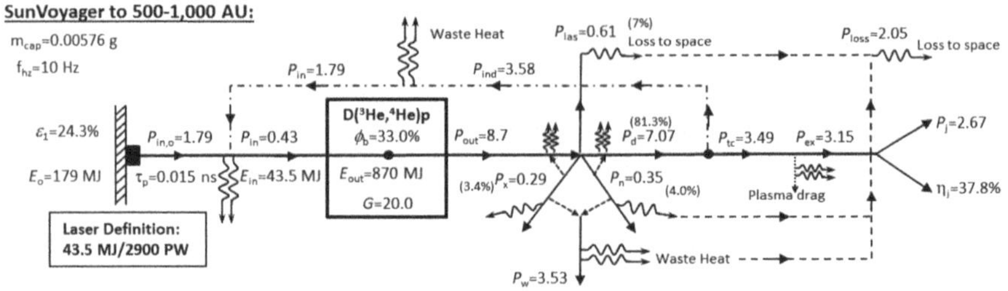

Fig. 10.5 Power flow map for engine operation of SunVoyager spacecraft from initial electrical energy generation through to exhaust power with bootstrap via induction for next cycle. Shown for a $m_{cap} = 0.00576$ g D^3He fuel capsule with energy gain $G = 20$, jet efficiency $\eta_j = 0.378$, laser efficiency $\epsilon_l = 24.3\%$[KFL]

For the VISTA 30 Hz high rep-rate model assuming an inductor system the power cycle as shown in the line map it can be seen that from the capsule gain due to thermonuclear energy release the output power is split into approximately $\sim 1/2th$ as emitted neutrons, $\sim 1/8th$ as radiation, $\sim 1/14th$ as energy loss and around $\sim 1/2th$ is released as the plasma debris. The model assumes 50% induction loss.

SunVoyager Power Cycle

It is also useful to look at the power cycle for SunVoyager which we discussed in Chap. 8, and a similar design configuration has been adopted. Since the design for VISTA is highly optimised, it is likely that for the SunVoyager the radiation may be somewhat higher than has been depicted in this analysis although this will suffice as a first estimate.

For the SunVoyager mission to 1,000 AU the power cycle line map is shown in Fig. 10.5 and Eq. (10.5). The power distribution is split as 7% space loss, 3.4% radiation loss, 4% neutron loss and then 81.3% to the plasma debris. We note that around 3.6 GW of power (assumed 50% induction loss) must then be bootstrapped via an induction line to power the next pulse cycle. The engine has a jet power of 2.67 GW and a jet efficiency of 37.8%. To drive this engine, it would need a laser characterised by 43.5 MJ driver energy and 2,900 PW peak power, all from an initial wall plug energy of 179 MJ. The stated laser efficiency of the laser at 24.3% comes out of the ignition physics model. Note however that constructing a credible power cycle around a gain of $G = 15.55$ proved to be very difficult and so the model shown in Fig. 10.5 is for a gain of $G = 20$. This is mainly down to the need for bootstrap power in the induction loop to power the next cycle.

$$P(I)^{24.3\%}_{1.79i} \times \bigodot^{10\,\text{Hz}}_{20G} \rightarrow P_{f,8.7} - 0.07P_l - 0.04P_n - 0.034P_x - 0.81P_d \rightarrow P^{37.8\%}_{j,2.67}. \tag{10.5}$$

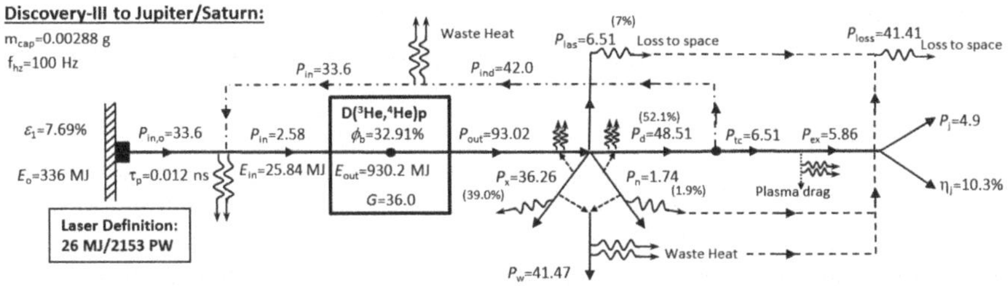

Fig. 10.6 Power flow map for engine operation of discovery-III spacecraft from initial electrical energy generation through to exhaust power with bootstrap via induction for next cycle. Shown for a $m_{cap} = 0.00288$ g D^3He fuel capsule with energy gain $G = 36$, jet efficiency $\eta_j = 0.103$, laser efficiency $\epsilon_1 = 7.69\%$[KFL]

Discovery-III Power Cycle

For the Discovery-III mission to Jupiter/Saturn also discussed in Chap. 8, the PCLM is shown in Fig. 10.6 and Eq. (10.6). The power distribution is split as 7% space loss, 39% radiation loss, 1.9% neutron loss and then 48% to the plasma debris. We note that around 42 GW of power must then be bootstrapped via an induction line to power the next pulse cycle. The engine has a jet power of 4.9 GW and a low jet efficiency of 10.3%, which is largely due to the high pulse frequency and low laser efficiency for this model. To drive this engine, it would need a laser characterised by 25.84 MJ driver energy and 2,153 PW peak power, all from an initial wall plug energy of 336 MJ. The stated laser efficiency of the laser at 7.69% comes out of the ignition physics model.

Once again constructing a viable power systems model at a gain of $G = 12.69$ proved to be difficult and so the model shown in Fig. 10.6 is for a gain of $G = 36$. This is mainly down to the low laser efficiency of 7.69% which significantly reduces the power available and so a significant gain improvement was required.

$$P(I)^{7.69\%}_{33.6i} \times \bigodot^{100\,\mathrm{Hz}}_{36G} \rightarrow P_{f,93} - 0.07P_l - 0.019P_n - 0.39P_x$$
$$- 0.48P_d \rightarrow P^{10.3\%}_{j,4.9}. \tag{10.6}$$

Note that for both the SunVoyager and the Discovery-III power cycles the power coefficients in the equations did not quite add up to unity (contrast this with VISTA where it did). For the SunVoyager this amounts to a missing power fraction of ~4.6%, so that the power fractions only come to 0.954. For the Discovery-III it amounts to a missing ~4.1%, so that the power only comes to 0.959. This is likely an artefact of the model calculations done at the time (particularly a very crude estimate for the neutron flux) and suggests there was additional waste heat, radiative loss, neutrons or induction overhead that was not fully accounted for. We will just label this additional term a miscellaneous power P_{misc} which would be required to close in order to give a full accounting of the power cycle.

Daedalus Power Cycle

The above models are somewhat different from the Project Daedalus second stage design [4, 5] in that for Daedalus it is an effectively perfect design in that no energy losses are defined for the system. This is because of the unique ICF capsule design in which it is assumed that all high energy neutrons and x-rays are absorbed into the outer shells of the material and so do not radiate the structure. Although the designers did accept that some radiation would escape at the outer edges, particularly when the mean free path of the high energy neutron exceeded the thickness of the shell. Yet, this was in the noise of the calculations and as a result the Daedalus is in the unique position of not requiring any waste heat radiator management system and any residual heat is assumed to be re-radiated by the internal reaction chamber.

We attempted to build our own model for the Daedalus second stage engine. However, this was within the optical laser paradigm, instead of the actual electron diodes assumed for the Daedalus study, since such a comparison would be more useful for this work. The engine stage involved a 288 mg pellet detonating at $f_{hz} = 250\,\text{Hz}$. The energy requirement was defined from an equation derived by this author which neatly fits the Daedalus theoretical work at high energy and simultaneously experimental results on the NIF at low energy. The adopted equation was described in Chap. 5 and in an earlier paper [6] and defines the minimum energy driver requirement as a function of the ICF capsule mass and burn-up fraction, under the assumption of a particular gain model $G = 4.385E_{dr}^{1/3}$.

For the Daedalus second stage engine design, calculations indicated a required power source electrical energy (equivalent to wall plug energy) of 404 MJ which would have to be accumulated from an initial reactor mass of 9.5 tons for a specification of 647 kWe and 68.1 We/kg (14.68 kg/kWe). For comparison the Daedalus first stage used a 20 tons reactor mass for a specification of 13 MWe and 650 We/kg (1.54 kg/kWe).

For the model developed as a part of this modern study, we assumed 1% efficiency loss to deliver a total of 400 MJ to the target in $\tau_p = 6.22\,\text{ps}$ or a power input of 101 GW. This is equivalent to ~64.3 PW peak power. Then assuming a burn-up fraction of $\phi = 13.3\%$, each pellet would produce an energy output of 13,320 MJ equivalent to an energy gain of $G = 33.3$, giving a power output of 3,330 GW.

For the Daedalus second stage engine the computed neutron flux was 4.613×10^{20} neutrons/pulse or 1.153×10^{23} neutrons/s. These neutrons are not produced directly from the D^3He reactions but from the side reactions, otherwise known as secondary reactions from the Deuterium reacting with itself and breeding Tritium. This includes DD $\rightarrow {}^3$He+n(2.45 MeV) and DD $\rightarrow$T+p(3.02 MeV) which leads to DT $\rightarrow {}^4$He+n(14 MeV).

Due to the smart design of the ICF pellet we can expect that all the 2.45 MeV will be absorbed into the outer dense fuel and so will not escape and will not contribute to any irradiation of the spacecraft structure. However, this will not be the case for a substantial portion of the 14 MeV neutrons which will have a longer mean free path. For those that originate from deep within the fuel they will be captured and absorbed, but for those further out towards the edge of the shell they will escape. If we assume that the neutrons produced from the secondary side reaction will be in a proportion of 50:50 in terms of 2.45:14 MeV respectively, then the question arises how much of the 14 MeV will escape. Without detailed modelling this is non-trivial to determine but for the purpose of this work we assumed that $\sim$1/10th of the total neutron flux escaped, which amounts to 10%.

If we accept that the majority of these would be 14 MeV neutrons (from DD side reactions), then each neutron would have an energy of around 2.2428×10^{-12} J, which implies a total neutron power of around 258.59 GW, and so 1/10th of this would be 25.86 GW. This is therefore the 14 MeV neutron flux we will assume escapes from each pellet to irradiate the spacecraft structure. We emphasise again however, that according to the Daedalus study, all these neutrons will be absorbed into the outer fuel of the pellet and so will not actually reach the spacecraft structure.

The model further indicated that 102 GW must be extracted from the plasma debris to power the next pulse cycle and with only an assumed 1% energy loss in the transmission this results in the 101 GW input power. The calculated power to the plasma debris was 3,300 GW, then 3,228 GW to the thrust chamber and 3,228 GW to the plasma exhaust if we assume no plasma drag. This would then give a model that produces 3,054 GW jet power and to close the power cycle would correspond to a jet efficiency of $\mu_j \approx 91\%$. This model is illustrated in Fig. 10.7 and the PCLM is given in Eq. (10.7). As can be seen, this is a highly idealised engine power flow.

Daedalus Second Stage Spacecraft

$$P(I)^{99\%}_{101i} \times \bigodot^{250\,\text{Hz}}_{33G} \rightarrow P_{f,3330} - 0.0P_l - 0.0P_n - 0.0P_x - 1.0P_d \rightarrow P^{91\%}_{j,3054}. \tag{10.7}$$

The value of $\eta_j = 91\%$ is consistent with other calculations by this author which were produced as an original derivation in earlier work [6] in which the jet efficiency is defined by an equation we described in Chap. 6 as follows

$$\eta_j \geq \frac{v_{ex}^2}{2\alpha\phi c^2}. \tag{10.8}$$

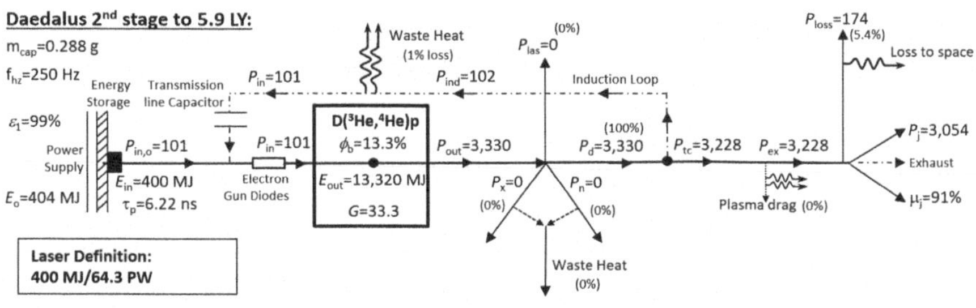

Fig. 10.7 Daedalus-like power cycle (GW) derived by this author[KFL]

So that for the Daedalus second stage engine we can see that

$$\eta_j = \frac{(9210 \times 10^3\ \text{m/s})^2}{2 \times 0.0039 \times 0.133 \times 299{,}792{,}458^2} \approx 0.909.$$

This is a very optimistic scenario with the assumption of minimal losses. For example, during the transmission-induction loop one might expect at least a ∼10% loss, if not much greater. The system also does not allow for the escape of x-rays and high energy neutrons from the pellets, due to the innovative design and properties of absorbance. This is despite the fact the Daedalus team admitted themselves that there will be some radiation near the edge of the pellet shell where the mean free path exceeds the remaining shell thickness. So, it would have been sensible to at least have designed in some redundancy for some minimal shielding.

We note that the Daedalus uses electron beam diodes. The author did attempt to construct an ICF model for the Daedalus second stage that depended on the use of optical lasers, and it was very difficult and relied on a laser efficiency assumption of $\epsilon = 81.2\%$ for the same energy gain of $G = 33$ and burn-up fraction of $\phi = 13.3\%$. Thus, even with this ideal design, the results are pointing towards a requirement for an ultra-efficient laser performance before such a mission would be possible.

Pegasus Power Cycle

We now apply the power cycle models to the Pegasus design and to illustrate the process we will show the calculations in more detail [7]. We have started from the Project Daedalus second stage design and scaled from there. The nuclear reactor output was derived to be a requirement for 15 MWe [8], and then assuming an output of 2 kWe/kg this results in a reactor mass of 7.5 tons, although in the mass budget we round it up to 10 tons for ancillaries. This energy is then stored (i.e. such as in supercapacitors) so that electricity can be released when required for the pulsed fusion engine and driving the lasers as a rapid discharge of high-powered energy. Since the required energy is 403.1 MJ this must be accumulated over 27.1 s per engine (108.4 s for all four engines). The total mass of the energy storage system m_{es} is then given by

$$m_{es} = \frac{P_r \times t_{acc}}{\epsilon_l \times E_{sp}}. \tag{10.9}$$

Where $P_r = 15$ MWe, $t_{acc} = 108.4$ s, the efficiency loss is 20% therefore $\epsilon_l = 80\%$, and the specific energy $E_{sp} = 0.02$ MJ/kg. This results in a mass of around 100 tons allocated for the energy storage system for all four engines

We have also designed the ICF pellets so that there should be minimal energy losses due to x-rays and neutrons as described in reference [9]. However, we think that is an unrealistic expectation for there to be zero losses, particularly due to the fact there is no true aneutronic system. For example, it has been estimated that high energy neutrons produced from DD self-burn reactions (secondary reactions) may contribute as much as ∼15% towards the total energy content. This author has independently calculated this and believes it may be as high as between ~17–20%, although that should be taken as a worst-case situation. It is necessary therefore to bound the performance between the two extremes of minimal energy losses and large energy losses to arrive at a reasonable value for the jet efficiency in the Pegasus spacecraft. We therefore adopt the assumption that some of that energy will escape and therefore some level of radiation shielding will be required in the vehicle.

Like the Daedalus study we have attempted to calculate the neutron flux levels associated with the engine operation. For both the first (acceleration) and second (deceleration) burn stages the neutron flux is computed to be 4.613×10^{20} neutrons/pulse or 4.613×10^{23} neutrons/s. Again, assuming the distribution is dominated by 14 MeV neutrons this would imply an output power of 1,034.6 GW, approximately four times that of the Daedalus. This is not a surprise since the pulse frequency is 1,000 Hz compared to the 250 Hz of the Daedalus and so is to be expected.

A question however is what happens to those neutrons since if they are not similarly absorbed into the outer fuel pellet then they will radiate the spacecraft structure with a significant dosage. The Pegasus pellets are indeed similarly designed as with the Daedalus, so will capture any neutrons from side reactions. However, as a policy of conservative redundancy, in our design model we assume that some fraction will escape and so this implies a need for both radiation shielding and heat radiator waste management. Assuming ~1/10th of the 14 MeV neutrons will escape from each pellet this amounts to 103.3 GW (per engine).

For the calculations now described, we assumed the values for the Pegasus model purely for the purpose of illustrating the method of analysis. The method for determining the power flow was to firstly define the minimum energy requirement from the laser drivers, where the pellet mass is $m_{cap} = 0.072$ g, For D^3He $\alpha = 0.0039$, $\phi = 0.133$, leading to an estimated minimum energy of $E_{dr} = 145.52$ MJ.

An ignition physics model was then constructed based on the premise of optical lasers. This led to a laser efficiency requirement of 36.1% and a total input energy of $E_o = 403.1$ MJ. With a pulse timing of $\tau_p = 0.16$ ns this implies an input power of $P_{in} = 145.52$ GW and a peak power of $P_p = 906.7$ PW. The ignition physics model led to a gain of $G = 23.06$ but it was found this was insufficient energy to close the power cycle with the induction loop. For this reason, the gain was multiplied by an *energy scaling factor* $\kappa = 1.0689$ producing a revised gain of $G = 24.65$, where $\kappa = G'/G$. The revised energy gain for Pegasus would correspond to an energy and power output of

$$E_{out} = \alpha \kappa \phi m_{cap} c^2, \tag{10.10}$$

$$P_{out} = E_{out} f_{hz}, \tag{10.11}$$

$$E_{out} = 0.0039 \times 0.133 \times 1.0687 \times 72 \times 10^{-6}\ \text{kg} \times c^2 = 3{,}587\ \text{MJ},$$

$$P_{out} = 3{,}586.92\ MJ \times 1,000\ \text{Hz} = 3{,}587\ \text{GW}.$$

This contrasts with the calculated jet power of

$$P_{jet} = \frac{1}{2}\dot{m}v_{ex}^2. \quad (10.12)$$

$$P_{jet} = 0.5 \times 0.072\ \text{kg/s} \times (8,897\ \text{km/s})^2 = 2{,}850\ \text{GW}.$$

Although note that the mass flow rate adopted here is for a single engine and with four engines the actual mass flow rate for the mission will be $\dot{m}_{tot} = 4\dot{m} = 0.288$ kg/s and this was mapped across from the Daedalus second stage engine in order to meet an approximate one century mission profile. This then results in a power difference between the output power and the jet power of

$$\Delta P = P_{out} - P_{jet}. \quad (10.13)$$

$$\Delta P = 3{,}587 - 2{,}850 = 737\ \text{GW}.$$

This then quantifies the total power loss which amounts to 20.5% loss of the total, or where 79.5% of the thermonuclear energy release is turned into useful work as jet power at the end of the propulsion cycle. The total loss can be distributed as power to several different loss mechanisms. This includes laser coupling loss P_{las}, which is the fractional loss in coupling any laser energy to x-rays such as associated with the use of a hohlraum in an indirect drive system. The VISTA design adopted a value of 7%. Although Pegasus is direct drive, we do associate some loss here since we have constructed the models based on some of the algorithms that originate from indirect drive experiments, and we assign a moderate 3%. Alternatively, we might associate this is with loss in the use of mirror reflectors to give a symmetric implosion by surrounding the capsule by a uniform radiation field.

The other losses include electromagnetic radiation loss in the form of x-rays P_x, high energy neutrons P_n. All the remainder goes into the plasma debris P_{pd}, which is what remains after the x-rays and neutrons have been extracted and is the principal kinetic energy of the exhaust via the debris ions from the expellant. It is from this plasma debris that the thrust is generated. The total loss is then

$$\Delta P = (P_{las} + P_x + P_n + P_{pd}) - P_j. \quad (10.14)$$

To put it another way

$$P_{out} = P_{las} + P_x + P_n + P_{pd}. \quad (10.15)$$

Where the plasma debris power breaks down into the sum of the inductance power P_{ind} and the thrust power P_{thr}, since the thrust power remains after the inductance power is extracted from the plasma debris

$$P_{pd} = P_{ind} + P_{thr}. \quad (10.16)$$

We can then estimate the laser plasma loss. For the VISTA study they assumed 7%, but for this vehicle design we assume 3%, so that the total laser plasma loss is given by

$$P_{las} = \epsilon_{loss} \times P_{out}. \tag{10.17}$$

$$P_{loss} = 0.03 \times 3{,}587 = 107.61 \text{ GW}.$$

The power from the neutron flux was calculated and we assume that most of this is absorbed into the fuel pellet, except for 1/10th in proportion which is assumed to be all 14 MeV neutrons resulting in $P_n = 103.46$ GW which is 2.88% of the output power. We can then calculate that

$$P_{out} - P_n - P_{las} = 3{,}587 - 103.46 - 107.61 = 3{,}376 \text{ GW}.$$

So it follows that

$$P_{pd} + P_x = 3{,}376 \text{ GW}.$$

Therefore

$$P_{ind} + P_{th} + P_x = 3{,}376 \text{ GW}.$$

And

$$P_x = 3{,}376 - P_{pd}.$$

For the power to the inductance, an assumption is required on the associated loss which is then emitted as waste heat. For the VISTA study [10] they assumed 50% but for this work we assume a moderate 10% (admittedly optimistic), which means that 90% of the power will be transmitted through the induction cycle, which results in

$$P_{ind} = \frac{P_{in}}{\epsilon_{ind}}. \tag{10.18}$$

So that we have an initial inductor requirement prior to waste heat loss of

$$P_{ind} = \frac{403.12 \text{ GW}}{0.9} = 447.91 \text{ GW}.$$

We can then define the jet efficiency by the ratio of jet power to the sum of inductive power and thrust power, a definition which was originally derived by Orth

[3] as follows

$$\eta_j = \frac{P_j}{P_{ind} + P_{thr}}. \quad (10.19)$$

Therefore

$$P_{thr} = \frac{P_j}{\eta_j} - P_{ind}. \quad (10.20)$$

On the assumption of a jet efficiency of 84.9%, as derived from the considerations of Eq. (10.8) this gives a value of

$$P_{thr} = \frac{2{,}850\ \text{GW}}{0.849} - 447.91\ \text{GW} = 2{,}910\ \text{GW}.$$

Then the plasma debris power is found from $P_{pd} = P_{ind} + P_{th}$ so that

$$P_{pd} = 447.91 + 2{,}910 = 3{,}357\ \text{GW}.$$

then the x-ray power is given by

$$P_x = 3{,}376 - 3{,}357 = 19.3\ \text{GW}(0.54\%).$$

Another factor we must consider is how well the expanded plasma debris will interact with the magnetic field and how much it deviates from an adiabatic expansion, and we refer to this as plasma drag. This depends on how conductive the plasma is so that the magnetic field can be expelled, and any exiting charged particles continue in a rearward direction consistent with the thrust direction vector. For those that do not, this will lead to some of the charged particles spiralling back towards the structure and this will be an effective loss from the potential thrust. So, the plasma drag essentially is a measure for how well the plasma decouples from the magnetic field. For the VISTA study they assumed a value of over 5%. For the Pegasus we have assumed an optimistic value for the plasma drag of 1% on the thrust power which amounts to

$$P_{drag} = 0.01 \times P_{th} = 0.01 \times 2{,}910\ \text{GW} = 29.1\ \text{GW}.$$

The exhaust power is then given by

$$P_{ex} = P_{thr} - P_{drag}. \quad (10.21)$$

$$P_{ex} = 2{,}910 - 29.1 = 2{,}879.5\ \text{GW}.$$

The jet power is then given by

$$P_j = P_{ex} - P_{loss}. \tag{10.22}$$

We assume a 1% for $P_{loss} = 28.79$ GW, therefore

$$P_j = 2{,}879.5 - 28.79 = 2{,}850 \text{ GW}.$$

Then we can confirm the jet efficiency via

$$\eta_j = \frac{P_j}{P_{ind} + P_{thr}}. \tag{10.23}$$

$$\eta_j = \frac{2{,}850\, GW}{447.9\, GW + 2{,}910\, GW} = 0.849.$$

This confirms the earlier calculated jet power and jet efficiency for the design based on the assumed parameters and so we can conclude we have a consistent and closed power flow model for the Pegasus. Note that all the calculations above were shown for the Pegasus model(a) which assumed a 36.1% laser efficiency.

However, a major problem with this model is that although it closed the power cycle and allowed for the induction, the excessive waste heat emission suggested the adoption of massive radiators that went beyond a practical limit. A problem with this is that the vehicle would become so heavy that it would not be possible to achieve the 100-year mission time and instead it would be accomplished over several centuries, depending on the specific radiator mass.

The main factor that causes the radiator mass to be so large is the poor efficiency of the laser driver. Although we should state that the assumed 36.1% efficiency is already highly optimistic and beyond the current state of art technological maturity. It was necessary therefore to increase the laser efficiency to a near perfect value and this meant a value, between $\epsilon_{laser} = 0.9$–0.99. However, this also would require a highly efficient radiator. For this reason, a laser efficiency of 99% was adopted for the model and this is therefore declared as a likely requirement before such a mission would be possible.

Table 10.1 shows the results of calculating the power flow for the different engines. Pegasus model (a) is the calculations shown above for $\epsilon_{laser} = 36.1\%$ and for an ICF capsule mass of 72 mg and a pulse frequency of $f_{hz} = 1{,}000$ Hz. Pegasus model (b) is for $\epsilon_{laser} = 32.49\%$ but for the capsule mass of 28.8 mg at 2,500 Hz. These two laser efficiencies were required in the representative ignition physics model discussed elsewhere [9]. Pegasus model (c) is essentially a re-calculation of model (a) but with the laser efficiency increased to $\epsilon_{laser} = 99\%$ and is the final model recommended for Project Icarus.

So, to summarise, this involved a pellet mass of $m_{cap} = 72$ mg grams detonating at $f_{hz} = 1{,}000$ Hz pulse frequency. It would require a wall plug electrical energy of 146.9 MJ operating at an efficiency of $\epsilon_1 = 99\%$ to deliver a total of 145.5 MJ to the

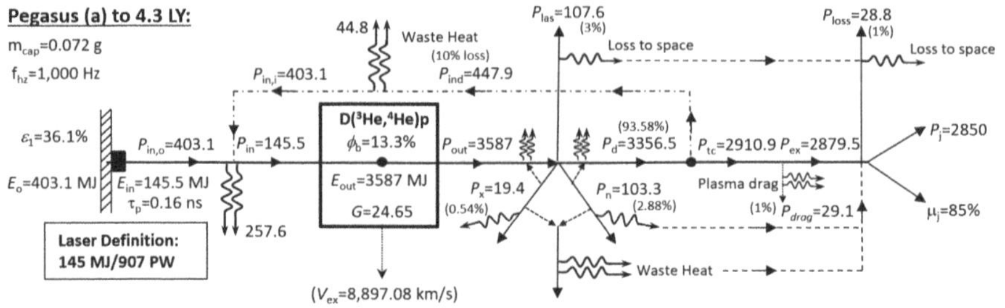

Fig. 10.8 Pegasus power cycle (GW) at 36.1% laser efficiency for a single engine in the propulsion system for Pegasus model(a)[KFL]

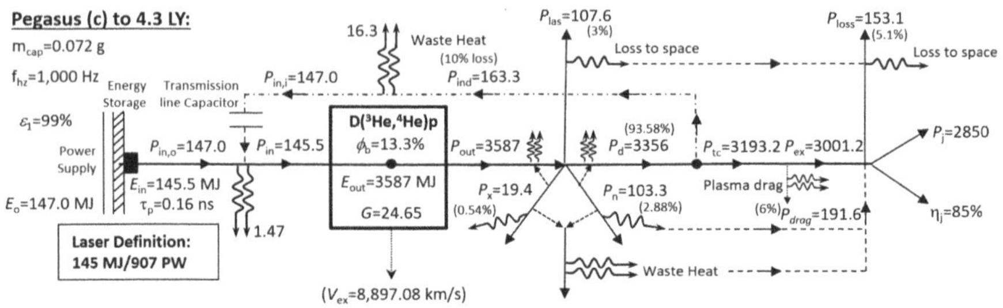

Fig. 10.9 Pegasus power cycle (GW) at 99% laser efficiency for a single engine in the propulsion system for Pegasus model(c)[KFL]

target in $\tau_p = 0.16$ ns or a power input of 146.9 GW. This is equivalent to 906.7 PW peak power of the laser system. Then assuming a burn-up fraction of $\phi = 13.3\%$ for the D^3He fuel, each pellet would produce an energy output of 3,586.9 MJ equivalent to an energy gain of $G = 24.65$. This gives a power output of 3,586.9 GW.

The PCLM equations below show the representative power split. The power cycle for a single engine is also shown in Figs. 10.8 and 10.9 as well as Eqs. (10.24) and (10.25). Examining the model, we see that from the energy output achieved, this is distributed as approximately 3% laser loss, just under 3% to high energy 14 MeV neutrons, around ~0.5% as x-rays. From all this there is then a remainder of over 93% for the plasma debris, and in model (c) just under ~5% of this is then extracted for the use of the induction loop to power the next cycle (~12% in model(a)). Remaining from that is the thrust power, and for which we assume a 1% plasma drag leading to the exhaust power.

So, whilst the Daedalus achieved a jet efficiency of around ~91%, and the Pegasus is at around ~85%. As discussed earlier, we have deliberately included losses to move away from the overly conservative assumptions of the Project Daedalus design. This has come with the penalty that it is much harder to achieve and results in significant system mass increase as a consequence; and we further note that even with the Pegasus it could be argued we have been overly optimistic in the system losses, and this author would be inclined to agree with such an opinion.

Yet, it was felt necessary to represent a more realistic and practical engineering design to illustrate how difficult it is to construct a fusion powered starship. Whilst the Daedalus design is not leaky, we argue that this is also not likely achievable. Although who is to say what may be possible in the future.

Pegasus Spacecraft Model(a)

$$P(I)^{36.1\%}_{447.9i} \times \bigodot^{1000\,\text{Hz}}_{24.65G} \rightarrow P_{f,3587} - 0.03P_l - 0.029P_n - 0.005P_x$$
$$- 0.936P_{pd} \rightarrow P^{85\%}_{j,2850}. \quad (10.24)$$

Pegasus Spacecraft Model(c)

$$P(I)^{99\%}_{163.3i} \times \bigodot^{1000\,\text{Hz}}_{24.65G} \rightarrow P_{f,3587} - 0.03P_l - 0.029P_n - 0.005P_x$$
$$- 0.936P_{pd} \rightarrow P^{85\%}_{j,2850}. \quad (10.25)$$

10.3 Thermal Heat Management

From an analysis of the power outflow through the waste heat management system, we can also estimate the mass of radiators required for the different systems. These are complex systems, and the reader is referred to a more comprehensive treatment of the subject [11]. For this section we will conduct what is a basic examination of the problem from the standpoint of simple radiator theory.

For this we need to know the Stefan-Boltzmann radiation law which describes the intensity of the thermal energy emitted as a function of the temperature. We then use the fact that the power radiated P_r is given as a function of the emissivity ϵ, area A, Stefan-Boltzmann constant $\sigma = 5.67 \times 10^{-8}$ W/m^{2}K^4, radiated temperature T

$$P_r = \epsilon A \sigma T^4. \quad (10.26)$$

The emissivity is defined by the surface emitting the radiation and it has a value between zero and unity, where the latter corresponds to a black body ($\epsilon = 1$). This is an idealized surface that absorbs all incident electromagnetic radiation—it absorbs all the colours of light and at all angles of incidence. The opposite to this is a white body ($\epsilon = 0$) is one that reflects all the incident electromagnetic radiation that falls upon it and it does so uniformly in all directions (isotropic). That fact that the radiation is a function of the temperature follows from Planck's law, when integrated over wavelength gives the Stefan-Boltzmann law. For our radiators in space, where we can approximate them as black body emission, we are interested in the temperature difference relative to the ambient temperature T_o as

$$P_r = \epsilon A\sigma(T^4 - T_o^4). \tag{10.27}$$

Essentially, within our propulsion system we are generating a lot of heat and whilst some of that will immediately go into space, a large part of it will circulate around the vehicle structure and so must be radiated away if it is not to damage the vehicle. In terms of the sorts of vehicles we are discussing in this book, the two dominant sources of heat energy are the laser system and the engines where the fusion detonation takes place. Therefore, we must use radiators in these areas, to manage this build-up of thermal energy. This requires some material that contains a coolant liquid, which can be heated and therefore circulate the heat prior to it being emitted by the radiator panels. If the amount of heat generated is large, then the radiators may also be large which then increases the mass of the vehicle and so will also affect the mission time. In designing spacecraft, we can rearrange this for the area and assume that each radiator is double sided and so the total area is halved for the size required which reduces the mass. We are assuming here that both sides of the radiator view deep space with the same F_{space} where they are deployed to avoid mutual occlusion.

In addition, we must assume a view factor f_v (e.g. ~5%) which determines how much of the radiation emitted from a surface will reach the other surfaces of the spacecraft. This means that in our case we are assuming that the larger portion (e.g. ~95%) of the heat emitted will be radiated into deep space efficiently. We therefore define the fraction that radiates into space from the other surfaces and does not hit the system we are calculating as $F_{space} \approx 1 - f_v$, where $F_{space} = 1 - 0.05 = 0.95$. Therefore, the radiation area is given by

$$A = \frac{P_r}{2F_{space}\epsilon\sigma(T^4 - T_o^4)}. \tag{10.28}$$

Since we may be dealing with different sets of radiators they also may emit at different temperatures. This is particularly the case for the radiators pertaining to the laser drivers and the fusion engines. For the Pegasus design for example, we assumed 900 K associated with the laser driver and induction system, and 2500 K associated with the engines. Both these were relative to a background temperature of 20 K, a number used at the time as an effective radiation temperature and the calculations were conducted on that basis. However, for any calculations the reader is advised to use a background temperature of 2.7 K, consistent with the Cosmic Microwave Background. Although, the difference is some small that there is a negligible effect on the radiator performance with either of these assumptions.

We then define an areal mass term $\rho\Delta x = \text{kg/m}^2$ which depends on the materials being used in the radiators. A sensible number might be 2 kg/m^2 but if this is hard to achieve then we must accept a higher value. Alternatively, if the radiators are still too heavy then we would need a value less than this, which then moves away from the presently practical and towards futuristic technology trends. Substituting

Table 10.2 Calculations of waste heat radiator power loss (GW per engine)

Parameter	Daedalus	Pegasus
Into structure from inductor	1.02	44.79
Into structure from driver	1.01	257.59
Into structure from neutrons	0.0	1.19
Into structure from x-rays	0.0	0.19
Into structure from plasma drag	0.0	29.09
Total to waste heat radiators	2.03	332.86
Fraction of total output	0.061	9.28

Table 10.3 Calculations of power loss (GW per engine)

Parameter	Daedalus	Pegasus
Into space from driver coupling	0.0	107.61
Into space from neutrons	0.0	103.46
Into space from x-rays	0.0	19.18
Into space from exhaust power	174.31	28.79
Into space from heat radiators	2.03	332.85
Total loss to space (excl.rad)	174.31	259.03
Fraction of total output	5.23	7.19

this into the equation for the area and rearranging we can find an equation for the mass of the radiator given by

$$m = \frac{\rho \Delta x P_r}{2F_{space}\epsilon\sigma(T^4 - T_o^4)}. \tag{10.29}$$

As an example, we can look at the Pegasus design. Calculating the radiator mass was a hard challenge since it has a strong dependence on the Power. Tables 10.2 and 10.3 show the results of calculating the amount of power generated and deposited into the different spacecraft structure from the different sources and in what proportion as waste heat. For these calculations we assumed $\epsilon \approx 0.9$. For comparison, we also show the results of calculating these values for a Daedalus-like second stage design These are the results for a single engine and so for the Pegasus would have to be multiplied up by the number of engines.

One of the biggest contributors to the radiator power is the power generated by the laser drivers and this is a strong function of the laser efficiency. For the nominal design for Pegasus a laser efficiency of 36.1% was adopted, based on what it took to get the representative ignition physics model to work and produce a descent gain that was usable. However, a problem with this value is that it results in an unacceptably high value for the radiator mass. As a result, it was necessary to specify a requirement that before the Pegasus could be constructed, incredibly efficient lasers would be required.

Tables 10.4 and 10.5 show examples for the radiator mass from the different contributions, as a function of areal mass (1–2 kg/m^2) and laser efficiency (36.1–99%). The nominal efficiency for the Pegasus laser driver was calculated to be

Table 10.4 Calculations of radiator mass (tons) assuming 1 kg/m^2

Parameter	$\epsilon = 36.1\%$	$\epsilon = 72.2\%$	$\epsilon = 80\%$	$\epsilon = 90\%$	$\epsilon = 99\%$
Mass from laser driver	4,049.4	880.8	571.9	254.2	23.1
Mass from induction cycle	704.1	352.1	317.7	2,82.4	256.8
Mass from plasma drag	7.7	7.7	7.7	7.7	7.7
Mass from neutrons	27.8	27.8	27.8	27.8	27.8
Mass from x-rays	5.1	5.1	5.1	5.1	5.1
Total mass (1e)	4,794.1	1,273.5	930.3	577.2	320.5
Total mass (4e)	19,176.4	5,094.1	3,721.1	2,309.0	1,282.0

Table 10.5 Calculations of radiator mass (tons) assuming 2 kg/m^2

Parameter	$\epsilon = 36.1\%$	$\epsilon = 72.2\%$	$\epsilon = 80\%$	$\epsilon = 90\%$	$\epsilon = 99\%$
Mass from laser driver	8,098.7	1,761.7	1,143.8	508.4	46.2
Mass from induction cycle	1,408.2	704.1	635.5	564.9	513.5
Mass from plasma drag	15.4	15.4	15.4	15.4	15.4
Mass from neutrons	55.7	55.7	55.7	55.7	55.7
Mass from x-rays	10.2	10.2	10.2	10.2	10.2
Total mass (1e)	9,588.2	2,547.1	1,860.6	1,154.6	641.0
Total mass (4e)	38,352.8	10,188.3	7,442.2	4,618.0	2,564.0

36.1%, but this would lead to a total radiator mass that was impractical. This ranges from 19,176 tons (at 1 kg/m^2) up to 76,705 tons (at 4 kg/m^2).

With these masses the spacecraft would take a total trip time on the order of several to many centuries to reach the destination. It is for this reason that as well as recommending a laser efficiency in the range $\sim$90–99% in the study it was also recommended to adopt a radiator rating of $\sim$0.5–1 kg/m^2. This takes us into futuristic technology extrapolation and is not possible today but may be in the future. Note that for these calculations (1e) refers to the mass for a single engine and (4e) refers to the combined mass of four engines.

For the Pegasus we had two different types of radiators. For the driver/inductor loads we assumed $T = 900\,\text{K}$ with $\epsilon = 0.9$, double-sided with $F_{space} = 0.95$, which gives an effective flux near $2\epsilon F_{space}\sigma T^4 \approx 63.61\,\text{kW/m}^2$ at this temperature. For the engines we assumed $T = 2500\,\text{K}$, $\epsilon = 0.9$, effective double-sided radiative flux to space of $2\epsilon F_{space}\sigma T^4 \approx 3.79\,\text{MW/m}^2$. That is a difference of a factor of $\sim$60 in flux and explains why the engine heat at high temperature radiates with far less area/mass than the driver/induction radiators, which heat at a moderate temperature in comparison. This also explains the differences in colour for the two types of radiators shown in the visual graphics of Chap. 15 of this book, consistent with emission temperature.

This ends our short discussion on radiator theory since the author is not an expert in this area. However, it is important to be able to at least get a first order of magnitude estimate of the thermal management mass requirement and to integrate that additional mass into the overall system mass and then re-calculate the mission.

Also, the thermal management issues are strongly coupled to the power cycle, especially when adopting an induction system. Yet the above should be sufficient to make first estimate calculations in advanced vehicle concept designs.

Problems of Chap. 10

10.1 A propulsion engine has a mass flow rate of 1.5 kg/s and an exhaust velocity of 100 km/s. The measured output power is 150 GW (i) Calculate the thrust and therefore jet power (ii) Calculate the ratio of useful work to total power losses to the system. Use the Gordinier method.

10.2 A space propulsion engine is described by the following Power Cycle Line Map shown below, where 5 GW is subtracted for the induction loop. (i) Interpret the meaning of each term (ii) If the induction waste heat loss was 40% what will be the initial induction power for the next cycle (iii) What is the input energy and the energy requirement at the initial reactor start up sequence (wall plug equivalent).

$$P(I)_{5i}^{10\%} \times \odot_{100G}^{150\,\text{Hz}} \rightarrow P_{f,300} - 0.05P_l - 0.8P_n - 0.15P_x - 0.20P_d \rightarrow P_{j,25.5}^{75\%}.$$

10.3 An advanced propulsion engine has radiators rated at 2.5 kg/m^2 receiving a thermal power dump of 25 GW. The radiators emit at a temperature of 1,000 K and with an efficiency of 0.8, and are characterised by a view factor of 7% (i) Calculate the total area required to radiate the heat and the area of each side (ii) Calculate the mass of the radiator and coolant, assuming the coolant will be 60% of the radiator mass (iii) Calculate the area of a single side assuming a double sided radiator, and therefore the length of the radiator.

References

1. Dyson FJ (1968) Interstellar transport. Phys Today 21(10):41–45
2. Gordinier MR, Davis JW, Schultz KR, Scott FR (2003) Nuclear fusion power. Encyclopedia of physical science and technology, 3rd edn, pp 671–699
3. Orth CD (2003) VISTA: a vehicle for interplanetary space transport application powered by inertial confinement fusion, LLNL, UCRL-TR-110500, 16 May
4. Bond A, Martin AR (1978) Project Daedalus: the mission profile. Final study report. JBIS Special Supplement:S37–S42
5. Martin AR, Bond A (1978) Project Daedalus: the propulsion system part 1: theoretical considerations and calculations. JBIS Special Supplement:S44–S62
6. Long KF (2024) Minimum condition requirements in laser driven ICF for advanced space propulsion applications. JBIS 77(1):29–32
7. Long KF (2025) Project Icarus: the Pegasus interstellar spacecraft propulsion power cycle. JBIS 78(10):330–333
8. Long KF (2025) Project Icarus: the Pegasus spacecraft concept design engineering systems description. JBIS 78(6):186–205
9. Long KF (2025) Project Icarus: the Pegasus spacecraft inertial confinement fusion ignition physics design. JBIS 78(10):344–368

10. Orth CD (2000) Parameter studies for the VISTA spacecraft concept, LLNL, UCRL-JC-141513. Presented at NASA advanced propulsion workshop in fusion propulsion, Huntsville, AL, 8–10 November
11. Lamontagne M (2018) Heat transfer in fusion starship radiation shielding systems. JBIS 71(12):450–457

Chapter 11
Large Crewed Vessels

The question, then, of whether starships must travel at almost the velocity of light or only about one-third that velocity depends greatly on a different discipline of physics (biology). Many analyses have been made with velocities about one per cent lower than the velocity of light. The conclusions are that the ships required would be ridiculous, and, therefore, interstellar travel even to the nearer stars is ridiculous. This represents a very naive system analysis of interstellar travel.

—Maxwell W Hunter II, 1966 [1]

11.1 Construction in Space

Currently, for any operations in space, the economics of the mission have to be considered in the context of our existing space launch capabilities. For rockets like those produced by the company SpaceX these are on the order of \$100 million per launch and once full reusability has been achieved the cost may come down even further. Whereas the current launch costs might be in the range \$10,000–25,000/kg, the goal would be to reduce it to \$1,000/kg or even much lower. Once this happens, trips into space will become routine and this is also likely to open space up to mass tourism—as we enter the age of space hotels. At this point, as illustrated in Fig. 11.1 the construction of large spacecraft will become feasible.

The main problem with any cost models for spacecraft is that we are extrapolating from existing numbers and from the assumption of an Earth dominated economy, where today 100% of resources are derived. Within decades this is likely to change so at first after the establishment of the initial lunar and Martian scientific stations we might see up to $\sim$5–10% of the resources used coming from non-Earth origin. Eventually this will be increased and the goal within a century or so would be for 95% of the resources used in space to have a non-Earth origin and perhaps only $\sim$5% comes from Earth such as for critical component technologies which might require high precision machining using rare metals.

At this point, commensurate with those space resources will be the establishment of an independent off world economy, [2–4] which is driven by space industrialisa-

K. F. Long, *Stellar Engines*, Springer Aerospace Technology,
https://doi.org/10.1007/978-3-032-18062-9_11

Fig. 11.1 Construction of a large starship in Earth orbit[KFL]

tion such as mining materials from asteroids as a key commercial incentive. By this time the construction of large spacecraft would not be undertaken from an Earth launch basis, but instead all the materials needed for it will derive from space. This especially includes strip mining of asteroids for their raw materials. An example for how such an operation might work in practice has been written up in a separate study [5].

It is constructive to consider asteroid mining by the derivation of a back-of-envelope approach to estimating how long it would take to produce the required material. The time τ_m to strip mine a proportion ϵ_p of an asteroid of object mass m_o is a function of the number of individual mining drones n, the rate at which material is removed from the surface through mining R_m and the efficiency ϵ_r of the average drone at converting raw source material into useable system mass at the factory, the latency time t_l between the asteroid raw material resource and the factory delivery point (which includes time for power charging), and the number of return trips per unit time t_f that an average drone can make from the asteroid to the factory. We wrap these together into a utilization factor $U = 1 - t_l t_f$ $(0 \leq U \leq 1)$. The mining

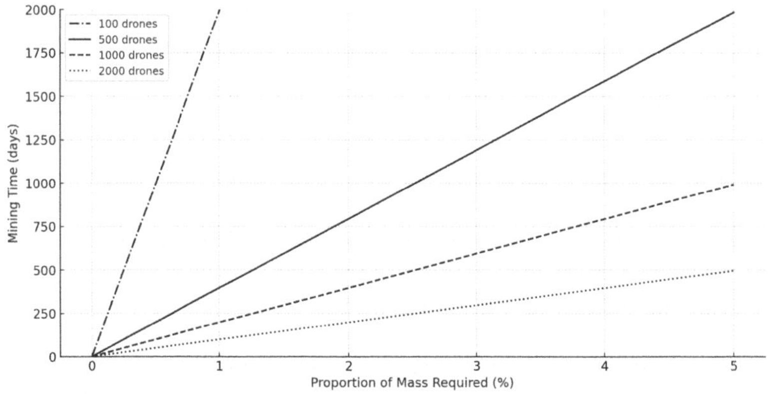

Fig. 11.2 Calculations for strip mining a 100,000 tons asteroid using robotic drones

time for a given object is then given by

$$\tau_m = \frac{\epsilon_p m_o}{n R_m \epsilon_r (1 - t_l t_f)} = \frac{\epsilon_p m_o}{n R_m \epsilon_r U}. \tag{11.1}$$

We can then also describe the throughput of the fleet of drones by the equation

$$\dot{m}_{use} = n R_m \epsilon_r U. \tag{11.2}$$

As an example, imagine we had a 100,000 tons asteroid and we require around 2,200 tons to be removed to construct a spacecraft using 1,000 drones that can remove mass at a rate of 0.1 g/s converting the mass to system mass with an efficiency of 70%, with each drone making an average of 1 trip/hour to the factory, assuming a latency time of 10 minutes. The total time for strip mining of this object would then be on the order of 10,476 hours or ~436 days. The throughput associated with this fleet would be 0.0583 kg/s. Figure 11.2 shows the results of calculating this problem for the same asteroid mass under the same assumptions but with different proportions mined and how long it would take.

Another factor that will likely reduce the cost is the economies of scale, where the cost per item is inversely proportional to the number that is constructed. In other words, if instead of building one large spacecraft, you build ten, the cost for each individual one will be lower. It would lower further if the raw materials mined from an asteroid are not just for a spacecraft but also for other purposes such as other types of spacecraft and space stations or habitats on moons or planets. This is likely to greatly impact the economics of building an interstellar probe in the future. This underscores the economic case for space industrialisation as we seek to expand humanity off world, out to the Solar System and beyond.

Living in space away from Earth will not be easy and it is unclear how this will affect human beings psychologically to be away from the environment from which we have evolved. Indeed, with different environmental conditions and perhaps a reduced gravity, we are likely to see a bifurcation of humanity into two separate species longer-term. Yet initially, the goal would be to generate environments that replicate Earth conditions as close as possible.

It is also likely to lead to massive amounts of jobs for humans that seek a new life or living and working in space. Although we might suppose that much of the work will be done by AI drones, and it will be, there will still be certain jobs that need a human eye or at the very least human supervision. Earth orbit and the surrounding planetary space is likely to become the stage of a new era of industrialisation that is surely to act as a catalyst for our progress in the exploration of space. It is likely that space industrialisation is likely to replicate some of the history of North America, where the Europeans were seeking to settle new lands and long railroad tracks were built off into the wilderness. The difference is that space is not a natural environment for humans and their animals and crops, and so we must create those environments artificially, either as a part of a large space habitat or on-board a large space station so that life can be sustained. For an excellent example on how such habitats can be achieved readers should consult the space habitat studies of Gerard O'Neill [6].

In the two sections that follow we will consider two scenarios where the construction of very large spacecraft becomes possible. Since they are so large, it would not be efficient, or ethical, to simply strip the Earth of its resources and to supply vast quantities of material into space. So instead, the assumption here is that most of the materials used for these large spacecraft comes from industrial mining operations of asteroids or even moons around other planets for which there is no life present. The sorts of vehicles considered have the capacity to carry hundreds or thousands of people which may be useful for colonisation strategies or just for holding large communities off world.

11.2 Enzmann Slow Boat

If humanity is to travel to the stars, and not just our robotic probes, then we must begin the process of considering the design of large vessels that can carry a crew on scientific exploration missions. These will also be the precursor to full colonisation missions of other planets and stars. Such an opportunity is offered in consideration of a concept known as a 'slow boat' for low travel speed (compared to relativistic flight) of less than 10% of the speed of light. Yet the interest of this work was to take the original concept a little further to consider it in the context of an ICF based propulsion system.

To facilitate this, we base the calculations on a concept known as the Enzmann spacecraft design, which was invented by Robert Duncan Enzmann over many years in a series of interesting studies [7–9]. The use of this vehicle architecture is not a statement that this is likely to be the best way to achieve the mission, but it is used as a case study for how such missions might look. In 2012 this author had published a paper with two other co-authors in which an initial appraisal of the Enzmann spacecraft concept was given [10]. The goal of this initial work was to determine a definition of the vehicle and its proposed architecture and then advance it to a revised concept [11].

Building on this we adopt one of the architectures that was chosen from that earlier study as being representative of the *Enzmann Slow Boat* concept. This involves a 30,000 tons vessel propelled by 3 million tons of fusion fuel to travel up to 9%c or 27,000 km/s and to complete the mission to the nearest stars in around 60 years. The initial population of the ship would be 200 people, and it would grow to 2,000 people throughout the journey. We make no comment here on the credibility of that population growth model but adopt it to remain consistent with the original design intent.

When Enzmann originally put together the spacecraft concept this was coming on the back of the Project Orion studies which proposed to use nuclear bomb technology to propel a large spacecraft [12]. However, Enzmann realised it may be better to employ smaller 'pulse units' contained within an internal reaction chamber instead of an external detonation and transfer of the momentum to a large pusher plate. However, the detonation of pulse units inside a reaction chamber is only really feasible up to a certain size or they risk destroying the thrust structure. Therefore, another type of ignition system is required that involves the detonation of lower mass pulse units.

In this case study we will adopt reaction chambers similar to those adopted on the Daedalus second engine stage which were on the order of 300 tons in mass [13, 14], but we further assume that in the future it may be possible to construct similar engines at 100 tons each and this would be a desirable goal. We will derive an acceptable vehicle configuration and mission profile based on these calculations. In addition, rather than just detonating single thermonuclear pulse units, we will adopt the method of thrust augmentation by mass flow rate increase introduced earlier in Chap. 8.

The original Enzmann design included a 30,000 tons dry mass vehicle with a length of 610 m and a spherical thermonuclear fuel tank diameter of 305 m. It contained three separate (dry) habitats for populations of people, which at launch was around 200. The propulsion system would have 24 engines (there were 8 engines in the earlier definition by Enzmann) to take it to 4.3 ly in around 60 years travel time.

In the Enzmann concept explored here we have extended the length of the vehicle beyond 610 m to what is now around 810 m, (depending on the engine and thermal radiator requirements). This is mainly due to the habitat sections which for the purpose of this study are given the dimensions of 100 m in length by 100 m in diameter, and there are three of them which are already 300 m. In addition to this will be the interlocking sections, the front propellant sphere, the command structure and the rear power and propulsion sections.

For the earlier Slow Boat configuration, the exhaust velocity was determined to be 11,700 km/s with a mass ratio per stage of around $R = 10.05$ and a total mass ratio $R_{tot} = 101.0$. Since the vehicle is to decelerate to orbital velocity the total velocity increment will be 54,000 km/s or $0.18c$ (for burn and deburn). The total

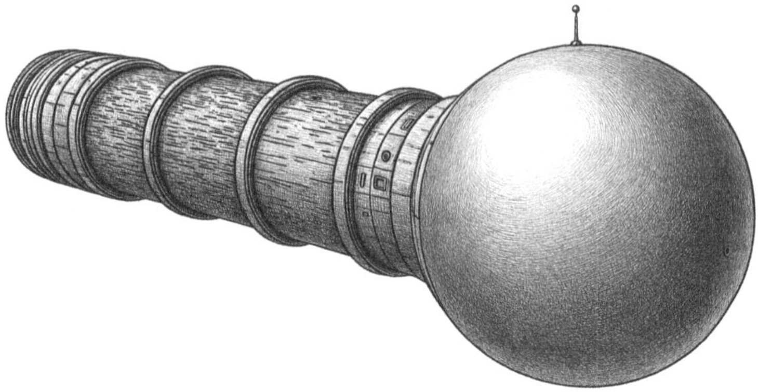

Fig. 11.3 Illustration of Enzmann spacecraft frontal view[KFL]

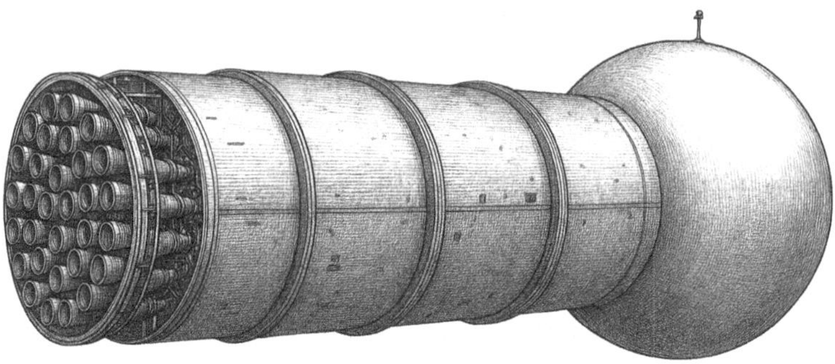

Fig. 11.4 Illustration of Enzmann spacecraft rearward view[KFL]

propellant allocated for the mission was to be $m_{prop} = 3$ million tons and with a mass flow rate of $\dot{m} = 5.02\,\text{kg/s}$ this would correspond to a total burn time on the order of $t_b = M_{prop}/\dot{m} = 18.95$ years for both acceleration and deceleration.

Figures 11.3 and 11.4 show the Enzmann spacecraft concept as envisaged by this author from a frontal and rearward perspective. The design includes the existence of the large metallic sphere at the front of the vessel to contain the frozen fusion fuel, as well as 24 engines at the rear (more are shown in the graphic). In addition, at a total length of around 810 m. The main difference between the Enzmann concept derived in section and the original concept is the move from internal reaction chambers for the pulse units to the use of external hybrid reaction chambers like the Daedalus study [13, 14]. It is the intention for this work that any pulse units, which we now term as ICF pellets, are moved rapidly into the target chamber and then hit by a large bank of lasers to compress and ignite them to a thermonuclear state. Any thrust would then be provided by the charged particles in the exhaust debris.

The Enzmann spacecraft in the configuration imagined is made of three main habitat cylindrical units which are illustrated in Fig. 11.5. Each habitat unit is further divided into four quarter subsections or zones which have the structure and motion

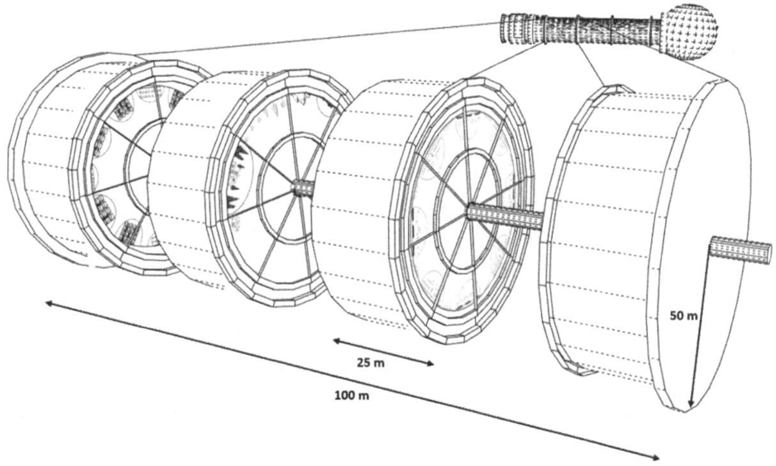

Fig. 11.5 Illustration of one of three Enzmann habitat sections divided into four subsection zones under rotation for artificial gravity[KFL]

like a rotating space station about a central axis point. Since the total length of each node is around 100 m in length, each subsection zone is 25 m in length, but wrapped around the 50 m radius cylinder so each section has a total circumference of around 314 m.

We will not describe the internal structure of the spacecraft in this book, in terms of the atmosphere, content, water, soil, all of which are discussed a little in the original paper [11]. We only state that each habitat section is split further into four subsections which contain a living zone, a working zone, recreational zone and a hydroponic zone. To generate an artificial gravity at the perimeter of around $1g_o = 9.81\,\text{m/s}^2$ the 50 m radius cylinder would have to rotate at 4.223 revolutions per minute or 0.07 revolutions per second, which implies that each rotation takes around 14.18 seconds.

Although the Enzmann envisaged a starting population of around 200 people, growing to 2,000 people by end of journey's end, since the mission is likely to exceed the original 60 years, we suggest that a starting population on the order of ~300–400 people would be preferred, growing to perhaps ~3,000–4,000 by journeys end.

Forward of the habitat structure but just aft of the large propellant sphere will be an operations management section. This will be the main command station for the vessel and contain the vital engineering stations. The other rooms will include a command-and-control bridge, meeting rooms, scientific analysis and diagnostics of all systems but also observations undertaken in space. Primary to this will be data management, which can be then fed to the working zone for further analysis. Any robotic subprobes for planetary or stellar surveys would also be deployed from this section and this may include human crewed 'yachts' to complete in-situ reconnaissance.

To the rear of the spacecraft will be the set of large engine reaction chambers which make up the main component of the thrust structure. This will also include nuclear reactors and laser systems for use in the propulsion system. The precise dimensions of this system would need to be determined but are vital for the estimation of the thermal waste heat requirements and the amount of surface area required for any external waste heat radiators to be wrapped around the cylinder. Alternatively, radiators could be constructed to the side of the cylinder giving the Enzmann the appearance of having rearward wings. A thermal heat assessment is not conducted within this preliminary study but would need to be undertaken if the design is to advance towards higher credibility. The remainder of this section will focus on the propulsion element of the vehicle design.

The spherical 'iceberg' would be contained with a 2 mm Titanium alloy skin with a density of 4,500 kg/m^3. At a radius of 152.5 m this would give it a total mass of just under 2,630 tons. Underneath this would be a sub-mm reflective plastic insulation blanket and for this we allocate a total mass of 370 tons.

Next comes what sort of mass is required for the capsules to drive the Enzmann spacecraft. We also note that for the declared 3 million tons propellant associated with the Enzmann spacecraft concept this was declared to be all thermonuclear fuel. We firstly assume an optimistic fuel performance using D^{3}He which is assumed to be mined from the gas giants [15], and this assumes a mass defect of $\alpha = 0.0039$. Alternatively, it could be assumed that this is using DD fuel, but it would require some way to enhance the performance such as by using the methods adopted within a catalysed DD stellarator reactor [16].

The exhaust velocity for the design was determined to be 11,700 km/s. For the model (a) assuming a mass flow rate of 5.02 kg/s, $\phi = 0.3$, This results in a total mission time of 55.93 years having achieved a cruise velocity of 29,655 km/s or 0.0989c. This performance required a pulse frequency of $f_{hz} = 100$ Hz and generated a thrust of 58.74 MN, a jet power of 343.61 TW.

The total propellant allocation of 3 million tons was split between 2,794,700 tons in the first acceleration stage during its 17.6-year burn time, followed by 205,300 tons during the 1.29 years deceleration time. This vehicle performance would suggest a jet efficiency for the propulsion performance of $\eta_j = 0.65$, which means that approximately 35% of the energy would be lost such as due to neutrons, x-rays and lack of jet collimation. A mission using D^{3}He would reach the destination in the required 60-years timeframe as defined by the original Enzmann concept. One assumes this was chosen as being within the plausible lifetime of a human being so that some of the crew that left Earth may also see the destination star system.

Next, the calculation was repeated using the same parameters but assuming a DD fuel mixture, which means using a value of $\alpha = 0.001$. This is conducted since the original Enzmann 'iceberg' was to be all Deuterium and so it is important to show the comparison. Assuming the same parameter value assumptions as above, this suggests an exhaust velocity of 6,966 km/s.

For the DD model, calculations suggested that the spacecraft now achieves a velocity of 17,656 km/s or 0.0589c. Using a mass flow rate of 5.02 kg/s it generates a total thrust of 34.97 MN, jet power of 121.8 TW and reaches the destination

Table 11.1 ICF pellet design for models with and without expellant assuming $n_e = 24$, $f_{hz} = 100\,\text{Hz}$, $\phi = 0.3$, $\dot{m} = 5.02\,\text{kg/s}$, $m_{TN} = 3$ million tons

Parameter	Enzmann	Model(a)	Model(b)
Fuel	DD	D^3He	DD
Exhaust velocity (km/s)	11,700	11,700	6,966
Thermonuclear fuel mass (g)	–	2.0917	2.0917
Mass defect α	0.001	0.0039	0.001
Jet efficiency η_j	–	0.651	0.9
Energy output E_{out} (GJ)	–	219.95	56.39
Energy driver E_{drive} (GJ)	–	3.352	1.208
Energy gain G	–	66	47
$N_{pellets}$	–	1.43×10^{12}	1.43×10^{12}
Thrust (MN)	58.74	58.734	34.971
Velocity (km/s)	26,981	29,655	17,656
Velocity (c)	0.09	0.0989	0.0589
Total burn time (Yrs)	18.94	18.93	18.93
Total trip time (Yrs)	60.0	55.93	85.91

target after a duration of 85.91 years. This exceeds the defined 60-year mission timeframe requirement and would certainly imply two generations of humans would be required to complete the mission. The main results of these two calculations are shown in Table 11.1.

For this study rather than allocating all the propellant to thermonuclear fuel we will instead split it between a minor thermonuclear fuel component and a largely hydrogen based expellant propellant. This is done for the purpose of increasing the mass flow rate and therefore thrust but for a minimised amount of thermonuclear fuel. The total mass of the propellant will increase however with the addition of the expellant and so this will impact the mission profile in terms of increased spacecraft mass and therefore a lower velocity attained for the same acceleration profile. This results in the total mission time being elongated compared to the non-expellant case.

Since we are using expellant propellant to reduce the quantity of thermonuclear fuel, the total mass flow rate is given by the combination of the thermonuclear capsule mass m_{cap} and the expellant mass m_{exp} along with the pulse frequency of detonation f_{hz} and the number of engines n_{eng} as follows

$$\dot{m} = (m_{cap} + m_{exp}) n_{eng} f_{hz}. \tag{11.3}$$

If we assume that the expellant mass is $m_{exp} = \theta m_{cap}$, where θ is the multiplication scale factor of the expellant mass relative to the thermonuclear capsule mass, and rearranging we find

$$m_{cap} = \frac{\dot{m}}{n_{eng} f_{hz} (1 + \theta)}. \tag{11.4}$$

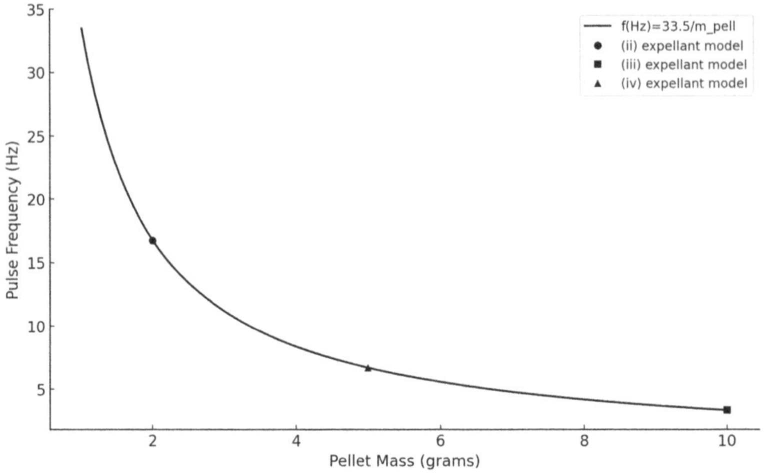

Fig. 11.6 Calculations of pulse frequency as a function of thermonuclear pellet mass

If we assume that $\theta = 11$, then $m_{exp} = 11m_{cap}$ which equates to $m_{cap} = 2.0917$ g and $m_{exp} = 23.008$ g. To put it another way $m_{cap}/m_{exp} = 0.09$. Now we must also determine the other propulsion parameters, so we are explicit in what has been assumed.

Figure 11.6 illustrates a set of calculations for the pulse frequency of detonation as a function for the thermonuclear pellet mass. These were conducted assuming the optimistic engine conditions which were based on the no-expellant mission (i) to match the mission profile. This includes the assumption of $\alpha = 0.0039$, $\epsilon_{mix} = 0.9$, $m_{exp}/m_{cap} = 11$, $\eta_j = 0.9$, $\phi = 0.7233$ leading to a total mass flow rate for the mission of $\dot{m} = 10.04$ kg/s.

It was found that a trend fit suggests an approximate relationship of $f_{hz}(Hz) \approx 33.494/m_{cap}(g)$. The expellant missions are then shown as missions (ii), (iii) and (iv) and these are shown in Table 11.2. Model (ii) represents a mission at high pulse frequency (16.667 Hz) but low pellet mass (2.09 g). Model (iii) represents a mission at high pellet mass (10.09 g) but low pulse frequency (3.5 Hz). Model (iv) is a compromise between them both at moderate pellet mass (5.09 g) and moderate pulse frequency (6.85 Hz). Table 12.2 shows some of the parameters for these different missions.

Table 11.3 shows the performance for the Enzmann spacecraft on its journey to the nearest stars. Examining the Model (iv) as the optimum expellant solution, the spacecraft accelerates at 0.024 m/s^2 for 8.6 years using a pulse detonation frequency of just under 7 Hz until it arrives at just over 3,900 AU from the point of origin, which is assumed to be at or near-Earth orbit. It achieves a cruise velocity of 6,616 km/s and continues to cruise over 270,472 AU or 4.28 ly. During this time there would be a change of generation on-board the vessel in that very few will

Table 11.2 ICF pellet design for D^3He ($\alpha = 0.0039$) models with and without expellant using 24 engines and assuming $\epsilon = 0.9$ (except for model (i)) and $m_{exp}/m_{cap} = 11$

Parameter	Model (i)	Model (ii)	Model (iii)	Model (iv)
Expellant	No	Yes	Yes	Yes
Exhaust velocity (km/s)	11,700	2,925	2,925	2,925
Thermonuclear fuel mass (g)	2.0917	2.0917	10.0917	5.0917
Expellant mass (g)	0.0	23.008	111.0084	56.008
M_{TN} (million tons)	3.0	0.25	0.25	0.25
M_{Aug} (million tons)	0.0	2.75	2.75	2.75
$N_{pellets}$	1.43×10^{12}	1.19×10^{11}	2.47×10^{10}	4.91×10^{10}
Pulse frequency (Hz)	100	16.67	3.45	6.85
Mass flow rate (kg/s) [$1/n_{eng}$]	0.2092	0.4183	0.4183	0.4183
Mass flow rate (kg/s)	5.02	10.04	10.04	10.04
Thrust (MN)	58.74	29.36	29.36	29.36
Jet efficiency	0.651	0.9	0.9	0.9
Burn-up fraction ϕ	0.3	0.72	0.72	0.72
Approx. burn time (Yrs)	18.94	9.47	9.47	9.47

remain from the original astronauts that set off on the journey and a mostly new crew will now be in place preparing to greet the new destination.

At a point of 270,494 AU or 4.28 ly distance the spacecraft will turn-around and operate under reverse engine thrust mode where it will decelerate for a period of 0.89 years over the remaining 1,640 AU distance. It will eventually fully slow down to near orbital velocity around any chosen destination target planet and arrival at this location would have occurred 203 years after leaving the home world.

Since we are discussing ICF pellet propulsion driven by lasers, it is constructive to describe some of the parameters associated with the pellet masses considered in this work in terms of the driver energy requirement and the intensity of the laser required and this is shown in Table 11.4. In addition to the thermonuclear capsule each detonation would be accompanied by injected expellant propellant. This helps to boost the mass flow rate of the exhaust products.

It is useful to describe these calculations a little and we can examine model (iv) with a 5.09 g capsule mass. The laser equations have been described in other publications by this author [17]. Assuming a highly optimistic laser efficiency of 20% this would require a wall plug energy of 90.28 GJ, leaving 18.06 GJ, for which we assume a further 30% loss due to frequency conversion leaving 12.64 GJ laser energy delivered to the target as a requirement to achieve the stated gain.

Further, assuming a 1 ns laser pulse this would correspond to a peak power output of 18,060 PW. If we then assume an average D^3He fuel density of 0.089 g/cm^3 this suggests a starting capsule radius of 23.99 mm. If we then assume that the laser illumination covers the entire surface of the sphere, we can estimate a spot size area of 0.396 cm^2 which then suggests a laser intensity requirement of 3.2×10^{17} W/cm^2, corresponding to an electric field amplitude of 1.5×10^{10} V/m. This is assuming a

Table 11.3 Spacecraft mission performance

Parameter	Model (i)a	Model (ii)	Model (iii)	Model (iv)
Expellant	No	Yes	Yes	Yes
Exhaust velocity (km/s)	11,700	2,925	2,925	2,925
$M_{TN,burn}$ (tons)	2,794,700	226,602	226,602	226,602
$M_{TN,deburn}$ (tons)	205,300	23,398	23,398	23,398
$M_{TN,tot}$ (tons)	3×10^6	2.5×10^5	2.5×10^5	2.5×10^5
$M_{aug,burn}$ (tons)	–	2,492,474	2,492,622	2,492,622
$M_{aug,deburn}$ (tons)	–	257,363	257,378	257,378
$M_{aug,tot}$ (tons)	–	2,749,837	2,750,000	2,750,000
Pulse frequency (Hz)	100	16.67	3.45	6.85
Mass flow rate (kg/s)	5.02	10.04	10.04	10.04
Acceleration (m/s^2)	0.05327	0.02443	0.02443	0.02443
Mass ratio burn	12.61	9.601	9.601	9.601
Mass ratio deburn	12.61	9.599	9.599	9.599
Mass ratio total	159.01	92.159	92.159	92.159
Thrust (MN)	58.74	29.365	29.367	29.367
Jet power (TW)	343.609	42.947	42.949	42.949
Specific power (MW/kg)	1.9875	0.248	0.248	0.248
Boost time (Yrs)	17.640	8.582	8.582	8.585
Boost distance (ly)	0.538	0.0617	0.0617	0.0617
Cruise speed (km/s)	29,655	6,616	6,616	6,616
Cruise speed (%c)	9.892	2.207	2.207	2.207
Cruise time (Yrs)	36.989	193.816	193.816	193.816
Cruise distance (ly)	3.659	4.0466	4.277	4.277
Deburn time (Yrs)	1.296	0.886	0.886	0.886
Deburn distance (ly)	0.168	0.0259	0.0259	0.0259
Total distance (ly)	4.365	4.365	4.365	4.365
Mission time (Yrs)	55.926	203.285	203.285	203.285

351 nm infrared laser which has been frequency tripled from 1,053 nm under direct drive.

In this section we have considered the nominal Enzmann design using DD fuel which reached its destination in around ~86 years. We then considered the adoption of D^3He fuel which allowed the spacecraft to reach the destination in around ~56 years. We then considered the adoption of expellant propellant as a means for reducing total thermonuclear fuel content, and under the different assumptions of pellet mass and the use of a factor $\times 11$ scaling on the expellant mass, the spacecraft reached the destination in around ~203 years.

Figure 11.7 illustrates what a revised Enzmann might look like to include a smaller 'iceberg' thermonuclear fuel section at the front and the addition of expellant tanks towards the rear. The tanks are held in place by structural support struts. The larger the mass of the expellant the larger these tanks or the more of

Table 11.4 Specification of energy driver requirements for thermonuclear capsule mass assuming laser energy conversion 20%, frequency conversion losses 30% with a laser pulse of 1 ns with 100 lasers per capsule

Parameter	(i)	(ii)	(iii)	(iv)
Thermonuclear fuel mass (g)	2.0917	2.0917	10.0917	5.0917
Wall plug energy E_{wp} (GJ)	23.9	46.28	150.78	90.28
Energy input E_{in} (GJ)	4.78	9.26	30.16	18.06
Energy driver E_{drive} (GJ)	3.35	6.48	21.11	12.64
Energy output E_{out} (GJ)	219.95	530.29	2,558.51	1290.88
Energy gain G	66	82	121	102
Peak power P_{peak} (PW)	4,780	9,260	30,160	18,060
Energy per beam E_b (MJ)	33.46	64.82	211.12	126.4
Power per beam (PW)	33.46	64.82	211.12	126.4
Radius capsule (mm)	17.75	17.84	29.99	23.88
Intensity I_L (W/cm^2)	8.45×10^{16}	1.64×10^{17}	5.33×10^{17}	3.19×10^{17}
Amplitude E_o (V/cm)	7.99×10^9	1.11×10^{10}	2.01×10^{10}	1.55×10^{10}

Fig. 11.7 Revised Enzmann with expellant propellant tanks[KFL]

them will be required. Since the overall spacecraft design has been elongated, and we have adopted expellant propellant, we refer to this design as the *Long-Enzmann Slow Boat* as compared to the nominal *Enzmann Slow Boat* shown earlier.

11.3 Asteroid Ship

In the previous section we have discussed the possibility of constructing vessels that can carrylarge populations of people. In the future, with access to robotics, automation and artificial intelligence, all of this should be possible. However, there is another option that is worthy of consideration. Instead of constructing a vessel, we might instead take an existing object, such as a moderate sized asteroid, and simply create a habitat within its internal structure, as depicted in Fig. 11.8. This would require hollowing out a large chamber and pumping in an atmosphere. Then

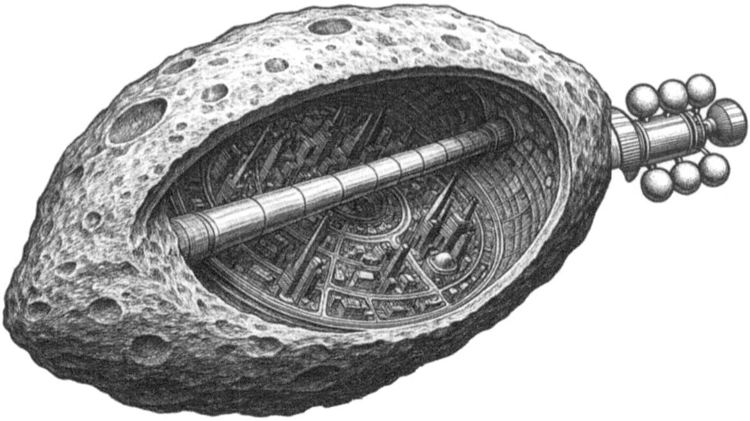

Fig. 11.8 Illustration of an asteroid starship under flight[KFL]

strapping engines to the back, it would be a self-contained structure that can travel the cosmos from star to star. Let us consider how this might work in practice.

One of the first to propose this type of space transportation system was Dandridge Cole, an American aerospace engineer from the 1960s [18] who referred to these as 'Bubble Worlds'. He proposed the selection of a metal rich asteroid, such as a nickel-iron object, and then hollowing it out using internal solar or nuclear heating. It would then be spun to provide artificial gravity for any crew. It would effectively become a generation ship capable of long duration space missions.

Cole was also building on the research of John Bernal [19] who had as early as 1929 proposed the construction of large hollow non-rotating spherical shells, now called Bernal Spheres, that could hold a population of 20,000–30,000 people. Bernal's research was also an influence on the space settlement studies at Stanford University in the 1970s [6]. The differences between this work and that of Cole's however was that he wanted to utilise the natural resources of space as colony ships.

In what follows, we will sketch out a concept for an asteroid starship making reasonable assumptions that the reader can follow. You are then invited to use this as a basis for designing an improved concept by using some of the calculations presented. We will start off with an assumed slightly oblate but spherical object, and since we will approximate it as a sphere, we will give it an outer diameter of 20 km (radius 10 km). If we assume a density of 3,000 kg/m^3 this would be a total volume of $V_o = 4.189 \times 10^{12}$ m^3, and a mass of $m_o = 1.257 \times 10^{16}$ kg. This is similar in scale to the mass of the Martian moon Phobos with its mass of 1.07×10^{16} kg and so this provides us with a visual comparison.

Now, let us further assume that a large group of autonomous drones (see Sect. 11.1) work fast and hollow out a large chamber within the asteroid, leaving behind a shell thickness of 100 m so that it has an inner diameter of 19.8 km (radius 9.9 km). Such an operation is depicted in Fig. 11.9. The total volume of this large cavity is 4.064×10^{12} m^3 so that the mass of asteroid left behind which forms the

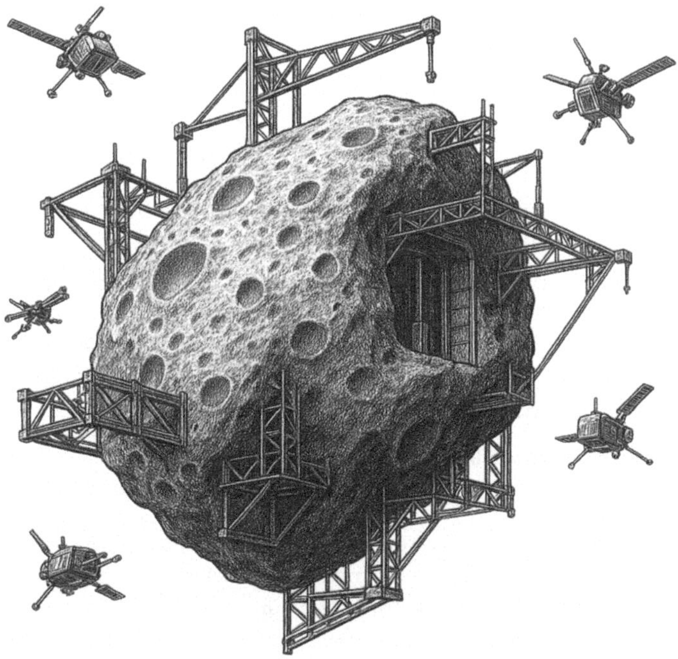

Fig. 11.9 Illustration of an asteroid starship under construction[KFL]

shell is 3.709×10^{14} kg, where the total mass removed is 1.219×10^{16} kg. This creates an open chamber area with a volume of 4.0644×10^{12} m^3.

Within this large cavity we will insert a large pressure vessel that is the artificial habitat, positioned at a much lower radius to facilitate space for mechanisms, engineering systems, structural stability and rotation. The contained internal city is assumed to be a sphere 10 km in diameter (radius 5 km), with a thickness on the order of a couple of metres but we will neglect that in the analysis. This is depicted in Fig. 11.10 and essentially, we have a rotating Bernal sphere contained within an asteroid. It would hold a population of order 10,000 people but allow margin for some minor population growth.

This internal habitat volume is essentially a self-contained pressure vessel pumped to 1 atmosphere or 101,325 Pa at a temperature of 20 °C (293.15 K) and with an air density of 1.204 kg/m^3. The air will be made up of 79.08% N_2, 20.95% O_2, 0.93% Ar and 0.04% CO_2. The volume of the air within this large pressure vessel will then be 5.2359×10^{11} m^3 corresponding to a mass of 6.304×10^{11} kg. Therefore, this breaks down into 4.922×10^{11} kg N_2, 1.321×10^{11} kg O_2, 5.863×10^9 kg Ar and 2.522×10^8 kg CO_2.

The reason we have set up the design this way is to avoid using the external rock itself as the outside surface of the pressure vessel, since we will need to rotate the habitat and it would not be a good idea to rely on the rock for tension and rotation.

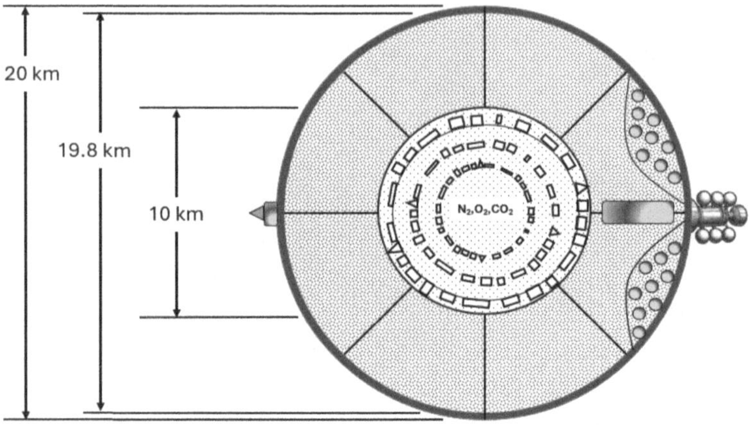

Fig. 11.10 Configuration of idealised asteroid starship geometry[KFL]

Instead, we are assuming the construction of a separate internal pressure vessel that will spin on large bearings for artificial gravity and then also use high strength metals or composites for the pressure structure. The outer asteroid surface will serve as a passive shielding material for radiation and micrometeoroids. The thickness of the outer shell will be given as a function of the internal pressure p, radius r and the allowable tensile strength σ as follows

$$\Delta x = \frac{P R_c}{2\sigma} \tag{11.5}$$

Although we have assumed 100 m thickness, we don't need this and a few meters of shielding material would have sufficed. But we are allowing for small impact erosion events and also sufficient material to position the structural ties and bearings that will be needed to hold the internal habitat in place. In addition, any large engineering systems can be placed within this cavity but also supplies. It would also be used to hold the propellant and engine systems. The extra-large thickness also facilitates local (and easily accessible) storage of a vital rock resource, which may be needed in the event of a construction project. The structural ties act as cable loops to carry the rotational loads efficiently.

In terms of the artificial gravity, we assume a revolutions per minute of 0.423 rpm so that there is an artificial gravity like the Earth at the outer edge of the habitat. This is partly chosen as it is well below the $\sim$1–2 rpm range where the Coriolis effect becomes noticeable to humans. We can then calculate the angular speed, acceleration at the outer rim of the habitat and the speed of rotation, as well as the rotational period as

$$\omega = \frac{2\pi \times 0.423\, rpm}{60} = 0.0443 \text{ rad/s}. \tag{11.6}$$

Then since

$$\omega = \sqrt{\frac{g}{r}}, \tag{11.7}$$

$$g_{rim} = \omega^2 R_c = (0.0443)^2 \times 5{,}000\ \text{m} = 9.81\ \text{m/s}^2 = 1g_o, \tag{11.8}$$

$$v_{rim} = \omega R = 0.0443 \times 5{,}000 = 221.48\ \text{m/s}, \tag{11.9}$$

$$T = \frac{60}{0.423\ \text{rpm}} = 141.84\ \text{s} = 2.36\ \text{minutes}. \tag{11.10}$$

These were calculated for the outer rim of the habitat. The gravity would reduce the closer you got to the center and so would the rotational velocity. At 1 km for example, the gravity would be 1.96 m/s^2 or 0.2g_o with a motion of 44.3 m/s. Since this gravity gradient might be a problem, it would be suitable to build structures in sets of two or three gravity rings so that when walking around the structure the gravity does not change, and it only changes when you walk along the radius of the spin axis.

Next, we must consider the Hoop stress (membrane pressure) on the pressure vessel that comprises the habitat at 1 atmosphere. For a spherical thin shell or order 1 m, the hoop stress is given as a function of the pressure, radius and thickness as

$$\sigma_p = \frac{pR_c}{2t}. \tag{11.11}$$

We therefore calculate a value per meter of wall thickness to be

$$\sigma_p = \frac{pR_c}{2} = \frac{101{,}325 \times 5{,}000}{2} = 253.31\ \text{MPa/m}. \tag{11.12}$$

We also calculate additional membrane stress associated with the rotation of the structure, and using a conservative thin-rim approximation this is given by

$$\sigma_{rot} \approx \rho_{mat}\omega^2 R_{rim}^2 = 78.49\ \text{MPa}. \tag{11.13}$$

Where we assume a shell material made up of Carbon Fiber Reinforced Polymer (CRFP) with a density of 1,600 kg/m^3. This justifies why it is better not to spin the entire rock but just an internal habitat section, since the tensile strength of natural rock is in the range ~2–20 MPa and so the rotation of the asteroid shell at the calculated membrane stress would cause it to be torn apart. Further, if we assume a habitat shell thickness of $R_{rim} - R_i = 0.5$ m then this computes to a shell mass of

$$m_{shell} = \frac{4\pi}{3}\rho(R_{rim}^3 - R_i^3) = 5.0255 \times 10^{11}\ \text{kg}. \tag{11.14}$$

Then if we total up the asteroid shell mass, atmosphere mass, habitat mass this comes to 3.72×10^{14} kg. Allowing for additional structures, fuel, water and all manner of things required by an independent space habitat we round this up to a total spacecraft mass of $m_{sp} = 3.8 \times 10^{14}$ kg, which leaves a total mass of 4.148×10^{12} kg for all additional systems. Next we must consider the propulsion of this large structure. We assume that the spacecraft is to move at a velocity of 100 km/s. The momentum changes for a mass of this size to achieve this velocity is given by

$$\Delta p = M_{sp} \times v = 3.8 \times 10^{19} \text{ Ns}. \tag{11.15}$$

Then the total impulse required will be a function of the momentum change and the time to accelerate up to the required speed. That for an acceleration time of 1 year it will be given by

$$F = \frac{\Delta p}{t_b} = \frac{3.8 \times 10^{19}}{3.156 \times 10^7} = 1.205 \times 10^{12} \text{ N}. \tag{11.16}$$

This will require an average acceleration of

$$a = \frac{v}{t_b} = 0.00317 \text{ m/s}^2. \tag{11.17}$$

We will now calculate some basic rocket propulsion requirements to get to the 100 km/s. For this we will assume an optimistic exhaust velocity of 10,000 km/s for a fusion fuel. The mass ratio is given from the ideal rocket equation as

$$\frac{M_o}{M_f} = e^{\Delta v/v_{ex}} = e^{10^5/10^7} = e^{0.01} = 1.01 \tag{11.18}$$

This suggests that to achieve the required velocity we will need $\sim$1% of the total spacecraft mass, which is

$$M_{prop} = (e^{0.01} - 1) \times M_{sp} = 3.8191 \times 10^{12} \text{ kg}. \tag{11.19}$$

This is a significant portion of our spacecraft mass. We might also ask how much energy it would cost for a mission like this, to accelerate the total mass of spacecraft at the required velocity. This is given by consideration of the kinetic energy relation

$$E = \frac{1}{2} M_{sp} \Delta v^2 = 1.9 \times 10^{24} \text{ J} \tag{11.20}$$

For interest, had we assumed a less optimistic exhaust velocity such as $v_{ex} = 1{,}000$ km/s, then we would have required $e^{0.1} = 1.105$ or $\sim$10.5% of the total spacecraft mass in propellant, which emphasises the advantage of a better fuel choice and a higher exhaust velocity. So we see that the key impact of the choice of exhaust velocity is the quantity of propellant required.

Let us consider how all this fits within a mission context. For this we examine how long it would take such a spacecraft to reach the nearest stars at 4.24 ly away for Proxima Centauri, which is a distance of 4.01×10^{13} km. For the speed of 100 km/s (0.03%c) this is equivalent to a travel time to the nearest stars of

$$t = \frac{d}{v} = \frac{4.01 \times 10^{13}}{100 \times 1 \text{ Year}} = 12{,}720 \text{ years.} \tag{11.21}$$

This is a long time and certainly a multi-generation ship with its low speed. Indeed, it is approximately equal to the timescale of known human history in terms of the start of our civilization. As we have seen, improving the exhaust velocity does not alter the trip time, but it does reduce the total propellant requirement for a constant mass mission. To reduce the trip time, we need to alter the velocity increment achieved (which does depend on the exhaust velocity). For example, if we aim for a velocity of 1,000 km/s, then this leads to a total trip time of ~1,272 years. If we can go as high as 3,000 km/s (~1%c) then the trip time drops to ~424 years. If the velocity was 10,000 km/s then the trip would take around ~127 years.

Next we consider the nuclear pulse specification. Since we have a defined total propellant mass and this is to be used in 1 year, this implies a mass flow rate of

$$\dot{m} = \frac{M_{prop}}{t_b} = \frac{3.8191 \times 10^{12} \text{ kg}}{3.1536 \times 10^7 \text{ s}} = 1.211 \times 10^5 \text{ kg/s} \tag{11.22}$$

This is ~121 tons/s which is large and represents some form of large detonation nuclear pulse engine, similar in scale to an Orion class propulsion system. The units are effectively hydrogen bombs of some description. Let us further assume that each unit has a mass of 1 ton, then we can estimate the required pulse frequency and the total number of units required for the burn

$$f_{hz} = \frac{\dot{m}}{m_{unit}} = \frac{1.211 \times 10^5 \text{ kg/s}}{1{,}000 \text{ kg}} \approx 121 \text{ Hz}, \tag{11.23}$$

$$N = t_b f_{hz} = 3.1536 \times 10^7 \text{ s} \times 121 \text{ Hz} = 3.819 \times 10^9 \text{ units.} \tag{11.24}$$

Table 11.5 shows alternative pulse frequencies that could be adopted in terms of the per unit mass, pulse frequency and acceleration times. There is substantial

Table 11.5 Specification of nuclear pulse propulsion frequency (f_{hz}) for the asteroid starship

Unit mass (kg)	t_b 0.5 Years	t_b 1 Years	t_b 2 Years	t_b 3 Years	t_b 4 Years	t_b 5 Years	t_b 10 Years
1,000	242	121	61	40	30	24	12
100	2,422	1,211	605	404	303	242	121
10	24,220	12,110	6,055	4,037	3,027	2,422	1,211
1	242,206	121,103	60,551	40,368	30,276	24,221	12,110

Table 11.6 Specification of asteroid starship to proxima centauri

Parameter	Value
Diameter asteroid (km)	20.0
Diameter internal habitat (km)	10.0
Spin (rpm)	0.423
Rotation time (min)	2.36
Gravity at rim (m/s^2)	9.81
Mass atmosphere (kg)	6.304×10^{11}
Mass rock shell (kg)	3.709×10^{14}
Mass habitat shell (kg)	5.025×10^{11}
Mass propellant (kg)	3.8191×10^{12}
Mass additional (kg)	4.148×10^{12}
Mass total (kg)	3.8×10^{14}
Hoop stress at rim (MPa)	78.49
Exhaust velocity (km/s)	10,000
Acceleration (m/s^2)	0.0032
Acceleration time t_b (Years)	1.0
Acceleration distance (AU)	10.54
Impulse (Ns)	3.8×10^{19}
Mass flow rate (kg/s)	1.211×10^{5}
Unit mass (kg)	1,000
Number units	3.819×10^{9}
Pulse frequency (Hz)	123
Thrust (TN)	1.211
Jet power (PW)	6,055
Velocity (km/s)	100
Mission distance (ly)	4.24
Total trip time (years)	12,689

scope for trading between parameters to help redefine the mission, and again this is left as an exercise for the reader who may want to attempt their own asteroid ship design. From the table we can see that there are no easy answers to making the propulsion environment less extreme. In some cases, the required pulse frequency is so high that it is tending towards a continuous propulsion engine. It is likely that for very large mass spacecraft, to minimise the per unit mass and pulse frequency it would therefore be suitable to parallise the thrust profile, which means more than one engine. We examine such an example in the next chapter. Note that in the above we have not considered deceleration and only looked at a straight-line flyby mission. It is left to the reader as an exercise to estimate the propellant mass required for a fully decelerated mission into orbital velocity around the destination star. We could also go much further in calculating the design using the knowledge from the various chapters of this book but again this could be an exercise for you. Table 11.6 shows a summary of this civilization-timescale interstellar spaceflight mission, where we also show other propulsion parameters relevant to the performance.

Problems of Chap. 11

11.1 In the first section of this chapter, we introduced the asteroid strip-mining equation and showed an example calculation. Imagine we want to strip mine material off the moon of Mars called Phobos which has a mass of 1.07×10^{16} kg, and we want to know how long it would take to strip mine 0.1% of its mass using 1 billion drones (i) Estimate the time in days to mine this object and also state your answer in years (ii) Estimate the throughput rate. In your calculations assume an efficiency of 90% but for all other numbers assume the values given in the example case study of Sect. 11.1.

11.2 An expellant based propulsion system has a capsule mass of 0.05 g and an expellant that is 50 times this (i) Assuming 8 engines and a pulse frequency of 100 Hz per engine what would be the mass flow rate per engine and total? (ii) If the boost time was 1 year for the acceleration of the vehicle, how much fusion fuel mass and how much expellant mass would be required?

11.3 In Sect. 11.3 we designed an asteroid starship and calculated the membrane stress associated with the rotation of the structure assuming CFRP. (i) Calculate the stress but assuming steel ($\rho = 7,850\,\mathrm{kg/m^3}$), Aluminium alloy ($\rho = 2,700\,\mathrm{kg/m^3}$ and Titanium alloy ($\rho = 4,430\,\mathrm{kg/m^3}$). (ii) Recalculate but assuming a larger radius of 10 km for the same rotational period (iii) For the case of CRFP, but assuming $\omega = 0.0886$ rad/s for a 10 km radius outer ring, what would be the new value of the membrane stress?

References

1. Hunter MW (1966) Thrust into space, Holt. Rinehart and Winston Inc., Austin
2. Ashworth S (2012) The emergence of the world-ship (I): the shift from planet-based to space-based civilisation. JBIS 65:140–154
3. Ashworth S (2012) The emergence of the world-ship (II): a development scenario. JBIS 65:155–175
4. Ceyssens F, Driesen M, Wouters K (2012) On the organization of world ships and other Gigascale interstellar space exploration projects. JBIS 65:4
5. Vermeulen A, et al (2019) Simulating the construction of conceptual space architectures to explore the potential of combined asteroid mining and space-based 3D manufacturing. In Proceedings of the international astronautical congress. IAC, New York City
6. O'Neill GK (2000) The high frontier: human colonies in space. Space Studies Institute, Space Frontier Foundation. Apogee Books, Burlington (originally published 1976)
7. Enzmann R (1966) Rocket science and fact-based science fiction. In Planetology and space mission planning conference. New York Academy of Sciences, New York
8. Stine GH (1973) A program for star flight. Analog XCII(2):10–27
9. Schroder T (1977) Slow boat to Centauri. Astronomy 6–15
10. Crowl A, Long KF, Obousy R (2012) The Enzmann starship: history & engineering appraisal. JBIS 65(6):185–199
11. Long KF (2024) Mission architecture calculations for the Enzmann interstellar spacecraft concept. JBIS 77(1):226–232
12. Dyson FJ (1968) Interstellar transport. Phys Today 21(10):41–45

13. Bond A, Martin AR (1978) Project Daedalus: the mission profile. Final study report. JBIS Special Supplement:S37–S42
14. Martin AR, Bond A (1978) Project Daedalus: the propulsion system part 1: theoretical considerations and calculations. JBIS Special Supplement:S44–S62
15. Palaszewski B (2012) Atmospheric mining in the outer solar system: resource capturing, storage and utilization, AIAA 2012–3742. In Presented 48th AIAA/ASME/SAE/ASEE joint propulsion conference & exhibit, Atlanta, GA, 30 July–01 August
16. Sheffield J, Spong DA (2016) Catalyzed D-D stellarator reactor. J Fus Sci Technol 70:36–53
17. Long KF (2022) On the application of pulse propulsion frequency in inertial fusion space mission design. JBIS 75(4):127–141
18. Cole DM, Cox DW (1964) Islands in space: the challenge of the planetoids. Chilton Books, Boston
19. Bernal JD (2018) The world, the flesh and the devil: an enquiry into the future of the three enemies of the rational soul. Verso Books, London (originally pulished in 1929)

Chapter 12
Massive World Ships

Imagination will often carry us to worlds that never were, but without it we go nowhere.

—Carl Sagan, 1980 [1]

12.1 Background to World Ships

If humanity is to truly become a starfaring species, we must consider the option to construct much larger vessels that can carry enormous populations of people. These are self-made worlds that are fully independent and sustainable. Such concepts were mentioned earlier in the world ship designs by Bond and Martin [2, 3]. This work has recently been advanced by this author in the publication of two papers which go into more detail in exploring these concepts but also within an ICF propulsion paradigm [4, 5].

A world ship is a massive spacecraft that can travel to the stars carrying a human crew in trip times which last centuries to millennia and in any case involve multiple generations of human beings [6]. The people live out their lives, perform some function on-board the ship, and will die on it. Most of the population that lives on the world ship during its journey to the stars will never see the planet of origin (Earth) or the planet of the destination (exoplanet) and so they must be content to live in this alien environment with stories of Earth history rapidly becoming ancient mythology.

A good introduction to the subject of generation ships in general is presented by others [7, 8]. A set of papers relevant to world ship studies was published by the European Space Agency covering subjects as diverse as artificial gravity, growing plants and life support systems [9–12]. This author had previously worked with students from the International Space University on the instigation of a world ship project called *Astra Planeta* which is worth a read [13].

In terms of the propulsion of a world ship there are many ideas by which this may be achieved. The most obvious approach is via nuclear pulse propulsion [14, 15]. But there are alternative ways such as in the use of non-nuclear Arks which have

K. F. Long, *Stellar Engines*, Springer Aerospace Technology,
https://doi.org/10.1007/978-3-032-18062-9_12

also been examined by the physicist Gregory Matloff [16–18]. In this chapter we are interested in the propulsion of a world ship via a nuclear fusion engine.

This involves a vessel travelling at 1,500 km/s on a 1,000-year trip time between stars. This will be a massive world ship driven by ICF engines. The main purpose of the study was to see if an architecture could be constructed that utilised this propulsion method and therefore what the likely mission parameters would be.

In earlier work Bond and Martin [2, 3] generated several world ship models for both dry and wet (internal ocean) vessels. These were essentially large habitats capable of sustaining a massive population of people over multiple generations. In this work we are attempting to replicate the Bond/Martin world ship dry model referred to as the Mark 2A. This was a giant cylinder 87.5 km in length and with a radius of 7.24 km. It would have a wall thickness of 2.88 m and a lateral surface area of 3,980 km^2.

Various other parameters for the Mark 2A design were defined including a wall mass of 1.04×10^{11} tons, regolith mass of 4.975×10^{10} tons, atmosphere mass of 1.403×10^{10} tons, making up the total habitat mass of 1.68×10^{11} tons. The design also specified a tankage mass of 3.09×10^{10} tons, propellant mass of 7.797×10^{11} tons, nuclear fuel mass of 3.696×10^{10} tons. This was obviously utilising nuclear pulse propulsion and so the nuclear fuel refers to the material associated with nuclear units, with each pulse unit having a staggering mass of around 2,600 tons.

Such a concept was explored as a part of the historical Project Orion which was conducted on a theoretical and experimental basis [14, 15]. For the calculations shown in this chapter we do not adopt this specific propulsion technology, although the technology would appear to be feasible for this application. Although we assume the same radius, we will assume that the world ship configurations generated in this work, which we call the Mk2A(ii) is 115 km in length: the main distinction being the propellant requirements and capsule mass size.

There are many issues with the idea of a world ship, such as the viability of population survival and genetic make-up, the sourcing and management of appropriate soils for plant growth and oxygen production [19]. There is also the potential for the break-out of a virus which must be contained, and others have written about this in the context of a potential meeting with extraterrestrials and a definition of protocols [20]. There is the policing of a social system that must balance requirements for democracy with the need for strong determined leadership with a clear chain of authority [21]. There is also the need to protect the human occupants from any excessive doses of cosmic rays and solar flares, and this usually implies the need for radiation shields utilising passive (mass shielding) and active (electromagnetic fields) methods [22].

The number of resources for each person required in a closed ecosystem is something that must be accounted for. This includes quantification of the atmosphere, animal and agricultural biomass, waste tanks and storage. In the studies by Richards [23] it was found that the minimum number was around $\sim$5 tons/person.

Numerical studies by others suggest that for a crew of 500 people living on an omnivorous balanced diet, around 0.45 km^2 of artificial land would be required to support all the necessary food growth and where the due to the genetic diversity the crew remains stable over time [24, 25]. This would represent around 0.001 km^2/person, or an agricultural area that is 30 m by 30 m each. For a world ship population of 1–10 million this would scale to a land area of 1,000–10,000 km^2 and so the total internal area of our world ship must exceed this size. The world ship for this study has an lateral surface area of $2\pi \times 7.24 \times 115 = 5{,}231\,km^2$ and an internal volume of $\pi \times 7.24^2 \times 115 = 18{,}938\,km$.

In studies for interstellar arks others found that for a crew of around 1,100 people they would require $\sim 10^8$ litres of Oxygen and $\sim 10^6$ litres of water per year [26]. For a population of 1–10 million over a 1,000-year timeline as proposed for the world ship studies here this scales to an Oxygen requirement of $\sim 10^{13} - 10^{14}$ kg and a water requirement of $\sim 10^{11} - 10^{12}$ kg. Given Oxygen is 21% of a breathable atmosphere this implies $\sim 2.1 \times 10^{12} - 10^{13}$ kg. For the Bond/Martin world ship which is the basis of our studies it had a total atmosphere requirement of $\sim 10^{13}$ kg [3].

In terms of the mission profile the Mark 2A would accelerate at 0.001 m/s^2 and reach a cruise velocity of 1,500 km/s or 0.5%. The total time spent on the acceleration and deceleration period was to be 47.56 years (<5% of the total mission time). The exhaust velocity would have been 1,880 km/s. The engine power was to be 2.6×10^6 TW and the propellant mass flow 5.205×10^5 kg/s. For an optimum mass ratio of $R = 4.9$ the total time to complete the mission would be around 1,000 years.

12.2 Scoping the Design

The Mk2A had a total length of 87.5 km and a cylindrical radius of 7.24 km. Although we assume the same radius we will assume that the world ship configurations generated in this chapter, which we call the Mk2A(ii) is 115 km in length: the main distinction being the propellant requirements and capsule mass size. The Mk2A(ii) will be discussed, and we call this vessel *Ra* after the Egyptian God of the Sun.

Whilst it may not be practical to completely match the original Mk2A design since the operation of the engine involves critical differences, the aim of this work is to reproduce the mission profile and assume elements of the vehicle structure so that an appropriate ‘apples to apples’ comparison can be made. This will help to highlight the utility (or lack of) for either the external nuclear pulse propulsion or ICF propulsion for this specific application. It was chosen to aim for the following requirements:

1. Near identical total dry spacecraft mass, m_{dry}.
2. Final cruise velocity on the order of 1,500 km/s (0.5%c), v_c.
3. Total trip time to not exceed 1,000 years between two stars on average, t_t.
4. Equivalent total burn/deburn time of around 50 years or less, t_b.
5. Equivalent exhaust velocity 1,880 km/s, v_{ex}.
6. Optimum total mass ratio 4.9 for minimum mass mission (burn + deburn), R.

The model was constructed by making some of the same assumptions of the original study so there is some level of consistency. Next, the goal was to re-configure the design to arrive at a more satisfactory vehicle architecture in terms of the propulsion requirements. Finally, the goal was to reach some conclusions about the engineering plausibility of using ICF technology to propel a world ship. This analysis will be conducted at the system level only and we do not go into sub-system engineering details which would require a more in-depth study. Although to illustrate potential areas where design improvements can be explored by others will be shown.

For the concept studied in this chapter we adopt hemispherical reaction chambers like that proposed for the Daedalus study [27]. These are surrounded by an array of energy drivers (i.e. lasers), and the capsules are fed into the target chamber from a containment tank. However, it will be apparent that one of these engines would facilitate too low a mass flow rate to produce adequate thrusting of the massive world ship in a timely manner.

To illustrate this some calculations were conducted to estimate using a single engine and with the Daedalus reaction chambers how high a pulse frequency would be required to generate the same thrust level of 9.7854×10^{11} N. In this calculation, following on from the world ship studies we assume a total thermonuclear fuel mass of 3.696×10^{13} kg and a total expellant propellant mass of 7.797×10^{14} kg, leading to a ratio of $M_{fuel}/M_{prop} = 21.09$. We also assume an exhaust velocity of 1,880 km/s.

For the Daedalus first stage reaction chamber this involved a capsule mass of 2.84 g, burn-up fraction of $\phi = 0.175$ and implied a jet efficiency of $\eta_j = 0.91$. The resulting pulse frequency requirement using a single engine would be 183 million Hz. For the Daedalus second stage reaction chamber this involved a capsule mass of 0.288 g, burn-up fraction of $\phi = 0.133$ and implied a jet efficiency of $\eta_j = 0.901$. The resulting pulse frequency was 1,800 million Hz. This is clearly not practical and is more like a continuous thrust engine rather than a pulse one.

A clear conclusion then is that if we are to push a world ship with ICF engines then many engines will be needed and in a parallel thrust arrangement. Layers of insulation and radiation shielding partitions can be constructed around the individual engines to mitigate any radiation dose. In addition, smart pellet designs can be adopted so that much of the emitted neutrons and x-rays are absorbed within a denser material layer at the outer shell of the thermonuclear fuel. The reaction chamber size is a function of the ICF pellet size and so a larger pellet would be accompanied by a larger reaction chamber, which then complicates calculations for mass budget.

Figure 12.1 illustrates what the reaction chambers might look like on the actual Mk2A(ii) *Ra* world ship, although missing from this image is a likely requirement for partition shielding between each chamber to mitigate a possible radiation coupling effect between each reaction chamber. Not including them in the design may result in a larger jet efficiency beyond the $\eta_j \approx 0.9$ assumed in this work, such as reducing it to $\eta_j \approx 0.8$ but we do not make this assumption and instead operate on the principle that appropriate radiation partitions are in place.

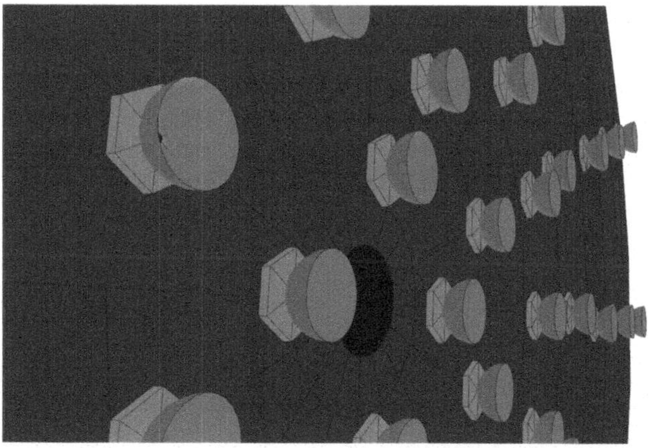

Fig. 12.1 Illustration of engine reaction chambers as adopted on the Mk2A(ii) *Ra* world ship[KFL]

We also note that each reaction chamber has a diameter of 50 m (slightly larger than the Daedalus second stage engines). The full diameter of the engine array would be 650 m, 1,050 m and 1,250 m for 16, 32 and 64 engines respectively. Yet, considering that the world ship cylinder has a radius of 7.24 km or a diameter of 14.48 km, this still only represents 4.5%, 7.2% and 8.6% of the total disc size for the three configurations respectively. The individual engines are spaced out and separated by insulation layer shields to minimise radiation coupling exposure due to the proximity of each engine. In particular, the radiation refers to the high energy neutrons and x-rays released from the capsule detonations.

However, as will be shown even this number of engines is not likely sufficient to push the world ship. Yet these engine configurations were discussed to illustrate the various issues with parallel thrust arrangements. There are essentially two options going forward: Firstly, one must use large capsule masses with a minimum number of parallel staged engines, but in the limit of very large capsule masses these tend towards a nuclear pulse engine like Orion. Second, instead one may go to much smaller capsule masses, but this implies a vast number of engines to match the thrust profile.

For the Bond/Martin world ships the assumption was that the pulse units were mostly contained of deuterium based fusion fuels. However, for the ICF driven world ship this is somewhat of a problem since the specific energy fraction of deuterium is very low, on the order of 0.792×10^{11} J/g for the D(D,^{3}He)n reaction ($\alpha = 0.0009$) and 0.977×10^{11} J/g for the D(D, T)p reaction ($\alpha = 0.0011$). This compares to the situation of using D(T,^{4}He)n at 3.408×10^{11} J/g ($\alpha = 0.0038$) or the mostly aneutronic reaction D(^{3}He, ^{4}He)p at 3.556×10^{11} J/g ($\alpha = 0.0039$). Whilst the DT reaction is the easiest to ignite in terms of reactivity as a function of temperature it must also be manufactured. The D^{3}He reaction is the hardest to ignite but produces the most energy, although it is less abundant as a resource.

In the calculations below we therefore assume that in the future it will be possible to manufacture high performance fusion fuel capsules that utilise an efficient energy conversion reaction on the order of $\alpha = 0.0039$. This may be a DD fuel that is mixed with a DT or D^3He fuel and it is likely manufactured in-situ on the world ship commensurate with its utility since otherwise any tritium will decay and release heat. This therefore also implies a relatively low to moderate pulse frequency of detonation.

In the calculations presented below an ICF capsule mass is derived to be on the order of 230 grams and this is based on an assumed $\alpha = 0.0039$. To illustrate, had we instead assumed a lower value of $\alpha = 0.00174$ (such as tending more towards a pure DD fuel) in order to get the same capsule mass this would require the impractical conditions of a unity burn-up fraction ($\phi = 1.0$), unity jet efficiency ($\eta_j = 1.0$) and a unity mix coupling efficiency ($\epsilon_{mix} = 1.0$).

We therefore imagine some very smart ICF capsule design configuration (i.e. utilising novel methods [28, 29]) which contains multiple shells which is made up of a mixture of different materials dominated by DD but to include layers of DT and D^3He such that a high specific energy fraction as assumed in this study is possible such that we can assume $\alpha = 0.0039$.

One of the methods by which this could be achieved is through catalysed D-D reactions, where the fusion products of tritium and helium-3 are recycled. Such a method was proposed in research for a stellarator reactor [30]. In this way a high abundance fuel of Deuterium is used but giving rise to similar output energies of the higher reactions such as expected with DT or D^3He. Another method may even include the possibility of seeding via antimatter catalysed fusion initiation as discussed by others [31, 32]. Had we not made this assumption then it would be difficult to achieve the required mission profile with such a low exhaust velocity.

Alternatively, a higher exhaust velocity can be assumed but this would suggest a lower degree of thermonuclear fuel to expellant mixture coupling and so therefore a reduced amount of mass flow rate augmentation and thrust. It was chosen to keep the exhaust velocity to around 1,880 km/s to facilitate a comparison to the Bond/Martin world ship studies [3].

We do not underestimate the vast challenge of mining such a large quantity of fusion fuel at a total mass of 3.696×10^{10} tons. Even if this was mined over 100 years prior to the trip it would have to be done so at a rate of 11.72 tons/s which clearly requires autonomous robotics operating on a large scale. This also implies that all destinations for the world ship should have a gas giant to facilitate mining for the next leg of the journey.

Since these calculations are novel, it is critical to derive a minimum mass requirement for the pulse units used in the world ship since up to now this has been a major uncertainty. On the one scale if they are too high in mass (i.e. >1 kg) then the system is tending towards a nuclear pulse engine and using very high laser drivers on such large capsules is unlikely to be as efficient as an internal detonation process. On the other scale, if they are too low in mass (i.e. $<$mg) then the energy generated will be inadequate to drive the vehicle in any timely manner, and particularly since the mass of the world ship is so large and therefore difficult to move without sufficient thrust.

Further, as shown earlier with very low mass capsules a vast number would be required and detonated at a very large pulse frequency (i.e. $>> 10^3$ Hz) which would tend towards a continuous pulse propulsion system; such a system would likely not be as efficient as say using a magnetic fusion engine on continuous mode. So, the interest here is in deriving an appropriate minimum mass that is mid-way between these two extremes.

To illustrate the problem let us assume that the capsule mass will be approximately in the range $1 \leq m_{cap}(g) \leq 1000$ and with an assumed the pulse detonation frequency of $\sim$1–100 Hz the mass flow rate of fusion fuel in the range $0.001 \leq \dot{m}_{fus}$ (kg/s) ≤ 100. The exhaust velocity is then approximated by $v_{ex} = (2\alpha\phi)^{1/2}c$.

So that on the assumption of a maximal burn-up fraction ($\phi = 1.0$) a maximum theoretical exhaust velocity range (neglecting losses) to be expected is on the order of 26,135 km/s (DT), 12,719–14,061 km/s (DD), 26,477 km/s (D^3He). Since the thrust $T = \dot{m}v_{ex}$, this implies a thrust in the range $0.001 \leq$ T (GN) ≤ 2.65; where we assumed that the burn-up fraction $\phi = 1.0$.

If a world ship had a dry mass on the order of $\sim 10^{11}$ tons, and taking the thrust of 2.65 GN, from $T = ma$, we find an acceleration of order $\sim 2.65 \times 10^{-5}$ m/s^2, which via $v = at$ would imply a minimum trip time to accelerate to 1,500 km/s on the order of of $\sim$1,796 years which exceeds a millennium.

These results can be combined and expressed as minimum world ship equations when driven by fusion propulsion for the minimum acceleration time t_b and the distance S achieved under acceleration given a ship mass M_{ship} and desired cruise velocity v_c as follows

$$t_b = \frac{v_c}{a} = \frac{v_c m_{ship}}{m_{cap} f_{hz} (2\alpha\phi)^{1/2} c}. \tag{12.1}$$

Thus, it can be observed that the boost period of the journey will be minimised with high capsule mass, high burn-up fraction, high pulse frequency. It will also be minimised the lower the ship mass and the lower the cruise speed one is trying to achieve; although a lower cruise speed comes at the cost of increased cruise time to the destination target since $t_{tot} = t_b + t_c$ where $t_c = S_c/v_c$. Alternatively, another way to write Eq. (12.1) is as a starship design function with the different design elements separated out for the capsule (cap), fuel (fuel), spacecraft (sc) and mission (mis)

$$t_b = \left[\frac{1}{m_{cap}\phi^{1/2}}\right]_{cap} \times \left[\frac{1}{(2\alpha)^{1/2}}\right]_{fuel} \times \left[\frac{m_{ship}}{f_{hz}}\right]_{sc} \times \left[\frac{v_c}{c}\right]_{mis}. \tag{12.2}$$

Similarly, one can derive the equation for the distance achieved under boost by substitution of Eq. (12.1) into $S = at_b^2/2$ so that

$$S_b = \frac{a}{4\alpha\phi f_{hz}^2}\left(\frac{m_{ship}}{m_{cap}}\right)^2\left(\frac{v_c}{c}\right)^2 = \frac{m_{ship} v_c^2}{2(2\alpha\phi)^{1/2} f_{hz} m_{cap} c}. \tag{12.3}$$

Table 12.1 Calculations of world ship burn time (yrs) and burn distance (ly) using Eqs. 12.2 and 12.3, under varying pulse frequency and capsule mass for an exhaust velocity v_{ex} = 1,880 km/s and $\alpha = 0.0039$

Capsule mass (kg)	1 Hz t_b (yrs)	1 Hz S_b (ly)	10 Hz t_b (yrs)	10 Hz S_b (ly)	100 Hz t_b (yrs)	100 Hz S_b (ly)
0.001	179,645,626	449,128	17,964,562	44,913	1,796,456	4,491
0.01	17,964,562	44,913	1,796,456	4,491	179,646	449
0.1	1,796,456	4,491	179,646	449	17,965	44.91
1.0	179,646	449	17,965	44.91	1,796	4.49
10.0	17,965	44.91	1,796	4.49	179.65	0.45
100.0	1,796	4.49	179.65	0.45	17.96	0.04
1000.0	180	0.45	17.96	0.04	1.79	0.004

Then by the end of this acceleration period the vessel would have reached around ~4.3 ly which is effectively the entire interstellar journey under acceleration. Yet, this was under the assumption of the maximum theoretical exhaust velocity and so in reality the journey would take much longer.

These equations are derived for the non-relativistic limit only and any significant velocity increase beyond the world ship regime (~0.005–0.03c) would require the velocity be written with relativistic corrections. In addition, the above equations assume a ship mass that is non-changing, when as fuel is used there would be mass loss. This assumption is appropriate since the vessel is so enormous in comparison to the mass loss throughout the burn period. However, a proper analysis of this would require the use of a logarithmic distance equation which takes account of the changing mass ratio (as described at the end of Chap. 1). Since this section is just an attempt to illustrate the problem with designing a suitable world ship driven by ICF, in this author's opinion this approximation is appropriate and serves the purpose.

The calculation results for a range of parameters around these assumptions is shown in Table 12.1. These calculations were performed for a 10^{14} kg ship mass, accelerating up to 1,500 km/s (0.005c) assuming $\alpha = 0.0039$. The calculations span the range of relatively low capsule mass (e.g. ICF) to very large capsule mass (e.g. nuclear pulse units), and from relatively low pulse frequency (e.g. 1 Hz) to moderately high pulse frequency (e.g. 100 Hz).

Examining the case study for 100 Hz with 10 kg capsule mass the calculated burn time is 180-years and the burn distance is 0.45 ly. This means that for a destination target of 4.3 ly, there would still be an additional 3.85 ly to travel at the cruise velocity of 0.005c which would take a further 770 years. Therefore, the total trip time would be 949 years. This is close to the goal of a 1,000-year trip time. The problem is this is on the unrealistic assumption of a 10 kg capsule mass which is not practical and likely to be more within the operational regime of nuclear pulse propulsion. This implies that if we are forced to use a lower capsule mass then the total trip time would be outside of the 1,000-year target. Overall, the results show that to achieve the 1,000-year mission with a cruise speed of 1,500 km/s, would likely require a capsule mass exceeding $\sim$100 g and at a pulse frequency approaching $\sim$100 Hz.

We highlight that in the calculations that follow we will adopt a capsule mass of around 230 g which is still quite a large mass but is much less than 10 kg. Yet, for a 100 Hz pulse frequency this would still correspond to a burn time of $t_b = 7{,}811$ years or a burn distance of $S_b = 195\,ly$ and so would not be practical for missions that are on the order of $5\,ly$ over a millennium. This simple analysis immediately suggests that the thrust must be augmented in some way to boost the mass flow rate and thereby allow for a much larger thrust. This would then have the effect of reducing the acceleration time to the required cruise velocity and therefore also the trip time. This will be addressed through two additions to the propulsion system.

The first is the need for many individual engines for the purpose of parallel thrust generation since the spacecraft is so massive. The second is the use of expellant propellant to augment the thrust by increasing the mass flow rate, where it is possible that the expellant may be an order of magnitude or more higher in mass than the thermonuclear capsule mass (in order to put it back into the mission profile regime of the mission scenario discussed above); although the injection of expellant mass comes at the cost of an increase in the mass fraction as well as a reduction in the exhaust velocity since it costs (thermonuclear) energy to heat up the expellant.

As an example, we can use this to compute the mass of the Daedalus minimum capsule mass required [27]. For the first stage we assume $\dot{m} = 0.71$ kg/s, $v_{ex} = 10,600$ km/s, $\phi = 0.175$. This leads to a minimum capsule mass on the order of 2.84 g. For the second stage we assume $\dot{m} = 0.072$ kg/s, $v_{ex} = 9,210$ km/s, $\phi = 0.133$. This leads to a minimum capsule mass on the order of 0.288 g. For both stages we assumed $n_{eng} = 1$, $\eta_j \approx 0.91$, $\alpha = 0.0039$ (for D^3He). It is also useful to note that the mass flow rate can also be written as the ratio of the total propellant mass m_{prop} to the burn time t_b as $\dot{m} = m_{prop}/t_b$.

On the assumption of a 10^{-3} m/s^2 acceleration as specified for the Mk2A world ship design, and a total mass on the order of 1.01×10^{12} tons this suggest a thrust requirement on the order of $\sim$1,000 GN. Multiplying this by the cruise velocity of 1,500 km/s wet get an estimate for the likely minimum power requirement which is 1.51×10^{18} W. However, if this calculation is based only on the 1.68×10^{11} tons habitat mass then it is reduced to a requirement of 2.52×10^{17} W.

As mentioned earlier, the original Bond/Martin concept employed a combination of fusion fuel and hydrogen propellant. This led to a combined mass flow rate of 5.205×10^5 kg/s, and with an exhaust velocity of 1,880 km/s implies a thrust of around 978.54 GN. This cannot be exactly replicated since we are looking at a different type of propulsion system, however we can adopt an analogy to this.

Since we are attempting to produce an ICF version of the Mk2A world ship, it is the intention to match the mission profile but also the mass flow rate, thrust, and overall systems masses, at least to a close approximation to facilitate a fair comparison. The mission profile is a cruise speed of 1,500 km/s (0.5%c) in 1,000 years with the acceleration and deceleration time being around 50 years or less (47.56 for the Bond/Martin world ship). The total mass flow rate was 520,500 kg/s with an exhaust velocity of 1,880 km/s leading to a thrust of 9.7854×10^{11} N.

The mass ratio was set to be around $R = 4.93$ based on the work of Shepherd [33] and accounting for both acceleration and deceleration. Since the cruise velocity to be attained is 0.5%c or 1,500 km/s this implies that for a decelerated rendezvous mission the total velocity increment to be achieved is 1%c or 3,000 km/s. So that indeed the mass ratio is given as $R = e^{(3{,}000/1{,}880)} = 4.93$.

We assume a total habitat mass of 1.6778×10^{11} tons. The nuclear component of the Bond/Martin design was 3.696×10^{10} tons and we assume this is equivalent to our thermonuclear fuel. The propellant component of the Bond/Martin world ship was 7.797×10^{11} tons and we assume this is equivalent to expellant propellant used for thrust augmentation. This is material injected into the exhaust stream with each thermonuclear capsule detonation and provided they are thermalised the products attain the same debris velocity.

The ratio of the expellant to thermonuclear mass will be such that $m_{exp} = 21.096 m_{cap}$ and we can then adopt this ratio in terms of how much quantity of expellant is to be injected per capsule. This helps to preserve a mass matching to the Bond/Martin world ship.

The capsule mass will be defined (total for all engines) by the chosen pulse frequency of detonation f_{hz} since $\dot{m} = (m_{cap} + m_{exp}) f_{hz}$. Therefore, if we assume the number of engines is $N_{eng} = 1{,}024$ (described below), then $\dot{m}/f_{hz} = 508.3$ kg/pulse. Then assuming $f_{hz} = 100$ Hz, this then implies that $m_{cap} = \dot{m}/[f_{hz}(1 + 21.096)] = 230$ g per engine. This then implies an injected expellant mass of 4.853 kg per pellet.

$$m_{cap} = \frac{\dot{m}}{22.097 n_{eng} f_{hz}} = 230 \text{ g}, \tag{12.4}$$

$$m_{exp} = 21.096 m_{cap} = 4{,}853 \text{ g}. \tag{12.5}$$

We also make the following assumptions. That the mixture coupling efficiency is close to ideal at $\epsilon_{mix} = 0.99$, the nozzle jet efficiency of each engine is $\eta_j = 0.901$ (based on the Daedalus second engine stage), the thermonuclear fuel fraction $\alpha = 0.0039$ (D^3He). This then implies an ICF capsule burn-up fraction of $\phi = 0.495$.

Based on the equation presented earlier in Chap. 5 this suggests that each pellet will require a driver energy of $E_{drive} = 165.7$ GJ and would correspond to an energy gain of $G = 239.9$. For all the 1,024 engines this would be a total driver input energy requirement of around 170 TJ. For each pellet detonated and with the assumed burn-up fraction given above the energy output would correspond to $E_{out} = 39.91$ TJ, which for all the engines combined would be a total energy output of 40,872 TJ.

For any laser driver system, even if it had a (fantastic) conversion efficiency of 25%, this would imply that the total laser power required for each cycle of the world ship would be on the order of 170 TJ/0.25 = 680 TJ. This is a phenomenal amount of energy and highlights the advantages of an external nuclear pulse engine.

Table 12.2 shows similar calculations to the one described above but assuming different numbers of engines. The required thermonuclear capsule mass and accompanying expellant propellant mass is also shown. The capsule mass requirement

Table 12.2 Calculations assuming $v_{ex} = 1{,}880\,\text{km/s}$, $m_{exp} = 21.09\,\text{m}_{cap}$, $f_{hz} = 100\,\text{Hz}$, $\epsilon_{mix} = 0.99$, $\eta_j = 0.909$, $\phi = 0.495$, $\alpha = 0.0039$ and where the total energy is to be multiplied by the number of engines

n_{eng}	$\dot{m}_{eng}/f_{hz}$ (kg)	m_{cap} (g)	m_{exp} (kg/shot)	E_{drive} (GJ/shot)	G	E_{out} (TJ/shot)
1	520,500	235,565	4,969	30,000	1,355	40,872
16	32,531	14,723	310	3,750	678	2,554
32	16,266	7,361	155	2,230	570	1,277
64	8,133	3,681	78	1,326	479	638.6
128	4,066	1,840	39	788	403	319.3
256	2,033	920	19.4	468	339	159.6
512	1,016	460	9.7	278	285	79.8
1024	508	230	4.85	166	240	39.9
2048	254	115	2.43	98	202	19.9

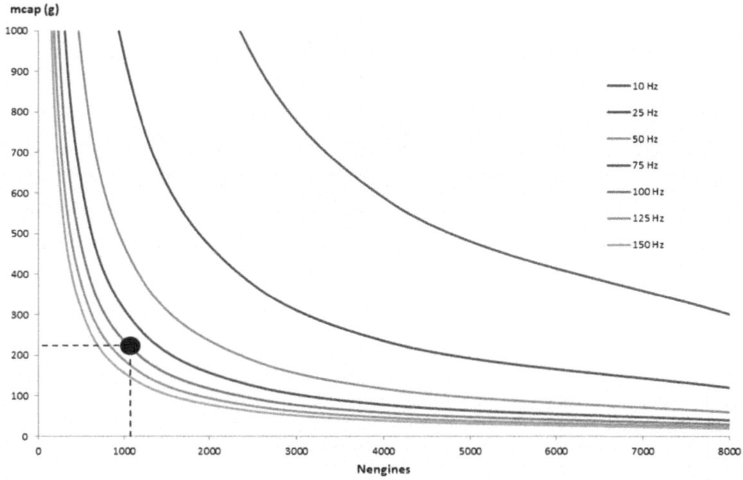

Fig. 12.2 Calculation of minimum capsule mass requirement as a function of the number of engines for different values of pulse detonation frequency

does drop the larger the number of engines used and in principle one could adopt thousands to tens of thousands of individual engines. However, each engine adds further mass to the system, and this must be kept in mind. Table 12.2 does not show calculations for the scenario of greater than 2,048 engines but the trend is well illustrated. The manufacture and emplacement of many engines would likely require an autonomous robotic replication technology to reduce the labour and time for construction. Alternatively, there may in the future be a way to synthetically grow engine nozzles according to a set replication algorithm, but this is speculation.

Figure 12.2 shows similar calculations to those presented in Table 12.2 but for different pulse frequency of detonation. The required capsule mass can be reduced the larger the number of engines but with increased system mass. For the remainder of the calculations presented in this work we focus on an apparent optimality point for 1,024 engines assuming an ICF capsule mass of 230 grams. It is not claimed that this is the best design point, but it seems a suitable compromise at 100 Hz pulse frequency from which to generate an initial concept design definition.

12.3 The World Ship Design

Figures 12.3 and 12.4 show illustrations of the world ship design as imagined. Although we have not represented the total 1,024 engine arrangement but a number somewhat less. In the concept shown here the world ship is made up of five separate habitats as living and working space for a large 1 million population of people, allowing for an expected growth rate of up to 10 million people [5].

The frontal habitat structure is the main engineering systems, command and control for the operation of the vessel. On top of this section is located a large drive motor which is the initiator for the spin of the world ship to provide for artificial gravity, also discussed earlier [5]. Each section is separated by a large structural partition ring which acts as a material strength component to prevent buckling under torsion but also as a radiation shield to protect individual habitats. To the front of the world ship is located a particle bombardment shield, forward sensors and a large high gain antenna for communication back to Earth or to other world ships.

To the rear of the world ship is the massive thrust structure which would contain the 1,024 individual reaction chamber engines. Since these would generate

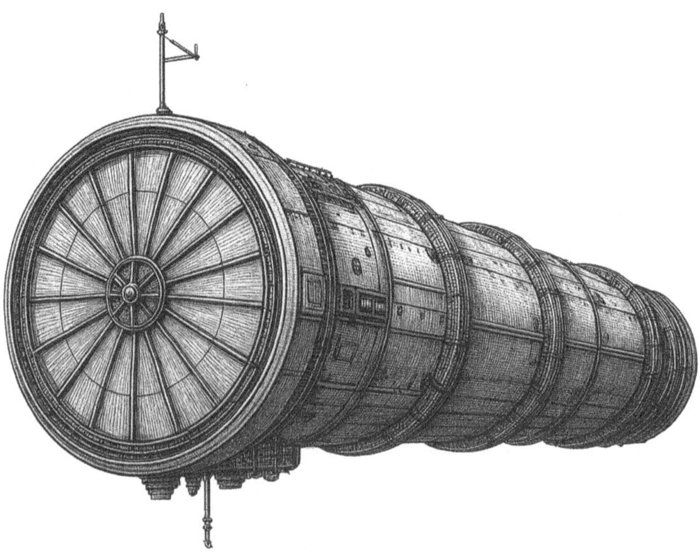

Fig. 12.3 World ship design concept front view[KFL]

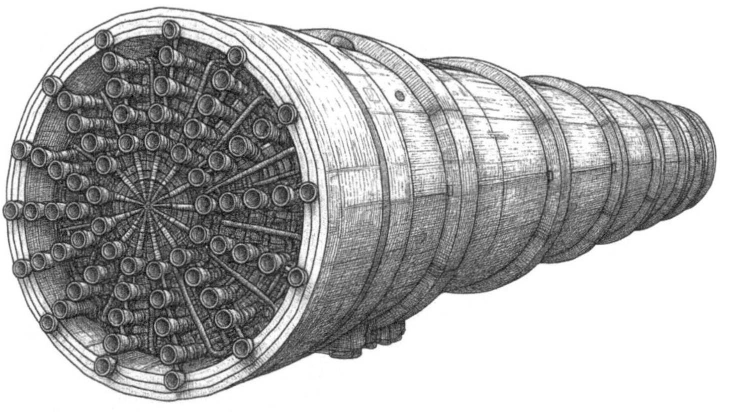

Fig. 12.4 World ship design concept rear view[KFL]

substantial radiation and waste heat, they would also be accompanied by large waste heat radiators. Any laser driver systems would also generate a large amount of heat and would need separate radiators. Since the world ship is contained in a large cylinder with thick walls it is critical that any heat is managed and is not deposited into the world ship unnecessarily and so heats up the internal volume. This may require waste heat radiators to be positioned along the length of the cylinder for each habitat to ensure proper thermal control within acceptable levels.

Inside of the thrust structure is the propellant bay housing the thermonuclear fuel and expellant propellant. Inside of this would be a large water bay which circulates around the vessel in a closed loop system. The fuel, expellant and water also acts to protect the habitat structures from any stellar flares or significant influx of radiation from space, where the vessel would face the engines towards the hazard and thereby minimise the dosage to the occupants.

It then cruises for just over 800 years over 264,088 AU or 4.176 ly. This would then be followed by a deceleration period of just under 15 years at a deceleration of 0.003 m/s^2 over 7,013 AU or 0.11 ly. The ship would reach the destination at 4.365 ly after 854 years total trip time.

Tables 12.3 and 12.4 show the calculation assumption, mission profile and performance for the model that matches the approximately 1,000-year total mission requirement and compared to the original 1984 model as published by Bond and Martin [3]. The Mk2A(ii) *Ra* model uses a pulse frequency of 100 Hz and this is assuming an exhaust velocity of 1,880 km/s using 1,024 engines in parallel thrust mode. For this mission the initial burn time is just under 35 years at an acceleration of 0.00142 m/s^2 upon which the ship would reach 5,720 AU or 0.09 ly. The cruise velocity will be 1,557 km/s which is equivalent to 328 AU/Year or 0.52 ly/Century.

There is a minor difference in the total burn time (acceleration and deceleration) between the two different models. For the 1984 Bond/Martin nuclear pulse engine the total burn time is less than $\sim$4.75% of the total mission time, whereas for the

Table 12.3 Design input for world ship models

Parameter	Nuclear pulse Mk2A (1984)	Laser fusion Mk2A(ii) (2023)
Number engines	1	1,024
Radius (km)	7.24	7.24
Length (km)	87.5	115
Burn mass ratio	2.22	2.29
Deburn mass ratio	2.22	2.29
Total mass ratio	4.93	5.24
Structure mass (tons)	1.68×10^{11}	1.67×10^{11}
Payload mass (tons)	–	1.6723×10^{6}
TN fuel mass (tons)	3.696×10^{10}	3.696×10^{10}
Propellant mass (tons)	7.797×10^{11}	7.797×10^{11}
Tankage mass (tons)	3.09×10^{10}	3.09×10^{10}
Exhaust velocity (km/s)	1,880	1,880
Energy per unit (J)	1.4×10^{19}	4.0×10^{16}
Pulse frequency (Hz)	0.2	100.0
Engine power (W)	2.6×10^{18}	4.0×10^{18}
Mass/unit (kg)	2.6×10^{6}	0.23
Expellant mass/unit (kg)	–	4.85

Table 12.4 Calculation mission profile and performance for world ship models

Parameter	Nuclear pulse Mk2A (1984)	Laser fusion Mk2A(ii) (2023)
Acceleration (m/s^2)	0.001	0.00142(+), 0.00307(−)
Mass flow rate (kg/s)	5.205×10^{5}	5.205×10^{5}
Burn time (Yrs)	–	34.81
Deburn time (Yrs)	–	14.91
Total boost time (Yrs)	47.56	49.73
Cruise time (Yrs)	952.44	804.19
Mission time (Years)	1,000.0	865
Cruise speed (km/s)	1,500 (0.5%c)	1,556.74 (0.519%c)
Thrust (GN)	978.52	978.52
Engine jet power (W)	9.19×10^{17}	9.19×10^{17}
Specific power (MW/kg)	–	8.16

2023 ICF design the total burn time is $\sim$5.82% of the mission time. Overall, the ICF engine appears to achieve the mission in a timescale that is a century and a half shorter than the nuclear pulse engine.

In this chapter we have considered the design of a world ship driven by ICF engines and compared it to the original 1984 Mk2A Bond/Martin design driven by nuclear pulse engines. There is no implication here that in making this comparison there is a statement of one being better than the other. Instead, it was the main purpose of this chapter to see what the propulsion parameters would look like when driven by an ICF engine and to gauge if ICF can be applied to this part of the interstellar design space.

That said the preliminary calculations from this work do suggest that an ICF driven world ship may arrive at the destination on the order of a century earlier than one driven by an external nuclear pulse engine. Although this result was using optimistic assumptions for the various propulsion parameters such as the jet efficiency and burn-up fraction and on this basis, it is too preliminary to reach any firm conclusions.

The calculated Mk2A(ii) *Ra* model mission meets the requirements to match the Mk2A mission profile under the input assumptions stated, except the burn/deburn time which has been relaxed a little and the slightly higher cruise velocity. However, next is the issue of whether this mission is practical and whether major design issues need to be addressed. In particular, the following issues are raised for further consideration in any future developments of the Mk2A(ii) *Ra* world ship.

The goal of producing an ICF driven world ship concept has been achieved although it is a matter of opinion on how plausible such a system would be. The main propulsion issues with the plausibility would appear to relate to (1) overly large capsule mass and the technological capability to drive them to an ignition state using lasers of very high efficiency (2) massive quantities of expellant mass per capsule and how this is coupled to the capsule thermonuclear ignition (3) very long acceleration times at low pulse frequency versus short acceleration times at very high pulse frequency and the appropriate selection of an optimum in between.

In this chapter the focus has been on the propulsion system, but this is not to say that there aren't significant other issues relating to the rest of the vehicle design which have not been addressed here. This includes the type of materials required for the different systems and in the habitat structure. The ability to maintain a large rigid cylinder structure under acceleration loads and rotational gravity spin. The distribution of the power and waste heat through the system in a way that maintains balance with power management at good efficiency. The appropriate mining, manufacture and storage of the fuel and propellant. The construction of such a large vessel is beyond our current technological capability and knowhow and instead it would firstly require the establishment of a solar system wide economy where easy and cheap access to the vast resources of space is made possible. It is worth making some observations about two of the essential ingredients, the atmosphere and the regolith.

For the atmosphere at 1.403×10^{10} tons this is not a small amount. This compares to the Earth's atmosphere which is around 5.148×10^{15} tons, around $\sim$0.003% of the Earth. It would be controversial to extract air from the Earth's atmosphere and arguably unethical and so this would have to originate from other sources. Since water is a source of oxygen, it can be found on the Moon at the poles or within lunar rock. It is also found on other moons in the Solar System but also on comets. So, any one of these may be the main source for generating such an atmosphere or alternatively mining the poles of Mars where hydrogen can be recovered by metabolic carbon vapour deposition. NASA is experimenting with such technology on a small scale for the use of space stations such as with the SpaceCraft Oxygen

Recovery (SCOR) project [34]. The creation of an atmosphere within a spacecraft has also been discussed by others [35].

Once the technology has been established to source and generate the required materials for an atmosphere it may also take some time to stabilise the atmosphere to the appropriate requirements for human respiration but also for essential agriculture on board; namely at 78% nitrogen, 20.9% oxygen, 1% argon and smaller traces of other elements like carbon dioxide, neon and methane. The vessel must also contain the right materials and structure to contain the atmosphere at the appropriate pressure. It is suggested that this process may take decades to a century, and even longer if the proposal is to generate the atmosphere using planted photosynthetic forests or hydroponic methods. Another way by which an atmosphere may be artificially generated is that the mass and spin of the world ship would attract individual particles in space through the inverse square law and over time this accumulation could build up to a substantial amount, but again this would likely take a long time unless it could be artificially accelerated. The depth of the atmosphere is also an important consideration in terms of protecting the occupants from cosmic rays, in addition to the shielding provided by the outer cylinder structure.

The Regolith is the bedrock of material that forms the foundations upon which the colony structures are built, and the agriculture must grow. This might be a combination of rocks, sand, dust or organic matter such as plants, trees, soils and fungi. Sourcing rocks is not such an issue since asteroids could in principle be captured and mined for this purpose, which may also provide a vital resource of minerals on-board for future use. Dust from surfaces like the moon or other moons may also be captured and deposited onto the internal surface of the world ship for use as a regolith material. The quantity required depends on the thickness but in line with the Bond/Martin studies we have assumed a total mass of 4.975×10^{10} tons to account for both acceleration and deceleration. The Earth has a total mass of 5.97×10^{24} kg and the amount of continental crust is on the order of 2.2×10^{22} kg with an average thickness that varies between 25–70 km. On the assumption that the soil is contained within the first few feet of this, then the total amount of soil would be on the order of 3.14×10^{17} (1/70,000) to 8.8×10^{17} kg (1/25,000). This means that the total regolith needed for our world ship would be on the order of 0.005–0.0015% of the total Earth regolith. Again, this is a question of ethics and especially if one is considering the construction of multiple world ships that usage fraction will increase.

Alternatively, it may be considered a feasibility to grow soils. The origin, development, evolution, formation of soils is known as Pedogenesis. Soils are complex constructs made of up a mixture of solids, liquids and gas, a mixture of complex minerals and organic matter. Air and water then fill the spaces in between in which the organisms can move. Soils will typically grow at a rate of around 0.1–0.2 mm/year which means that after a century of growth you will have $\sim$1–2 cm of soil. This implies that on a 1,000-year journey over 5 ly travelling at 0.5%c the expected soil formation would be in the range 10–20 cm. The growth of the soil depends upon the type of climate and the extremes of weathering and temperature, the sorts of living organisms within it, the topographical relief or shape of the landscape it is growing in. It is also a strong function of the parent material such as rocks, boulder clay or a composite amalgam.

A soil will be made up of several mineral components which vary in size fraction. This includes sand (0.02–2 mm) which is mostly composed of quartz grains, silt (0.002–0.02 mm) which is mostly unaltered compared to the parent material, and clay (<0.002 mm) where the weathering process changes their physical and chemical composition into secondary materials. The typical minerals from which the soils are formed include quartz (SiO_2), Calcite ($CaCO_3$), FeldSpar ($KAlSi_3O_8$) and Mica ($K(Hg,Fe)_3AlSi_3O_{10}(OH)_2$). Another factor in soil growth is the age of the landscape it is growing in.

A soil texture will then be defined by its relative percentage of sand, silt and clay. The texture is important since it will determine how much water a soil can hold. A sandy soil will drain rapidly and hold little water and nutrients due to the large size of its particles. Conversely, a clay soil has much smaller particles and so can hold a lot more water and will drain slowly. Soils store approximately 0.001% of all the water on the planet in the form of moisture. For the growing of plants, an ideal soil loam might be $\sim$40% silt, $\sim$40% sand and $\sim$20% clay. All this points towards a slow growth rate for soils within a world ship and certainly a very careful stewardship of the land within the habitats.

Finally, Table 12.5 shows examples of the journey times of the Mk2A(ii) *Ra* World Ship to alternative destinations within the nearby stars out to around $\sim 20\,ly$ distance. For these the propulsion parameters are kept constant as are the assumptions on fuel and expellant and so they are not optimised. The acceleration and deceleration time is also the same for each mission at 34.81 years and 14.91 years respectively, all using 1,024 engines in parallel thrust for a cruise velocity of 1,557 km/s or 0.52%c.

Also shown in Table 12.5 is the number of generations of humans throughout the voyage on the assumption that an average human lifespan is 50 years, 75 years or 100 years respectively. These calculations neglect the possibility of life extension ideas through hibernation, organ replacement, genetic engineering or indeed consciousness upload. Similarly, we neglect the possibility of life reduction

Table 12.5 Mission profiles for different interstellar destinations using the Mk2A(ii) *Ra* world ship assuming an equivalent fuel loading and propulsion values

Target star	S (ly)	t_c (yrs)	t_{tot} (yrs)	N_{gen} (50 yrs)	N_{gen} (75 yrs)	N_{gen} (100 yrs)
Proxima Centauri	4.365	804	865	17	11	9
Barnard's star	5.963	1,112	1,162	23	15	12
Wolf 359	7.856	1,477	1,526	31	20	15
Ross 154	9.706	1,833	1,883	38	25	19
Epsilon Eridani	10.475	1,981	2,031	41	27	20
Tau Ceti	11.912	2,258	2,307	46	31	23
Gliese 674	14.849	2,823	2,873	57	38	29
82 G Eridani	19.705	3,758	3,808	76	51	38

such as through genetic decay due to exposure to radioactivity in deep space or bone loss due to a low gravity environment. Yet the estimates illustrate that the nth generation will be the one that arrives at the target destination.

The reality is that around ~5% or less (depending on the distance and average age) will be a part of the arrival party. The other ~95% or more will never see the destination star and its associated planets other than through telescopes. Similarly, they will never see the Earth home world of their origin. This brings into question the entire point of making such a journey, although for the *Star Children* born on the vessel they would never have known any different and stories of Earth would be but legend and ancient history.

Problems of Chap. 12

12.1 The Moon has a mass of 7.35×10^{22} kg and has an orbital speed around the Earth of 1.022 km/s at an average distance of 384,400 km which it completes in approximately 27.32 days (i) If the orbital speed was increased by 0.5 km/s but in the same sidereal period what would be the new average distance from the Earth? (ii) If the change in orbital speed was to occur over 1,000 years what would be the force required for this change? (iii) Assuming giant engines were fitted to the surface of the Moon, and each produced 1 TN of thrust, how many engines would have to be assembled on the Moon's surface to make this change?

12.2 The Earth has a mass of 5.972×10^{24} kg and an orbital period around the Sun of 365.256 days (one sidereal year). It has an average orbital radius of 1 AU $=$ 1.496×10^{11} m (i) If the orbital period of the Earth was changed to 360 days, what would be the revised average orbital radius? (ii) How much closer to the Sun is this new radius as a percentage? (iii) The current orbital speed of the Earth around the Sun is 29.78 km/s, how would this change given the new orbital period?

12.3 Carrying on from the previous question, we can consider the Earth to be equivalent to a giant world ship. (i) If the velocity change was to be imparted onto the Earth over a period of 10,000 years how much force (thrust) would this require (ii) If the Saturn V rocket during Project Apollo generated around 34 million Newtons of Thrust, how many Saturn V equivalent rockets would we need to impart this velocity change on the Earth? (iii) If we assume there are around 200 countries in the world and the total number of rockets is to be constructed equally among all nations at a cost of $1.5 billion per rocket, how much would each nation have to spend and what is the total global construction cost?

References

1. Sagan C (1980) Cosmos. Random House, New York
2. Martin AR (1984) World ships - concept, cause, cost, construction and colonisation. JBIS 37(6):243–253
3. Bond A, Martin AR (1984) World ships - an assessment of the engineering feasibility. JBIS 37(6):254–266

4. Long KF (2023) Population demographics and other issues for the massive Ra world ship model - part 1. JBIS 76(11):262–272
5. Long KF (2024) Inertial confinement fusion propulsion for the massive Ra world ship model - part 2. JBIS 77(4):119–137
6. Long KF, Galea P, Mann A (2011) Brave new future worlds. Futuristic journeys to distant stars. Spaceflight 53(11):428–432
7. Kondo Y, Bruhweiler FC, Moore J, Sheffield C (eds) (1971) Interstellar travel and multi-generation space ships. Apogee Books Publication, Burlington
8. Caroti S (2011) The generation starship in science fiction: a critical history, 1934–2001. McFarland & Company, Inc., Jefferson
9. Hall TW (2020) Artificial gravity in interstellar travel. Acta Futura 105–121
10. Medina FJ (2020) Growing plants in human space exploration enterprises. Acta Futura 151–163
11. Volponi M, Lasseur C (2020) Considerations on life support systems for interstellar travel: a regenerative story. Acta Futura 133–149
12. Hein AM, Smith C, Marin F, Staats K (2020) World ships: feasibility and rationale. Acta Futura 75–104
13. Acierno K, Leblanc O, Bevington J, Liu Y, Bhattacharjee S, Manoharan H, Ragala H, Chen G, Ramos B, Jong DD, Rococo J, Geng L, Russell-Conway F, Ghosh A, Shar M, Gujarati K, Sherriff A, Klettke M, Wilder A, Lazar F, Zuo P (2015) Astra planeta final report, International space university masters of space studies team project, 20th March
14. Dyson FJ (1968) Interstellar transport. Phys Today 21(10):41–45
15. Dyson G (2002) Project orion - the atomic spaceship 1957–1965. The Penguin Press, London
16. Matloff GL (1985) On the potential performance of non-nuclear interstellar arks. JBIS 38(3):113–119
17. Matloff GL (1986) Faster non-nuclear world ships. JBIS 39(11):475–485
18. Matloff GL (2012) World ships: the solar-photon sail option. JBIS 65(4/5):114–118
19. Armstrong R (2017) Star ark: A living self-sustaining spaceship. Springer-Praxis, Berlin
20. Traphagan JW, Wisian K (2020) Protocols for encounter with extraterrestrials: lessons from the covid-19 pandemic. JBIS 73(7):234–238
21. Holmes DL (1984) Worldships: a sociological view. JBIS 37(6):296–304
22. Birch R (1982) Radiation shields for ships and settlements. JBIS 35(11):515–519
23. Richards IR (1981) A closed ecosystem for space colonies. JBIS 34(9):392–399
24. Marin F, Beluffi C, Raylor R, Grau L (2018) Numerical constraints on the size of generation ships from total energy expenditure on board. Annual food production and spacefarming techniques. JBIS 71(10):382–393
25. Marin F, Beluffi C, Fischer F (2021) Genetic evolution of a multi-generational population in the context of interstellar space travel. JBIS 74(5):162–184
26. Marin F, Beluffi C (2020) Water and air consumption aboard interstellar arks. JBIS 73(7):239–247
27. Martin AR, Bond A (1978) Project Daedalus: the propulsion system part 1: theoretical considerations and calculations. JBIS Special Supplement:S44–S62
28. Tabak M, Clark DS, Hatchett SP (2005) Review of progress in fast ignition. Phys Plasmas 12:057305
29. Perkins LJ (2009) Shock ignition: a new approach to high gain inertial confinement fusion on the national ignition facility. Phys Rev Lett 103:045004
30. Sheffield J, Spong DA (2016) Catalyzed D-D stellarator reactor. J Fus Sci Technol 70(1):36–53
31. Gaidos G, Lewis RA, Meyer K, Schmidt T, Smith GA (1999) AIMSTAR: antimatter initiated microfusion for precursor interstellar missions. Acta Astronautica 44(2–4):183–186
32. Kammash T (2002) Antiproton driven magnetically insulated inertial confinement fusion (MICF) propulsion system. NIAC 98–02 final report

33. Shepherd LR (1952) Interstellar flight. JBIS 11(4):149–167
34. Stanley CM (2023) Spacecraft oxygen recovery project carbon analyses. NASA/TM-20220017369
35. Strogonova LB, Sorokin AE, Vasin YuA, Eelyavskii AE (2019) Creating an atmosphere within spacecraft. Russ Eng Res 39:813–815

Chapter 13
Fantastic Propulsion Concepts

In principal, it looked like this solved the problem from the standpoint of the physics....Of course, the engineering is virtually impossible today, but give us a 100 years of advanced engineering and it may be possible.

—Robert Bussard [1]

13.1 Interstellar Ramjet

One of the major issues with reaction engine-based rocket systems for space propulsion is the need to carry your own fuel. You can use more efficient fuels which gives you more energy per unit mass of material, but you still must carry along some fuel which adds mass to the vehicle. However, when we look at fusion engines, we note that one of the principal ingredients for a fuel source is hydrogen or hydrogen isotopes such as deuterium and tritium. An ionised hydrogen is just a proton. It is a curious occurrence that in between the stars happens to be copious amounts of hydrogen (and other materials but at lower abundance) which makes up what is known as the interstellar medium.

Imagine a scenario where you don't have to carry your fuel, but you are simply mining it on route to your destination. This is the interstellar ramjet, which was first conceived in 1960 by the physicist Robert Bussard [2]. Such a system requires the use of a large magnetic funnel at the front of the vehicle so that the charged protons can spiral down the field lines and into the propulsion system. A concept for such a vessel is shown in Fig. 13.1.

It is called a ramjet since it uses a large magnetic scoop which generates an electromagnetic field and so scoops up the interstellar hydrogen, which might be perhaps 1 atom/cm^3 over thousands of km in area. Once the protons are passed through the propulsion system, in principle they would be fused and any products from the reaction would be expelled as exhaust material to produce continuous thrust. This happens via the proton-proton chain and the Deuterium burning chain for example, which are shown below. Firstly two sets of proton reactions need to

K. F. Long, *Stellar Engines*, Springer Aerospace Technology,
https://doi.org/10.1007/978-3-032-18062-9_13

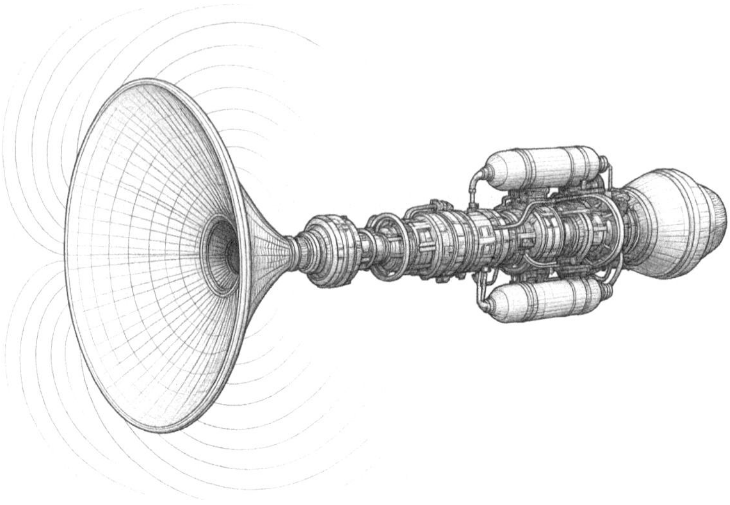

Fig. 13.1 A concept for an interstellar ramjet spacecraft[KFL]

happen so that we begin with 4p as follows

$$p + p \rightarrow^2 H + e^+ + \nu_e,$$

$$p + p \rightarrow^2 H + e^+ + \nu_e.$$

This then forms two Deuterons ^{2}H which is one neutron and one proton (but no electron so is positively charged). It is essentially just the nucleus of a Deuterium atom D. The pair of resulting reactions from this are

$$^2H + p \rightarrow {}^3He + \gamma,$$

$$^2H + p \rightarrow {}^3He + \gamma.$$

Both of these reactions have then produced ^{3}He and these combine to form ^{4}He

$$^3He + {}^3He \rightarrow {}^4He + 2p.$$

Where ^4He has been generated. The total net reaction chain can be described as follows

$$4p \rightarrow {}^4He + 2e^+ + 2\nu_e + \gamma.$$

Where the additional two protons 2p have been returned into the reaction system to generate more reactions. From this we can calculate the mass defect, where $m_H =$

$1.007825u$ and $m_{4He} = 4.002603u$, where u is in atomic mass units, and $1u = 931.5\,\text{MeV/c}^2$. Then estimating the mass difference between the starting atoms and the finishing ones we find

$$\Delta m = m_{initial} - m_{final} = 4m_H - m_{4He} = 0.028697u. \tag{13.1}$$

Then the energy release is calculated to be

$$Q = \Delta mc^2 = 0.028697u \times 931.5\frac{MeV}{u} = 26.73\text{ MeV}. \tag{13.2}$$

This is also equal to around 6.5×10^{14} J/kg.

There are some issues with the interstellar ramjet however, which have discouraged researchers from attempting to refine the design ideas. The first is that proton-proton fusion is extremely hard to achieve due to its very small cross section. Thus, to increase the probability of two particles colliding this suggests very long confinement times. Alternatively, one may find a way to turn the hydrogen into other particles such as deuterium, helium-3 or tritium to facilitate D-^{3}He or D-T fusion. Any of these options, long confinement times or conversion to other fuels, takes energy which suggests a lack of efficiency and therefore a greatly reduced energy gain. The interstellar ramjet teaches us that in using a fusion fuel one likely needs to first assemble the material such as inside some sort of inertial capsule—hence ICF.

Another issue is that some calculations suggest that the drag generated by the large magnetic scoop may exceed any thrust generated by the engines [3]. This is especially so at low speed and in realistic ISM densities ($\sim$0.1–1 cm^{-3}). The drag scales with the rate that the spacecraft removes energy from the intercepted flow of the interstellar medium. We can see this by looking at the mass collection rate $\dot{m}_{scoop}$ from an ram scoop of effective area A which is generated by the magnetic field, onto a mass density stream of $\rho = nm_p$ where m_p is proton mass, and n is the number density of protons in the interstellar medium. In the frame of the ramjet the Lorentz-contracted collection mass rate is then given by

$$\dot{m}_{scoop} = \gamma\rho Av. \tag{13.3}$$

Where $\gamma = 1/\sqrt{1-\beta^2}$ and $\beta = v/c$. For each unit of mass that enters the incoming stream it will have a kinetic energy which acts to heat up the scoop of the ramjet and any surrounding plasmas or magnetic fields. It is given by

$$\epsilon = (\gamma - 1)c^2. \tag{13.4}$$

The drag power on the ramjet will then be

$$P_d = \dot{m}_{scoop}(\gamma - 1)c^2. \tag{13.5}$$

Therefore the drag on the interstellar ramjet will be given as a function of the relativistic velocity

$$P_d = \gamma\rho A v(\gamma - 1)c^2 = \gamma n m_p A v(\gamma - 1)c^2. \tag{13.6}$$

Note, that we can also look at the non-relativistic case for when $v << c$, where $\gamma \approx 1+\beta^2/2$ so that we get an equation that is similar to those we find in aeronautics where drag coefficient $C_D \sim 1$ for conventional flight

$$P_d \approx \rho A v\left(\frac{v^2}{2c^2}\right)c^2 = \frac{1}{2}\rho A v^3. \tag{13.7}$$

Therefore, any spacecraft must exceed the power of this drag through its thrust to undergo net acceleration. This drag will be rather large when we consider the enormous size of the magnetic scoop which may be thousands of km across. As the hydrogen fuel builds up in the funnel this may be difficult to manage, and the entire system will be hot due to the emission of Bremsstrahlung and synchrotron radiation losses. Any incoming protons would have to be moderated down to a lower energy before they could be used for thrust generation. This is because they would have too high a kinetic energy and so would just blast through the spacecraft structure causing damage and would not be usable. Yet to slow these high energy particles down would take significant energy, perhaps more than is generated from the fusion process one is trying to achieve. Another way to say this is that the interstellar protons are effectively too hot and we need to significantly slow them down (moderate) into a cold and confined state if we are to use them in a fusion reaction.

To deal with the slowness of the proton-proton chain, another approach has been investigated for the interstellar ramjet. This involves the idea of using heavy fusion catalysts such as the isotopes of carbon ^{12}C and Neon ^{20}Ne, where they are used to speed up the conversion of hydrogen to helium and then release energy [4]. This approach is known as Catalytic hydrogen burning via the CNO cycle and is much faster. For example:

$$^{12}C + p \rightarrow {}^{13}N + \gamma,$$

$$^{13}N \rightarrow {}^{13}C + e^+ + \nu,$$

$$^{13}C + p \rightarrow {}^{14}N + \gamma,$$

$$^{14}N + p \rightarrow {}^{15}O + \gamma,$$

$$^{15}O \rightarrow {}^{15}N + e^+ + \nu,$$

$$^{15}N + p \rightarrow {}^{12}C + {}^{4}He.$$

Another approach involves the use of a Ram Augmented Interstellar Ramjet (RAIR) and was developed by Alan Bond [5] in which there are two propellant streams operating through the engine. In one interstellar hydrogen is still converted to helium via fusion reactions but it is sourced from on-board propellant tanks. In the other, interstellar hydrogen is collected from the incoming stream, thus providing for additional reaction mass by the collected un-fused protons. So the collected protons only act as the momentum reaction mass to augment the thrust.

In some ways, this method is similar to the injection of expellant into the exhaust stream that have been examined in the other concepts of this book. Both of the alternative approaches are highly novel and deserve considerable attention if we are to ever make the interstellar ramjet a reality.

In an interstellar ramjet the fusion power can be described by the burn-up fraction ϕ, the specific energy of the fusion reaction chain E_{sp} and the rate of collection of hydrogen into the magnetic scoop dm/dt so that

$$P_{fus} = \phi E_{sp} \frac{dm}{dt}. \tag{13.8}$$

In addition to the fusion power, the thrust of the ramjet will be given as a function of the jet efficiency η_j and exhaust velocity v_{ex} so that

$$T = \frac{2\eta_j P_{fus}}{v_{ex}}. \tag{13.9}$$

Therefore combining the two equations

$$T = \frac{2\eta_j \phi E_{sp}}{v_{ex}} \frac{dm}{dt}. \tag{13.10}$$

To illustrate with an example, let us say we have a spacecraft with an exhaust velocity $v_{ex} = 10{,}000$ km/s, that is characterised by a burn-up fraction of $\phi = 0.33$, and a jet efficiency of $\eta_j = 0.9$. We further assume that the ramscoop can funnel in a mass flow rate of hydrogen into the reactor of $dm/dt = 10$ kg/s. The specific energy associated with the p-p chain is $E_{pp} = 6.4 \times 10^{14}$ J/kg. This results in a fusion power of $P_{fus} = 2.112 \times 10^{15}$ W and therefore a jet power of $P_{jet} = \eta_j P_{fus} = 1.9 \times 10^{15}$ W. The thrust will then be $T = 380.2$ MN.

One of the benefits of such a system is that provided the efficiency of the particle capture and energy conversion to thrust is high then such a vehicle may achieve speeds close to the speed of light, that is relativistic flight and perhaps near optical velocities ($v \rightarrow c$). As it continues to accelerate, the speed of the vehicle will also increase and so also will the amount of hydrogen mass that is scooped up and so this would need active management such as by changing the field configuration or geometry of the large scoop structure.

We can compute this from a consideration of the relativistic energy equation where the total energy is given as a function of the proton rest mass energy m

(938.27 MeV/c^2), the speed of light c and the Lorentz factor γ

$$E_{tot} = \gamma mc^2. \tag{13.11}$$

If we examine this case of $v = 0.9c$ then

$$\gamma = \frac{1}{\sqrt{1-(0.9)^2}} = 2.294. \tag{13.12}$$

Then the total energy is given by

$$E_{tot} = 2.294 \times 938.27\ \text{MeV} = 2,152\ \text{MeV}. \tag{13.13}$$

Where the rest mass energy is given by

$$E_o = mc^2. \tag{13.14}$$

Then subtracting the rest mass energy from the total energy, we find the kinetic energy

$$E_{kin} = E_{tot} - mc^2 = 2,152 - 938.27 = 1,214\ \text{MeV}. \tag{13.15}$$

At $0.1c$ a proton will have a kinetic energy of around 4.73 MeV. At $0.3c$ it will have a kinetic energy of 45.3 MeV and at $0.5c$ it will be 145.15 MeV. At the much higher speed of $0.9c$ it will have an energy of 1,214 MeV. Therefore, the faster the ship is travelling the more energetic the protons will be, and relativistic. Figure 13.2 shows the kinetic energy for protons over a range of vehicle velocities. Figure 13.3 shows the same calculation but for electrons for comparison. The kinetic energy of

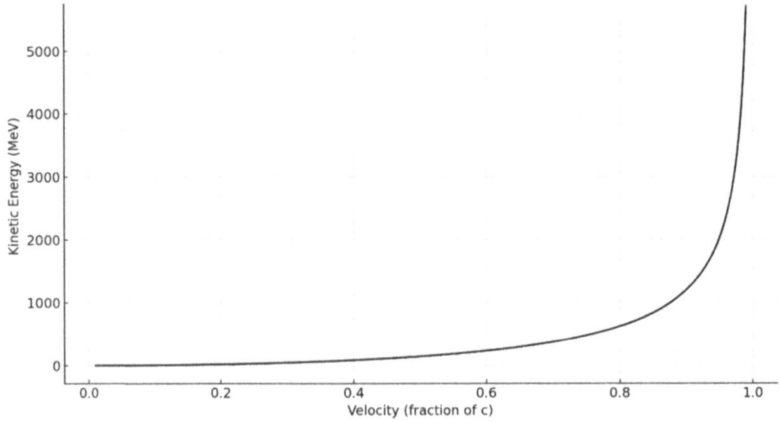

Fig. 13.2 Calculated kinetic energy for a proton

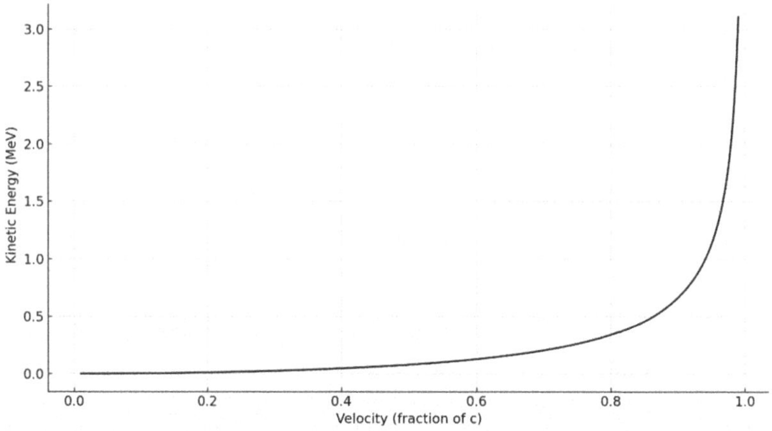

Fig. 13.3 Calculated kinetic energy for an electron

the electron rises much more slowly due to its much smaller mass compared to the proton. By implication the proton would need much more energy to reach relativistic speeds. The electron calculation is shown in case one of the readers would like to have a think about designing an electron based interstellar ramjet and this is left for you as a challenge.

It was previously shown by Anthony Martin [6] that the rate of mass collection for an interstellar ramjet is given as a function of the hydrogen mass m, the speed of light c, the Lorentz factor γ, the relativistic beta β, the average interstellar medium hydrogen density n and the collection area A_o so that the collection rate in the units of kg/s is given by

$$\left(\frac{dm}{dt}\right)_c = mc\gamma\beta n A_o = 5.019 \times 10^{-19}\gamma\beta. \tag{13.16}$$

Following on from the same equation, if we then assume a collector radius of 1,000 km, an interstellar medium hydrogen density ranging from $10^6 - 10^9 m^{-3}$, a velocity range of 0.5–0.9c, then we find a collection rate range of 1.8–3,614 kg/s. If we assumed this was all ionised hydrogen where $m = 1.6726 \times 10^{-27}$ kg, then this collection rate range would be the equivalent of $\sim 1.1 \times 10^{27} - 2.2 \times 10^{30}$ protons/s. However, if the velocity gets even closer to the speed of light, then the collection rate rises even further, such as at 0.99c it will be around 11,167 kg/s which would be the equivalent of 6.67×10^{30} protons/s.

In this chapter it is not the intention to have a significant discussion on the interstellar ramjet, but rather to examine one potential solution to making it viable, that has not been previously considered by others and so this is an original idea. This is a fantastic solution however, hence the chapter name, and there is no claim here that such a solution is practical. Yet it is shown since it is the aim of this book to expand the imagination of the reader and to think outside the box in terms of creative innovations. We might refer to what follows as an extremely advanced propulsion engine and highly speculative. Yet the aim in presenting this is to stretch the imagination of the reader as to the wider possibilities.

13.2 Black Hole Engine

We now arrive at what will be the most fantastic concept presented in this book. This is shown not because the idea is currently credible, but to further illustrate the process of designing interstellar machines through creative imagination. The idea goes back to the time when the Large Hadron Collider (LHC) at CERN in Geneva, Switzerland, was being switched on. This was occurring to the background of developments in theoretical physics where it was believed from string theory there are likely to be more than three spatial dimensions. On the basis this was true, the collision of TeV proton particles might result in them immediately collapsing into their Schwarzschild radius via a process called Hawking evaporation [7]. Some physicists proposed that, in certain extra-dimensional models, LHC energies could produce microscopic black holes that would immediately evaporate via Hawking radiation [8, 9].

It was noted by this author that this may present an opportunity for addressing some of the technical challenges associated with the interstellar ramjet. The idea was originally written up in a draft paper in 2009 [10] but was never submitted for publication and only circulated around a few colleagues. This was mainly due to the uncertainty involved in predicting the charged particle distribution from Hawking evaporation. The idea was developed later however, when the author worked with a student Andrew Alexander, who completed a study of the concept as a part of his master's thesis at the International Space University [11]. The sort of concept envisaged is illustrated in Fig. 13.4.

Black holes were first proposed as a solution to the collapse of a massive star in 1939 by Robert Oppenheimer [12]. His equations described a spherically symmetric sphere of stellar matter collapsing into a singularity, from which the escape velocity was so great that not even light could escape. Then, it was in 1974 that the physicist Stephen Hawking first predicted that black holes are not entirely black, but in fact will emit radiation due to quantum mechanical effects which occur near the event horizon. If this emission continues to occur over time, then this will lead to energy loss and therefore mass loss of the black hole, so that eventually it will completely evaporate. This occurs because near the event horizon of a black hole quantum fluctuations constantly produce virtual particle-antiparticle pairs which rapidly annihilate soon after they are created. Yet if one of these particle pairs falls into the black hole and the other escapes at the horizon, then the escaping one will be observed as radiation emission. The source of power behind this radiation is the mass loss of the stellar object, since $E = mc^2$ and for a non-rotating, uncharged

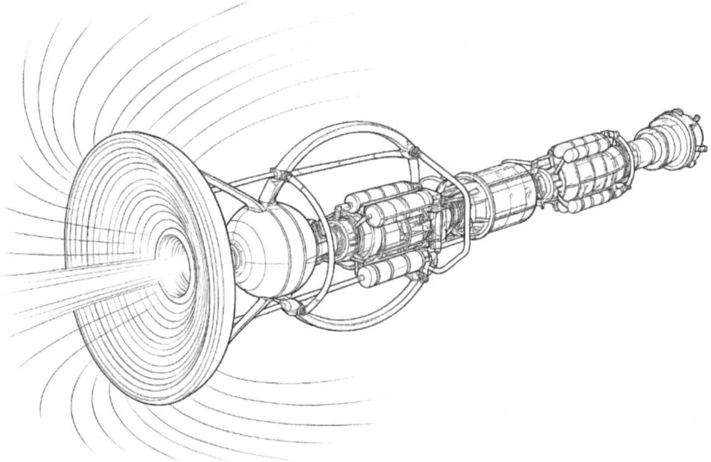

Fig. 13.4 A micro-black hole engine[KFL]

black hole the power is inversely proportional to the mass through

$$P \approx \sigma A T_H^4 \propto M^2\left(\frac{1}{M}\right)^4 = \frac{1}{M^2}. \tag{13.17}$$

A particular observation of Hawking was that the mass will decrease in inverse proportion to the temperature at the horizon, which is known as the Hawking temperature and is given by

$$T_H = \frac{\hbar c^3}{8\pi G M k_B}. \tag{13.18}$$

Therefore, whilst a large black hole will evaporate very slowly, a much smaller mass black hole will evaporate much faster. As the mass shrinks further, the evaporation process will accelerate and so end in a violent burst of energy. The evaporation time will be given by

$$\tau_H = \frac{5{,}120\pi G^2 M^3}{\hbar c^4}. \tag{13.19}$$

For a 1 solar mass black hole the evaporation time will be around $\tau_H = 2.09 \times 10^{67}$ years. However, for the purpose of propulsion, it was not a macroscopic or stellar black hole that was of interest, but much smaller black holes that approach the atomic mass of subatomic particles. Using the same equation, the evaporation of two colliding protons, such as in a p-p reaction, if they were able to form a micro-black hole, would be around $\tau_H = 3.14 \times 10^{-96}$ seconds and so is essentially an instantaneous evaporation. However, near such tiny masses the semiclassical approximation is no longer trustworthy so this is purely illustrative.

We can also take the Stefan-Boltzmann radiation law and from this write out an associated power for the radiation emission for an area A of radius R_s, where $\sigma = 5.6704 \times 10^{-8}\, \text{W/m}^2\text{K}^4$,

$$P = A\sigma T_H^4 = 4\pi R_s^2 \sigma T_H^4. \tag{13.20}$$

So, you might ask, well how does this help us? In classical physics we consider a four-dimensional space time, where there are 3 dimensions of space and 1 dimension of time. Within this framework, as a solution to the Schwarzschild equation from the general theory of relativity, the radius for gravitational collapse of an object is described as a function of its mass

$$R_s = \frac{2GM}{c^2}. \tag{13.21}$$

If we calculate this for a single proton at 7 TeV energy, where $1\,\text{eV} = 1.602 \times 10^{-19}$ J the mass equivalent is given by

$$M = \frac{E}{c^2} = \frac{1.1215 \times 10^{-6} J}{c^2} = 1.248 \times 10^{-23}\ \text{kg}. \tag{13.22}$$

Where for this mass $T_H = 9.854 \times 10^{45}$ K and $\tau_H = 1.635 \times 10^{-85}$ s which is effectively instantaneous. Then the radius for collapse is very small at around $R_s = 1.853 \times 10^{-50}$ m and so we can conclude that quantum effects will dominate. Under these assumptions, you will never get to gravitational collapse of a proton since any gravitational forces are completely negligible compared to quantum forces. Yet, things get very interesting when we consider quantum descriptions of gravity and in particular models of higher dimensional spaces.

One of these is called the ADD model (Arkani-Hamed et al. [13, 14] which claims to describe why gravity is so much weaker than the other fundamental forces—also known as the Hierarchy problem. Within this framework, our observed 3-dimensional reality is called a 'brane' which is embedded within a higher dimensional space known as the 'bulk'. Whilst the other fundamental forces are confined to the brane, gravity can leak into the extra dimensions; it is dispersed and spread out. There is also an alternative model to ADD which is known as the Randall-Sundrum (RS) model, which has one extra dimension of finite size. Whilst in the ADD model the extra dimension is flat; within the RS model it is warped. We will not examine the RS model further here and readers should consult some of the research papers for more information [15, 16].

Going back to the ADD model, whilst at long distances gravity behaves according to Newton's inverse square law $F \propto 1/R^2$, at short distances it is much stronger, typically at distances less than the size of the extra dimension. This can be represented by considering the Planck mass M_{Pl} as proportional to the mass at the true fundamental Planck scale M_D in $4 + n$ dimensions, and the size of the extra dimension R so that

$$M_{Pl}^2 \approx M_D^{2+n} \times R^n. \tag{13.23}$$

Where n is the number of extra dimensions in the model, so that if $n \geq 2$ but R is small and finite, then the true Planck scale M_D could be as low as a few TeV. This is within the scale of the Large Hadron Collider (LHC) at CERN and this is what caught the interest of this author many years ago as the LHC was about to go operational. In this model, gravity would deviate from Newton's law at the sub-millimeter scale and so for colliders like the LHC which operate on the TeV scale, it might be possible for colliding protons to produce tiny black holes that evaporate via Hawking radiation.

The size of the extra dimension is given by rearranging the above equation to show that larger values of n will allow for smaller extra dimensions, as follows

$$R = \left(\frac{M_{Pl}^2}{M_D^{2+n}}\right)^{1/n}. \tag{13.24}$$

Where $M_{Pl} = 1.22 \times 10^{19}$ GeV and $M_D = 1$ TeV $= 10^3$ GeV.

$$R = \left(\frac{(1.22 \times 10^{19})^2}{(10^3)^{2+n}}\right)^{1/n} = 10^{\frac{30}{n}-18.7}. \tag{13.25}$$

Where we approximated $1GeV^{-1} \approx 1.97 \times 10^{-16}$ m which implies $Log_{10}(1.97 \times 10^{-16}) \approx -15.7$ and so

$$Log_{10}\left(\frac{M_{Pl}^2}{M_D^{2+n}}\right) \approx 30. \tag{13.26}$$

Therefore for $n = 2$, we find that $R = 0.1$ mm with this approximation, but without using the logarithm and using the precise values the radius is computed to be $R = 2.44$ mm. For interest, for $n = 3, 4$, we find that $R = 1.044 \times 10^{-8}$ m and $R = 2.1759 \times 10^{-11}$ m respectively. Some higher dimensional models explore the possibility of $n = 10, 11$, in which case $R = 3.2489 \times 10^{-16}$ m and $R = 1.6567 \times 10^{-16}$ m respectively.

A key result that comes out of adopting a higher dimensional model, is that the total energy density in a collision will exceed the modified Planck scale, so that if the particles come within the effective Schwarzschild radius in $4 + n$ dimensions, then they will be trapped within the event horizon and therefore create a miniature black hole. This object would have an effective radius which is given as a function of the fundamental Planck scale mass M_* at the TeV level up to order unity n dependent factors so that

$$R_s^{(4+n)} \approx \left(\frac{E}{M_*}\right)^{1/(n+1)} \times \frac{1}{M_*}. \tag{13.27}$$

The evaporation time would then be given by

$$\tau_H \approx \frac{1}{M_D}\left(\frac{M}{M_D}\right)^{\frac{(n+3)}{(n+1)}}. \tag{13.28}$$

So that with $M_D = 1\,\text{TeV}$, $M = 7\,\text{TeV}$, $n = 6$

$$\tau_H \approx \frac{1}{10^{12}\,eV}\left(\frac{7}{1}\right)^{(6+3)/(6+1)} \approx 10^{-26}\ \text{s}. \tag{13.29}$$

For $n = 2, 3$, we find that $\tau_H = 1.69 \times 10^{-26}$ s and 1.22×10^{-26} s respectively. From this we can see that with fewer dimensions the rate of evaporation is slowed down, since the radiation is confined to a lower number of dimensions. For completeness if we assume $n = 10, 11$ then we find $\tau_H = 6.56 \times 10^{-27}$ s and 6.37×10^{-27} s respectively. We see that with a higher number of dimensions the evaporation time is decreased.

The idea was to use the physics of higher dimensional quantum gravity and the evaporation of 7 TeV protons as a way of solving several of the problems associated with the interstellar ramjet. Since the proton-proton chain has a very small cross section for fusion, we don't need to worry about that. Instead, we simply allow the protons to be attracted into the magnetic scoop, where they are then passed down into the spacecraft and travel through a collider ring, where they can meet within a reaction chamber. Since there will be many protons incoming into the scoop simultaneously, we can envisage that there will be many reactions occurring in the reaction chamber. Although any formed black hole will be electrically neutral since it would shred its charge during the collapse in what is known as the Balding Phase, and any charged particles will be emitted during the Evaporation phase.

The radiation associated with the Hawking evaporation is described when we approximate a black body emission as a function of the Hawking temperature T_H and $\Gamma_i(E, M)$ which is a grey body factor to account for the amount of transmission probability through a potential barrier and it depends on the black hole mass as well as the particle type and energy. For a real model we might expect that $\Gamma_i(E, M) \leq 1$. The ± 1 is to allow for either fermions $(+)$ or boson $(-)$ particles. Since the black hole will radiate all particles where the Compton wavelength fits within its horizon, in principle we can expect any of the particles from the standard model to be emitted via a spectrum

$$\frac{dN_i}{dt dE} = \frac{\Gamma_i(E, M)}{2\pi\hbar}\frac{E^2}{e^{E/k_B T_H} \pm 1}. \tag{13.30}$$

Where + is for fermions and − is for Bosons. Note also that for $\Gamma_i \leq 1$ this is strongly energy and spin-dependent. Taking $\Gamma_i = 1$ corresponds to an upper bound. To illustrate how we might use this equation, let us imagine we have a particle of 7 TeV energy which is equivalent to $M = 1.25 \times 10^{-23}$ kg, so that the temperature will compute to $T_H = 9.828 \times 10^{45}$ K. Since this is an approximation, we can just assume an optimistic maximum grey body transmission probability of $\Gamma_i(E, M) = 1$, so that 100% of the radiation is assumed to escape the potential barrier. Hypothetically, if the required emission rate was $E = 1$ TeV, which is equal to 1.6×10^{-7} J. Therefore, the derivative will be

$$\frac{dN_i}{dtdE} = \frac{1}{2\pi \times 1.054 \times 10^{-34}} \times \frac{(1.602 \times 10^{-7})^2}{e^{(1.602 \times 10^{-7})/(9.498 \times 10^{23})} + 1}. \tag{13.31}$$

This computes to a value of 1.938×10^{19} particles/s/J.

To derive thrust from this process we need there to be charged particles that come out of the Hawking evaporation emission. But there are many types of particles that might be emitted during the Hawking evaporation. Table 13.1 gives a summary of the sorts of particles that might be emitted, and which ones might be suitable for thrust generation. We emphasise that this is an approximation and not a normalised

Table 13.1 Approximate order-of-magnitude particles emitted in Hawking evaporation[KFL]

Particle	Mass (MeV)	Charge	Thrust	Amount	Notes
Protons	938	+1	Yes	~5–8%	Magnetic control, high momentum
Antiprotons	938	−1	Yes	~5–8%	Same as protons, but annihilation issues
Electrons	0.511	−1	Yes	~10–15%	Magnetic control, lower momentum
Positrons	0.511	+1	Yes	~10–15%	Same as electrons, annihilation issues
Muons	105.7	±1	Yes	~10%	Heavier, but short life 2.2 μs, rapid acc'n
Pions ($\pi\pm$)	139.6	±1	Yes	~15–20%	Decays to pions to muons, use promptly
Kaons ($K\pm$)	493.7	±1	Yes	~15–20%	Magnetic control, but decay quick
$W^{\pm}$ Bosons	80,400	±1	No	<5%	Very massive, decay quickly
Quarks	2.2–4,180	±1/3–2/3	Yes	~20–30%	Hadronize quickly
Top quarks	173,000	+2/3	No	~0	Massive, decay too fast
Gravitons	0	0	No	~0	Extremely weak interaction with matter
Gluons	0	0	No	~10%	Immediate hadronize
Photons	0	0	No	~5–8%	No magnetic field control, neutral
Neutrinos	~0	0	No	~20–25%	Non-interacting, neutral

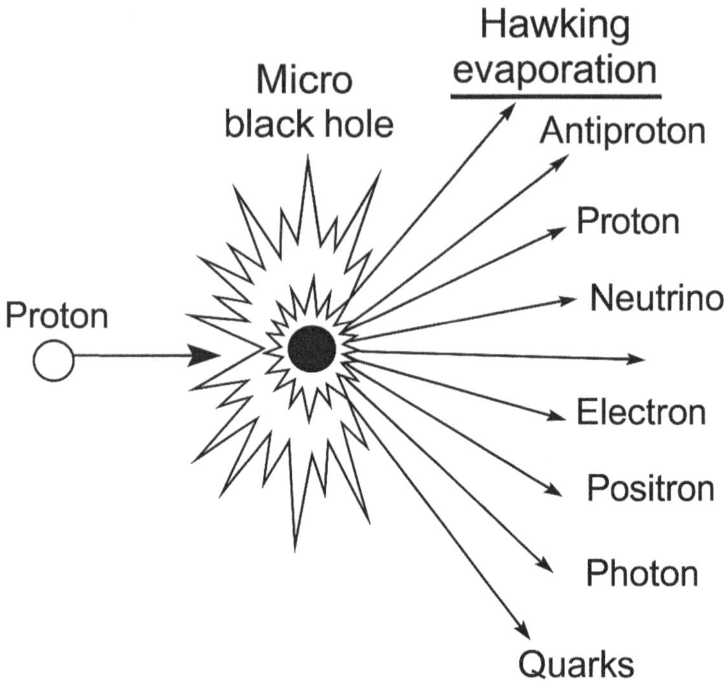

Fig. 13.5 Illustration of black hole evaporation

spectrum and so the quoted percentages are merely indicative of what we might expect rather than precise values.

It will be clear that the ideal particles that come out of the Hawking emission process are those that are charged, contain sufficient mass for momentum, and are reasonably well long lived so that it is practical to magnetically control them before they decay. As illustrated in Fig. 13.5 the primary particle candidates then are the electrons, muons and charged hadrons, such as protons and Pions (π^+, π^-) or Kaons (K^+, K^-) particles, which emerge from the quark hadronization process, which is where they turn into other particles within 10^{-24} s such as Pions, Protons and Kaons. Since they have mass and charge these hadronized particles will carry momentum which can be used for thrusting. The exception to this is the Top Quark which is too massive and then decays rapidly. Other particles either do not meet the criteria or are simply lost to space.

From the Hawking evaporation, it is important to be able to estimate the number of particles that are emitted. This is referred to as the multiplicity N which is given by the ratio of the total energy available to the average energy per particle

$$N = \frac{Mc^2}{k_B T_H}. \tag{13.32}$$

Calculating this for a 7 TeV proton or 1.121×10^{-6} J, which has an equivalent mass of 1.247×10^{-23} kg, where the Planck scale in the ADD model we have $M_* = 1\,\text{TeV} = 1.602 \times 10^{-7}$ J. The Hawking temperature assuming $n = 6$ dimensions is then given by

$$k_B T_H = M_* \left(\frac{M_*}{M_{BH}}\right)^{1/(n+1)} = 1.602 \times 10^{-7} \left(\frac{1.602 \times 10^{-7}}{1.1214 \times 10^{-6}}\right)^{1/7}$$
$$= 1.213 \times 10^{-7} \text{ J}. \tag{13.33}$$

And the temperature by

$$T_H = \frac{k_B T_H}{k_B} = 8.7875 \times 10^{15} \text{ K} \tag{13.34}$$

Note that in the above we have used the temperature conversion

$$k_B T_H \sim M_* \left(\frac{M_*}{M_{BH}}\right)^{1/(n+1)}. \tag{13.35}$$

So that to get temperature from energy we have

$$T(K) = \frac{k_B T(J)}{k_B}. \tag{13.36}$$

Therefore

$$N = \frac{M_{BH} c^2}{k_B T_H} = \frac{1.2477 \times 10^{-23} \times c^2}{(1.3806 \times 10^{-23})(8.7875 \times 10^{15})}. \tag{13.37}$$

Alternatively in natural units of energy, where $\hbar = c = k_B = 1$ we find

$$N = \frac{1.1214 \times 10^{-6}}{1.2132 \times 10^{-7}} = 9.243 \text{ particles}. \tag{13.38}$$

Table 13.2 shows the results of calculating this for the different number of dimensions. We see that within the ADD gravity model when the black hole mass is held fixed, increasing the number of extra-dimensions raises the Hawking temperature and reduces the multiplicity, as well as shortening the lifetime of the evaporation. The number of particles emitted may not seem very large but recall this is just from a single evaporation event. There will be many such events occurring simultaneously, depending on the density of the interstellar medium and the speed of the ship.

Now that we have shown the likely emission spectrum from a micro-black hole evaporation, we can consider how this can be used for the purpose of propulsion. This is achieved by guiding the protons to collide, collapse inside

Table 13.2 Black hole temperature, lifetime and multiplicity vs extra dimensions

n	$k_B T_H$ (J)	τ_H (s)	N (particles)	$dN_i/dtdE$ (particles/s/J) [$\times 10^{18}$]
2	8.3746×10^{-8}	1.6853×10^{-26}	13.3905	3.9418
3	9.8489×10^{-8}	1.2185×10^{-26}	11.3860	5.1407
4	1.0855×10^{-7}	1.0030×10^{-26}	10.3304	5.9928
5	1.1583×10^{-7}	8.8099×10^{-27}	9.6816	6.6190
6	1.2132×10^{-7}	8.0303×10^{-27}	9.2433	7.0950
7	1.2561×10^{-7}	7.4912×10^{-27}	8.9276	7.4677
8	1.2905×10^{-7}	7.0970×10^{-27}	8.6895	7.7667
9	1.3187×10^{-7}	6.7967×10^{-27}	8.5037	8.0115
10	1.3423×10^{-7}	6.5604×10^{-27}	8.3546	8.2155
11	1.3622×10^{-7}	6.3698×10^{-27}	8.2323	8.3881

their Schwarzschild radius, and then Hawking evaporate leading to the emission of charged particles for thrust generation using a magnetic nozzle. This will result in an energetic and directionally controllable exhaust. This avoids the need to moderate the protons as is required in the interstellar ramjet and uses the natural frame of the interstellar protons since they are already moving at fast speed. This is therefore a potentially relativistic flight concept, made especially possible, since it would not have to carry any of its own fuel, except perhaps for the initial acceleration phase to bring the vehicle up to a reasonable velocity where the incoming proton flux is sufficient.

The main charged particles associated with the thrust generation will be electrons, muons, pions and quarks that decay to hadrons. These are then deflected using magnetic fields and through the thrust chamber. Some degree of proton focussing on the inlet may be required to ensure that the plasma density is controlled and manageable.

13.3 Black Hole Performance

Figure 13.6 shows a more detailed concept layout for such a vehicle. The payload, which may be robotic or crewed, is hidden behind a large radiation shield which also doubles as a heat management system. This protects the payload from Hawking radiation. A large CERN style collider ring is used to help channel the incoming protons so that they can collide within the reaction chamber.

Let us make some reasonable assumptions to gauge the potential performance of such a machine. Since the Hawking radiation produces ultra-relativistic particles, we can approximate the exhaust velocity as being close to the speed of light $v_{ex} \approx c$ and we will assume $0.99c$ for simplicity. This means that the average particle velocity

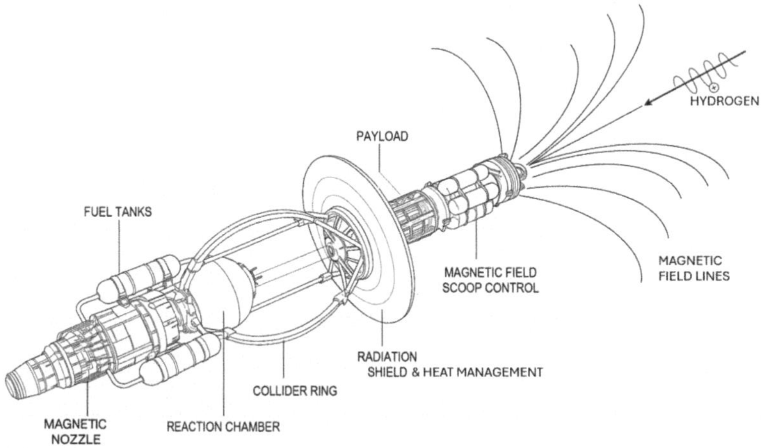

Fig. 13.6 A concept for a black hole augmented interstellar ramjet (BAIR) engine[KFL]

from the spectrum will be given by

$$v_{ex} \approx \frac{pc^2}{E} = 2.9679 \times 10^8 \text{ m/s.} \tag{13.39}$$

From the particles emitted in the Hawking evaporation, we might assume that around $\eta \sim 60\%$ of the mass-energy goes into charged particles that we can use for thrusting. If we then assume that the emission corresponds to around 1 microgram per second evaporated, then this gives a mass flow rate of $\dot{m} = 10^{-9}$ kg/s or 0.001 mg/s. The thrust will then be

$$T = \eta \dot{m} v_{ex} = 0.6 \times 10^{-9} \times 2.9679 \times 10^8 = 0.178 \text{ N.} \tag{13.40}$$

This will correspond to a jet power of

$$P_j = \frac{1}{2}\eta \dot{m} v_{ex}^2 = \frac{0.6}{2} \times 10^{-9}(2.9679 \times 10^8)^2 = 26.43 \text{ MW.} \tag{13.41}$$

The exhaust velocity is also equivalent to a specific impulse of

$$I_{sp} = \frac{v_{ex}}{g_o} = \frac{2.9679 \times 10^8}{9.81} = 3.025 \times 10^7 \text{ s.} \tag{13.42}$$

So with ~0.001 mg/s flow rate the thrust is 0.178 N. If we assumed a total spacecraft mass of around 1,000 tons, then from this we can compute the required acceleration $a = F/m = 1.782 \times 10^{-7}$ m/s^2. Then using the equation of motion discussed in Chap. 1 for distance and rearranging

$$t = (\frac{2S}{a})^{1/2}. \tag{13.43}$$

This results in a trip time of around 30,312 years to the nearest stars at 4.3 light years. This assumes a constant acceleration halfway followed by deceleration for the second half. If instead we increased the mass flow rate to 1 mg/s the thrust would be 178 N and the trip time 958 years. Yet to construct a vessel that can carry humans we need to go faster. If we increased the mass flow rate to 10 mg/s, 100 mg/s or 1,000 mg/s, then the trip times would be of 303 years, 96 years and 30 years respectively. As can be seen in Table 13.3 with a mass flow rate exceeding 100 g/s it may be possible to reach the nearest stars in a trip time of less than a human lifetime.

For the high velocity missions at 28.4%c, 89.8% this corresponds to a Lorentz factor of $\gamma = 1.043$ and $\gamma = 2.27$ respectively. For the other missions the relativistic effects are negligible. Although these calculations are somewhat speculative, they do demonstrate the potential of such a fantastic engine. Note that the velocities shown in Table 13.3 are (non-relativistic) kinematic peak velocities based on $v_{peak} = at/2$ and not a cruise velocity. The mission profiles also assume constant acceleration and a linear thrust with a continuous mass flow rate evaporation of black holes.

Shown also in Table 13.3 is the number of 7 TeV black holes evaporations required per second $\dot{N}$ to match the mass flow rate specified. This is estimated from

$$\dot{N} = \frac{\dot{m}}{M_{BH}} \tag{13.44}$$

So in the case of a 7 TeV black hole evaporation, which as we have seen will have a mass of 1.246×10^{-23} kg, for a mass flow rate of 1 mg/s, this suggests we would need 8×10^{16} black holes evaporation events per second. This would inform the required size and magnetic field strength of the proton ramscoop positioned at the front of the spacecraft.

The BAIR concept offers the potential to solve the energetic proton issue at low capture cross section so associated with the interstellar ramjet, which stands as a technical bottleneck to its fruition. Although it does depend on the existence of TeV scale black hole formation. In addition, when we follow the physics its leads to extreme aerospace engineering, as expected, so that such a machine would not be technologically possible today, even if the physics worked.

The BAIR concept combines elements of quantum gravity, starship engineering and plasma propulsion. In effect, it uses protons from interstellar hydrogen but not in a fusion reaction and so is not a fusion engine; although one may be required at the beginning to bring the vehicle up to high velocities where the incoming proton flux is sufficient to enable black hole formation. Its performance depends on the precise particle emission rates from the Hawking evaporation and how the mass and charged particles are distributed across a spectrum. Whilst the thrust associated with BAIR may end up being very small, the specific impulse is potentially enormous which

Table 13.3 Potential mission profiles for black hole evaporator engine

$\dot{m}$ (mg/s)	T (N)	P_j (MW)	a (m/s^2)	v_{peak} (km/s)	v/c (%)	t_{tot} (Yrs)	$\dot{N}$ (7 TeV BHs/s)	γ
0.001	0.17	26.43	1.78×10^{-7}	85.11	0.028	30,312	8.03×10^{13}	$\sim$1.0
0.01	1.781	264.3	1.78×10^{-6}	269.15	0.089	9,585	8.03×10^{14}	$\sim$1.0
0.1	17.81	2,642	1.78×10^{-5}	851.13	0.28	3,032	8.03×10^{15}	$\sim$1.0
1	178.1	26,426	1.78×10^{-4}	2, 691.5	0.89	958	8.02×10^{16}	$\sim$1.0
10	1,780	264,261	1.78×10^{-3}	8, 511.28	2.84	303	8.025×10^{17}	1.0004
100	17,807	2,642,609	0.0178	26, 915.03	8.98	96	8.025×10^{18}	1.0041
1,000	178,076	2.64×10^{6}	0.178	85, 112.79	28.39	30	8.025×10^{19}	1.0429
10,000	1.78×10^{6}	2.64×10^{8}	1.781	269, 150.29	89.78	9.6	8.025×10^{20}	2.2705

makes it ideal for a propellant efficient deep space exploration vessel; although how to practically collect the fuel is yet a separate challenge. Compared to the interstellar ramjet, the proton flux may be a lot less, but this makes the propulsion system more manageable, since what we need is a steady influx of particles and not a chaotic tsunami which is too much to handle.

The probability of black hole formation is likely to be a strong function of the proton density within the magnetic scoop collection. We can estimate the probability of black hole formation by formulating an equation for the number of black holes formed per second $\dot{N}_{BH}$, which depends on the number of proton collisions per second $\dot{N}_p$, the effective interaction length L of the scoop collision zone, and the cross section for black hole formation σ_{BH}, defined by the hydrogen density in the interstellar medium n, the speed of the spacecraft v and the scoop area A as follows

$$\dot{N}_p = nAv, \tag{13.45}$$

$$\dot{N}_{BH} = \dot{N}_p n \sigma_{BH} L, \tag{13.46}$$

$$\dot{N}_{BH} = n^2 A v \sigma_{BH} L. \tag{13.47}$$

The probability per proton pair is then given by

$$P_{BH} = \frac{\sigma_{BH}}{\sigma_{tot}}. \tag{13.48}$$

Where σ_{tot} is the total inelastic proton-proton cross section at the energy, which might be around ~60–100 mB at TeV scale energies. So for example using a density of $n = 0.3\,\text{cm}^{-3}$, $v = 0.9c$, scoop area $A = 10,000\,\text{km}^2$, $L = 10\,\text{m}$, $\sigma_{BH} = 10^{-34}\,\text{m}^2$, this computes to an estimated 0.000243 black holes/second formed, or 1 black hole every 4,115 seconds, which is likely inadequate for a good acceleration profile. In the earlier Table 13.3 we calculated a required production rate of order $\sim 10^{20}$ black holes/s.

Table 13.4 shows three examples of how the production rate would change if we increased several of the parameters. In Model (b) we still only get to $\dot{N}_{BH} \approx 9 \times 10^8$ black holes/s. In model (c) we also significantly increase the density, which might occur for a very dense molecular cloud, and then we obtain something like $\dot{N}_{BH} \approx 10^{20}$ black holes/s. In model (d) we reduce the density by an order of magnitude but increase the length and area of the scoop, to also give us like $\dot{N}_{BH} \approx 10^{20}$. Readers are invited to play around with these parameters to design a better machine. These models do not represent realistic engineering designs but merely illustrate the sensitivity of the black hole formation rate to the different parameters of the design.

To get anything approaching the required black hole rate shown in Table 13.3 would imply a very long interaction length, significantly enlarged area and a much higher proton collision rate. This suggests that in its current form the black hole evaporator engine would not be sufficiently efficient to convert protons into black hole evaporators and then into charged particles, to enable the required

Table 13.4 Hypothetical Ramscoop parameters for black hole engine

Parameter	Model (a)	Model (b)	Model (c)	Model (d)
L (km)	0.01	100	100	1,000
R (km)	56.42	10,852	10,852	34,318
A (km^2)	1×10^4	3.7×10^8	3.7×10^8	3.7×10^9
n (m^{-3})	3×10^5	3×10^6	3×10^{12}	1×10^{11}
σ_{tot} (m^2)	1×10^{-34}	1×10^{-32}	1×10^{-32}	1×10^{-32}
$\dot{N}_{BH}$ (s^{-1})	2.43×10^{-4}	8.98×10^8	9.98×10^{19}	9.98×10^{19}

thrust. Another option would be to carry a substantial portion of fuel on-board the spacecraft so that the number of protons present in the system can be artificially augmented, but that is another problem altogether. As mentioned there are also alternative models to the ADD one, which could be explored to see if the constraints become easier. The black hole engine would certainly make a good thesis project for a student to extend the work presented here towards a more feasible concept.

Whilst we have presented a concept for the BAIR, we must also acknowledge some of the difficulties. Firstly, this is a speculative idea based on the assumption of large extra dimensions for which physics does not yet have a solid experimental foundation. We have then used the ADD model to demonstrate the potential, and this is just one version of quantum gravity that we might have adopted. As well as the charged particles there will be a substantial amount of radiation associated with each evaporation in the form of x-rays or gamma-rays and so this suggests a need for a high degree of shielding.

Not to mention that the particle density of the ISM is extremely sparse and this is the same problem for the interstellar ramjet. We might be looking at values of $n = 0.1-1$ particle/cm^3 which is equivalent to $10^5 - 10^6$ particles/m^3. Flying through this medium at relativistic speed, the number of particles collected per second would then be given by $\dot{N} = nAv$, where v is the spacecraft velocity and A is the scoop area.

If instead we were to fly the vehicle at very low speeds, even if we use a very large scoop the mass flow rate is likely to be very small, and insufficient material would be collected for a high collision probability. There is likely a threshold performance, below which the BAIR becomes non-viable and it is defined by these parameters. Similarly, it may be that even at very high relativistic speeds it becomes a difficult engineering challenge to manage the number of black hole detonations. This then suggests there is likely a sweet spot in performance for the operability of the BAIR. This might suggest that integrating a BAIR engine with an interstellar ramjet at high velocity and a propellant carrying fusion engine at low velocity, would represent a good hybrid design solution.

In addition, we must acknowledge that so far, the LHC at CERN has not observed TeV scale black hole production which would be a key signature of extra-spatial dimensions as predicted by the ADD model explored in this chapter. Current results appear to exclude semiclassical black holes with masses below ~9 TeV and quantum black holes below ~5–6 TeV, across 2–6 extra dimensions [17]. So far the experiments seem to rule out scenarios with M_{*} ~1 TeV from ADD-like models that would predict black hole production within the collider. If extra-dimensions do exist, they are so small and so few that we are struggling to detect them and for now we must be content with the Planck scale which is much higher than the energies accessible to the LHC, in the regime of ~10–100 TeV. However, such ideas are certainly fun to think about, and it is through exercises like this that innovative machines to the stars can one day be designed and constructed. To go far, we must think big and we must have the courage to compute the seemingly impossible if our imagination is to become reality.

Problems of Chap. 13

13.1 Assume an interstellar ramjet travelling at a speed of ~2/3rd$c$ that collects hydrogen at a rate of 1.2536×10^{26} protons per second though an ISM density of 10^{7} m^{-3}. What would be the radius of the ramjets collector? For the same speed, what would be the kinetic energy of a helium-4 particle as it approaches the large ramscoop?

13.2 Assuming classical physics (no higher dimensions) calculate the Schwarzschild radius of a single proton? Then estimate the Hawking temperature, emitted radiation power and evaporation time?

13.3 (i) Calculate the thrust of a BAIR engine assuming the mass flow rate into the engine 50 mg/s and an exhaust velocity that is 10% less than the speed of light. If the spacecraft has a total mass of 10,000 tons, how long would it take to reach Alpha Centauri at 4.3 light years distance? (ii) What was the cruise velocity achieved as a percentage of light speed? (iii) Calculate the ISM proton density required for a BAIR engine which has a magnetic scoop size of 10,000 km^{2} and is travelling at $0.99c$, to detonate 100 black holes per second. For the calculation assume $L = 10$ m and a cross section of $\sigma_{BH} = 100$ pb, where 1 pb $= 10^{-12}$ barns and 1 barn $= 10^{-28}$ m^{2}.

References

1. Ventura T, Robert W (2007) Bussard's last interview. American antigravity, Transcript by Mark Duncan, 14 October 2007
2. Bussard RW (1960) Galactic matter and interstellar flight. Astronaut Acta 6:170–194. Fasc. 4
3. Zubrin RM, Andrews DG (1990) Magnetic sails and interstellar travel. JBIS 43:265–272
4. Whitmire DP (1975) Relativistic spaceflight and the catalytic nuclear ramjet. Acta Astronaut 2:497–509
5. Bond A (1974) An analysis of the potential performance of the ram augmented interstellar ramjet. JBIS 27(9):674–685
6. Martin R (1973) Magnetic intake limitations on interstellar ramjets. Astronaut Acta 18:1–10
7. Khachatryan V, Sirunyan AM, Tumasyan A et al (2011) Search for microscopic black hole signatures at the large Hadron Collider. Phys Lett B 697(5):434–453
8. Dimopoulos S, Landsberg G (2001) Black holes at the LHC. Phys Rev Lett 87:161602
9. Bleicher M (2007) How to create black holes on Earth? Euro J Phys 28(3):509

10. Long KF (2009) The black hole evaporator engine: a theoretical propulsion scheme using TeV scale quantum gravity. Paper in draft
11. Alexander A (2014) Project BAIR: the black hole augmented interstellar rocket, Msc Thesis. International Space University, Strasbourg (co-supervised by K. F. Long and C. Welch)
12. Oppenheimer JR, Snyder H (1939) On continued gravitational contraction. Phys Rev 56(5): 455–459
13. Arkani-Hamed N, Dimopoulos S, Dvali G (1998) The hierarchy problem and new dimensions at a millimeter. Phys Lett B 429(3/4):263–272
14. Arkani-Hamed N, Dimopoulos S, Dvali G (1999) Phenomenology, astrophysics, and cosmology of theories with submillimeter dimensions and TeV scale quantum gravity. Phys Rev D 59(8):086004
15. Randall L, Sundrum R (1999) Large mass hierarchy from a small extra dimension. Phys Rev Lett 83(17):3370–3373
16. Randall L, Sundrum R (1999) An alternative to compactification. Phys Rev Lett 83(23):4690–4693
17. CMS Collaboration (2017) Search for black holes and other new phenomena in high-multiplicity final states in proton-proton collisions at Sqrt(s) = 13 TeV. Phys Lett B 774:279–307

Chapter 14
The Starship Engineer

With proper planning in the coming years, it would be hoped that the political need, when it comes, can be met with a minimum disruption to our then ongoing interplanetary space programmes and our life here at home on Earth. With planning, we can produce an interstellar exploration programme that will not only send us to the stars once, but will produce a strong technology base for our continued exploration of the Universe and the determination of our place within it.

— Robert Forward, 1976 [1]

14.1 Advanced Propulsion Commentary

Propulsion Synthesis

When examining the different spacecraft concepts described in this book that utilise a nuclear pulse engine, it will be clear that there is a range of parameters which are strongly dependent on the spacecraft design but also the type of mission. This also depends on the mission requirement such as a 100-days Mars mission versus a 100-year interstellar mission, and so it should be no surprise that the engine performance is also different. In general, an interstellar mission will be many orders of magnitude higher in terms of propulsion requirements when discussing performance. This is why interstellar flight is so difficult compared to interplanetary flight.

We can consider a range of operational performance values for the different mission concepts. For the interplanetary missions (out to $\sim$100s AU) data is included for the VISTA Mars mission and the IFR Pluto mission and the HWN concept. For the interstellar missions (out to $\sim$100,000s AU) data is included for Project Daedalus, Resolution, Endeavour, Pegasus and the studies by Halyard. We do not include the data from the Longshot study due to issues of credibility of design and we do not include the NASA Schulze work since in this author's opinion the study leaned on VISTA for guidance and so does not add new information. In between these two regimes will be interstellar precursor missions but these are not considered here [2]. We have also excluded the external nuclear pulse Orion from

K. F. Long, *Stellar Engines*, Springer Aerospace Technology,
https://doi.org/10.1007/978-3-032-18062-9_14

Table 14.1 Estimates for mission performance parameters for spacecraft concept designs applied to interplanetary and interstellar missions

Mission parameter	Interplanetary	Interstellar
Exhaust velocity (km/s)	152–3,260	4,540–10,600
Payload mass (tons)	50–1,500	100–450
Propulsion mass (tons)	300–1,000	100–2,660
Propellant mass (tons)	3.72–11	2,000–256,000
Cruise velocity (km/s)	110–1,542	13,557–37,153
Trip time (years)	0.137–0.197	49–1,404
Pulse frequency (Hz)	1–500	13.7–2,500
Capsule mass (g)	0.00288–0.046	0.00288–23
Driver energy (MJ)	33–83.3	30–2,700
burn-up fraction (%)	30–47.6	4.2–17.5
Capsule gain	12.7–1,500	18–66
Mass flow rate (kg/s)	0.00015–1.5	0.0048–0.713
Thrust (MN)	0.024–0.24	0.0218–7.54
Jet power (GW)	2.67–53.8	1,220–40,050
Specific power (kW/kg)	16.1–257	2,860–40,500
Jet efficiency (%)	10.3–42	81–98

this data and of course the fantastic propulsion concepts which are not sufficiently mature in definition.

Table 14.1 shows the results of collating this data for spacecraft propulsion design parameters that are important in any assessment of nuclear pulse methods. Once again, we do not observe any strong trends in terms of designs which implies non-optimal solutions and a lack of rigour in design selection. We can make the following statements however (1) interstellar designs require a much higher value of exhaust velocity, jet power and specific power when compared to interplanetary designs, in order to achieve the higher cruise velocity (2) the quantity of propellant required for interstellar missions greatly exceeds that required for interplanetary missions by several orders of magnitude (3) a very high jet efficiency seems to suggest a very low burn-up fraction and in order to achieve a very high burn-up fraction in fusion engines it is likely that the jet efficiency would be low.

Table 14.2 shows an example of pulse frequencies and pellet masses adopted for various space propulsion engines. There is a diverse range of values although the pulse frequencies were chosen to fit the acceleration profile of the mission. It is clear however that bar some exceptions (i.e. VISTA) few others have an insight into what sort of pellet mass and pulse frequency to use in a space propulsion mission, and this was a part of the motivation for deriving some of the equations that have featured in this book. Indeed, it points towards the non-optimality of the designs and the need for a consistent modelling tool. This author has previously attempted such an effort with some success [3].

These range of values also contrast with those used for the NIF which may utilise $\sim$mg capsule masses at $<<1$ Hz pulse frequency as an experimental laboratory [4, 5]. Future facilities can be expected to operate at 10s Hz. Apart from VISTA, Daedalus and Pegasus most of these design concepts had pellet parameters (mass,

Table 14.2 Pellet parameter space for different spacecraft concept designs

Concept	Fuel	m_{cap} (mg)	f_{hz} (Hz)	ϕ	G
Exp. Lab.	DT	≤ 1	$<< 1$	~0.05	1.5–3
Inert.React.	DT	≤ 1	1–50	0.1–0.15	10
HWN	DD	15	500	0.30	25
IFR	DD	15	100	0.64	1,000
VISTA	DT	46	1–30	0.476	1500
Daedalus 1st	D^3He	2,840	250	0.175	66.6
Daedalus 2nd	D^3He	288	250	0.133	33.2
Halyard	DD	23,000	15	0.042	7.0
Pegasus (a)	D^3He	72	1,000	0.133	23.1
Pegasus (b)	D^3He	28.8	2,500	0.133	18.8

pulse frequency, burn-up fraction, gain) which were not based on any rigorous reasoning.

In reference [6] it was observed that for a given capsule mass, an increase in the propellant mass will require a higher pulse frequency to burn the capsules for the given mission parameters. This is also evident in the equation

$$f_{hz} = \frac{N_{cap}^{tot}}{t_b} = \frac{m_{prop}}{m_{cap} t_b}. \tag{14.1}$$

For a constant m_{prop}, as $m_{cap} \to 0$, $f_{hz} \to \infty$, which tends towards a state of non-pulsing propulsion also known as continuous propulsion such as proposed for the NASA Discovery-II design concept [7], which has the purpose of minimising re-circulating power fraction loss and maximising the fraction of useful charged power output, whilst minimising radiator mass. Similarly, for the same constant, it is observed that as $m_{cap} \to \infty$, $f_{hz} \to 0$.

In the limit of infinitesimal capsule mass (tending towards a single particle) and infinite pulse frequency this is an analogue of an interstellar ramjet [8]. Such a spacecraft would not have any significant quantities of propellant on-board other than for the initial start-up of the fusion engine, but instead the fuel would be mined from the interstellar medium as the spacecraft moved at high velocity. The interstellar medium is mostly hydrogen, although proton-proton fusion has an incredibly small cross section. The reason the pulse frequency would be so high is due to the fact the spacecraft would be moving at extremely high velocity typical of speeds approaching that of light, and so the mass flux of particles coming into the ramjet will also be high. Any fusion reactor must be able to fusion this mass flow rate rapidly or moderate the protons down to a lower energy before using them in any fusion reaction.

In the scenario where $m_{cap} \to m_{prop}$ and the pulse frequency is minimised. This may represent very few pulse detonations per second, where this can be more appropriately described as a set of nuclear explosions than a controlled set of pulse detonations since the thermonuclear masses are so large. These sorts of scenarios then move away from being described as an ICF engine to a nuclear bomb engine such as was proposed for the Project Orion study [9], a massive 400,000 tons spacecraft of which 300,000 tons was the fuel mass.

A real-world example of this technology may only involve say 10 detonations per second with the explosive products transferring momentum to a large pusher plate which rests on shock absorbers. Although the pulse frequency would be low due to the nature of the bomb units the energy coming out of each one would be substantial resulting in high acceleration over a short duration up to a cruise speed of around $\sim 0.05c$ (atomic bombs) and $\sim 0.1c$ (hydrogen bombs) [9]. So, an ICF engine is really a form of nuclear pulse rocket resulting from design evolution, but in the limit of small capsule masses and higher pulse frequency which changes the mode of engine operation (i.e. laser target reaction chamber instead of a pusher plate).

In the scenario where $m_{cap} = m_{prop}$ for a single detonation but in the limit of a very small m_{prop} this would be representative of an inertial fusion laboratory experiment where say only a single shot is fired in a day allowing time for analysis of the data. In the limit of extremely large m_{prop} tending towards an astrophysical scale in size and for a single event, this would be representative of the detonation of a single star by nuclear fusion in a supernova, leaving only a minor core remnant behind; although it is important to point out that whilst all the other systems are confined by the inertia of the material, in the case of a star it is confined by gravity. Yet what we are trying to visualise here is a kind-of continuum of events over a large range when taking in the appropriate limits.

The different scenarios by which thermonuclear ignition may be used are depicted in Table 14.3 and Fig. 14.1. The approximate values shown for the pulse frequency and capsule mass should not be considered absolute limits but illustrative of the likely domain of applicability. It is acknowledged that the operation principles

Table 14.3 Illustrative synthesis for the range of fusion detonation types from the lowest mass to the highest mass and with the expected detonation frequency

Interstellar propulsion	*Relativistic ramjets*
$(v \sim 0.01–0.15c)$	$(v \sim 0.5–0.99c)$
$f_{hz} \sim$ 100–1,000s Hz	$f_{hz} >$ 100s–1,000s Hz
$m_{cap} \sim$ 1–10,000s mg	$m_{cap} <$ mg (effective), but high $\dot{m}$
Interplanetary propulsion	*Non-relativistic world ships*
$(v \sim 0.001–0.01c)$	$(v \sim 0.01–0.03c)$
$f_{hz} \sim$ 10–100 Hz	$f_{hz} \sim$ 1–50 Hz
$m_{cap} \sim$ 1–100 mg	$m_{cap} \sim 1–10^7$ mg
Inertial fusion energy	*Nuclear (bomb) propulsion*
$f_{hz} \sim$ 1–50 Hz	$(v \sim 0.001–0.05c)$
$m_{cap} \sim$ 0.01–1 mg	$f_{hz} \sim$ 1–50 Hz
	$m_{cap} \sim 10^7 - 10^{10}$ mg
Laboratory experiments	*Stellar supernova*
$f_{hz} <<$ 1 Hz	$f_{hz} <<$ 1 Hz (single shot)
$m_{cap} \sim$ 0.001–1 mg	$m_{cap} \sim 10^{33+} mg$

of each engine are different, but what is being presented here is a synthesis on how they are all connected. By taking the mathematical limits of the capsule mass or the pulse frequency and everything in between, one can envisage a vast scale of different types of systems. This helps to illustrate for example how a nuclear pulse engine when taken to one limit ends up as an ICF engine, and then how this engine when taken to another limit ends up as a (quasi) continuous propulsion engine and eventually analogous to an interstellar ramjet. The effective pulse frequency for a ramjet represents the limiting case of very small effective unit mass at ultra-high event rate and this is distinguished from actual capsules. Note that in Fig. 14.1 the mass flow rates neglect the possible use of multiple engine stages which would increase the total the mass flow rate.

Finally, a comment on the idea of self-similarity as applied to inertial fusion systems; although we exclude the stellar supernova from this observation since its engine operates at a vastly different scale. The above discussion pertains to a range of pellet masses, driver energies, efficiencies, plasma densities, material response, hydrodynamics, transport and sizes, different types of reactions and associated particle emissions. In considering these there is an implied understanding that the capsule operation will all be the same provided the design configurations are also similar. This would be an observation of self-similarity, where no matter the scale of the operation and physics response should be the same. This is likely over a sub-set of ranges, but at some point, this assumption becomes invalid and particularly when we are dealing with ns $\rightarrow$ μs $\rightarrow$s timescales and sub-mm $\rightarrow$cm spatial scales over energy ranges of sub-MJ $\rightarrow$100s MJ.

We note that even between the ICF capsule masses at $\sim$1 mg discussed for non-relativistic world ships [10, 11], this is still many orders of magnitude difference in mass for the next one up which is for the nuclear bomb propulsion systems which stated here begin at around $\sim$10s–100s kg and effectively would represent small mini-hydrogen bombs. The transition between the two regimes would likely be manifest of a different regime of physics operation and therefore the assumption of self-similarity may break down. It may therefore not be appropriate to assume that under variation in drive conditions over a range of capsule masses and pulse frequencies the physics operation of the implosion engine will be characterised by *Scale Invariance*.

Yet in the gain scaling algorithms discussed in Chap. 5 this is exactly what has been done using a power-law gain scaling like $G \propto E_{dr}^{1/3}$ for example, as a heuristic interpolated across nearby operating points. This also highlights the limits to our discussion of a synthesis. This may also be the reason why the ignition physics models for the designs were increasingly problematic and required even thicker ablator shells to generate the implosion pressure. This also brings into question the validity of the $G = 1{,}500$ assumption of VISTA, and whilst the Project Daedalus utilised $G = 33 \rightarrow 66$ this is arguably more conservative as a design solution despite the more ambitious mission.

As an aside this author previously worked on the Breakthrough Initiatives Project Starshot advisory committee and made this exact point in relation to the use of

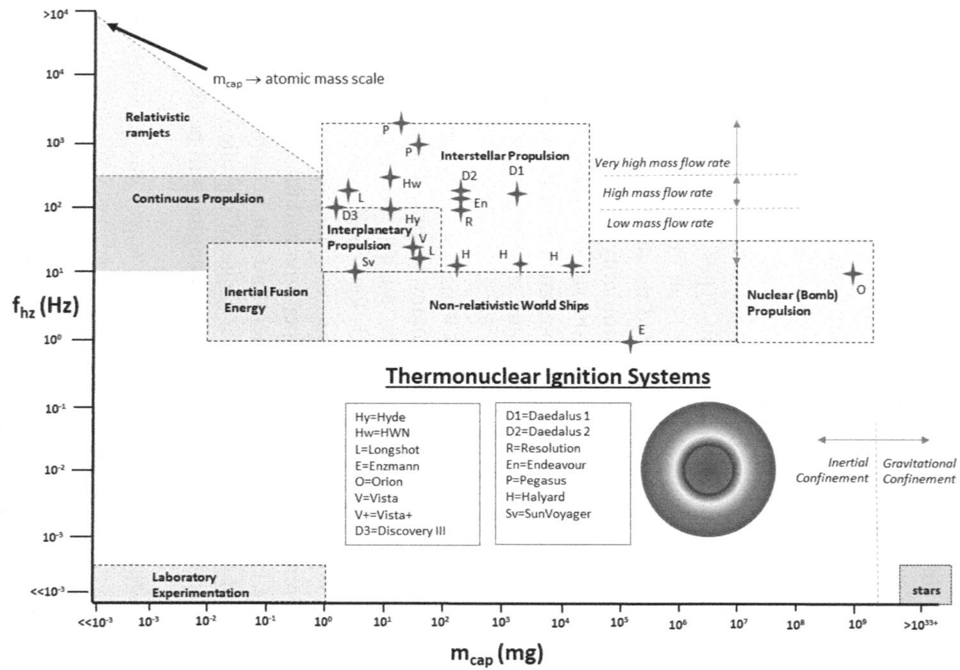

Fig. 14.1 Illustrative synthesis (parameter map) of different uses for fusion energy with approximate domains of application to include design points from the literature[KFL]

laser-beaming sails. The project was looking at the use of ~100–200 GW beamers projected onto a ~3–4 m sail in orbit, to be accelerated up to 60,000 km/s or 0.2c. Yet the prediction of the system performance would be based on experiments and material response at much lower powers. This emphasises the importance of experimental validation and prototype testing in the support of theoretical models.

Getting back to the nuclear pulse propulsion, a question arises as to the limits of pulse frequency that may be possible. For a space propulsion system, one might imagine that the minimum would be around 1 Hz; although frequencies less than this can be envisaged for nuclear pulse Orion type systems, and values less than this are representative of inertial fusion reactors or laboratory experiments. One can imagine a situation however when the first space propulsion systems are built where capsules are detonated at say once per minute or 0.016 Hz. That said, if the capsule masses were sufficiently large, perhaps as a transition between ICF propulsion and nuclear pulse, this might require very low pulse frequencies. Examples might include one detonation every six minutes or 1/360 s (0.0027 Hz) or at an extreme even one detonation every hour or 1/3600 s (0.00027 Hz). The choice of a low pulse frequency would be driven by the requirements for the mission and the engineering practicalities of the engine operation and how rapidly each cycle can be repeated. Going beyond this would suggest an engine performance that is not ready for the applications of space propulsion and that it is still in the experimental demonstration phase. Utilising too low a pulse frequency will result in very long acceleration burn periods and so elongate the mission profile, unless the average energy released is the same as over many smaller detonations.

The upper boundary on the pulse frequency will depend on how rapidly the spacecraft can fire a laser or other energy driver necessary to create the fusion reaction. A pulse frequency of ~100s Hz is the typical wing flapping frequency of small insects. It is feasible that with sufficient technological developments it may be possible to reproduce this within our engineering state of art. A 100 Hz pulse frequency is the equivalent of firing a laser 100 times per second or every 0.01 s or 10^{-2} s (1 centisecond). Going beyond this will take the frequency into 10^{-3} s (1 milli-s), 10^{-6} s (1 μs) or even 10^{-9} s (1 ns). As the frequency gets smaller, the time-scale eventually becomes faster than the physics processes at work. This includes the physics associated with the hydrodynamic implosion, ignition, thermonuclear burn, capsule disassembly all of which takes place over the $ns - \mu s$ range. There is also the plasma expansion within the nozzle, structural material response time and the energy bootstrap time to power the next pulse cycle, all of which occur over longer timescales than nuclear reactions.

14.1.1 Hypothetical ISS-NIF Spacecraft

Since in this book we have speculated on some fantastic machines, it is constructive to end this analysis on a sober footing of where we are today. For this we will simply look at NIF and imagine if it was a space propulsion engine and then speculate

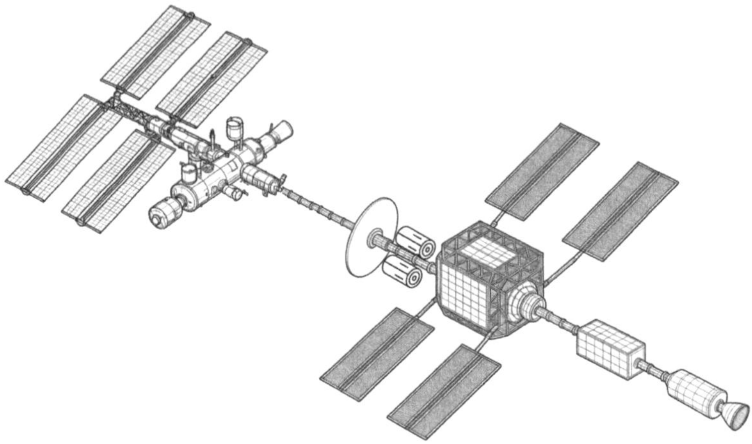

Fig. 14.2 Illustration of ISS-NIF-like spacecraft to Mars[KFL]

on its likely performance for a mission to Mars. For this we will assume a total spacecraft dry mass equivalent to what we can build in space today, which is the ISS at around $m_o = 450$ tons. This would include the propellant, engines, structure, lasers, payload and any other systems.

Such a concept is illustrated in Fig. 14.2, although this should not be taken too seriously since there are many reasons why this representation may not be practical. Instead, this should be considered as a thought experiment so we can gauge the technological status today compared to where we need to be, to build the sorts of visionary machines discussed in this book.

Note also that that NIF is highly inefficient (at <1%) which would suggest that any such machine would need a substantial nuclear reactor power source, as opposed to having access to a national electrical grid. But we can assume that in the future this efficiency will improve at least. For this exercise, in the interest of simplicity we will assume a perfect nozzle collimation and a unity jet efficiency.

We will adopt the ICF capsule designs shown in Fig. 5.4 of Chap. 5 which includes a fuel mass in the of order 0.227–0.4 mg, as representative of what we might be able to test with an outer radius not exceeding 0.82–1.04 mm. The capsule is driven by a 2.05 MJ laser driver (351 nm ultra-violet light using 192 beam lines). To be optimistic, we will assume a high burn-up fraction of $\phi = 10$–20% is achievable after say ten years of experiments and using a DT fuel with $\alpha = 0.0038$. This would imply a potential exhaust velocity of $v_{ex} = \sqrt{2\alpha\phi}c = 8{,}264$–$11{,}688$ km/s. This would also correspond to an energy output of $E_{out} = 7.75$–27.32 MJ which is a gain of $G = 3.8$–13.3.

It goes without saying that the shot rates at NIF are very low since this is an experimental facility. Currently a shot might be repeated on a timescale on the order of every ~4–8 hours which is equivalent to a pulse frequency of 0.0000694–0.0000347 Hz. If they were able to increase this the time between shots to around

1 hour then this would be equivalent to 0.000277 Hz. Taking the mass flow rate $\dot{m} = m_{pell} f_{hz}$ this gives us a value of $\dot{m} = 6.305 \times 10^{-8}$ kg/s $= 63.05$ ng/s. Then from this we compute the thrust $T = \dot{m} v_{ex} = 9.509 \times 10^{-4}$ N, which is a minuscule value. If we make the assumption that $m_f = m_o - m_{prop}$, then from the ideal rocket equation and rearranging the mass of the propellant is then given by

$$m_{prop} = m_o(1 - e^{-\Delta v / v_{ex}}). \tag{14.2}$$

If we launch from Earth orbit at LEO this requires around ~9.5 km/s. For a Hohmann transfer from Earth orbit to Mars this would require a hyperbolic excess velocity of ~2.9 km/s. However, the required injection velocity for Mars would be ~3.7 km/s . Therefore we specify this as a minimum velocity requirement to aim for.

For a mass of 0.227 mg for each pellet, this would compute to 110 kg of fuel and would suggest we need of order $N = 4.86 \times 10^8$ pellets. However, the problem with this set up is that it would require an acceleration time that is of order thousands of years, so this won't work as it is. Instead, we increase the pulse frequency to $f_{hz} = 30$ Hz, similar to VISTA, and with a total fuel mass of 201.4 kg this leads to a cruise velocity of 3.7 km/s, after 342 days acceleration. This would be followed by 72 days at the cruise velocity so that the total number of days for the mission would be 414 days.

Table 14.4 shows a set of calculations with some of the assumptions altered, and this includes for total velocity increments of Δv, $2\Delta v$ and $3\Delta v$. The aim was to at least get to a mission architecture that was of order ~100–150 days in duration and this is shown in the final two rows with a fuel mass of order ~0.5 mg and a (Project Daedalus-like) pulse frequency of 250 Hz. Such missions would require ~427–604 kg of DT fuel. The different calculations are shown and a mission to Mars is possible in around ~100–110 days. We note that in the above analysis we neglected fuel mass and time for deceleration which is left as an exercise for the reader.

It is left for others to play around with these numbers, but it seems to this author that there is a set of parameters we can imagine for a NIF-ISS type scenario where the mission becomes feasible in the coming decades. Critical to any such system will be to significantly increase the laser efficiency which will reduce the power demand and the total system mass, which is significantly dominated by the waste heat radiators, as well as the laser driver and engines. Yet, this was a fun exercise in seeing what might be feasible on paper.

14.2 Starship Design Philosophy

In this section we are going to discuss the subject of design philosophy when it comes to starship engineering. There are many similarities with conventional aerospace engineering, but since it is speculative and typically at the boundaries of extrapolation capacity, there are also key differences that rely on our creative

Table 14.4 Calculations for a NIF-ISS propulsion parameters assuming 2.05 MJ driving energy using DT fuel for a direct flyby mission to Mars

m_{cap} (mg)	N_{shots} ($\times 10^9$)	f_{hz} (Hz)	$\dot{m}$ (g/s)	T (N)	v_{ex} (km/s)	ϕ (%)	E_{out} (MJ)	G	M_{fuel} (kg)	Δv (km/s)	T (Days)
0.227	0.8872	30	0.0068	56.3	8,264.7	10.0	7.75	3.78	201.41	3.7	415
0.227	0.6274	30	0.0068	79.6	11,688.1	20.0	15.51	7.56	142.43	3.7	364
0.4	0.5035	30	0.012	99.2	8,264.7	10.0	13.66	6.67	201.41	3.7	340
0.4	0.3561	30	0.012	140.3	11,688.1	20.0	27.32	13.33	142.43	3.7	312
0.4	1.0068	100	0.04	330.6	8,264.7	10.0	13.66	6.66	402.74	7.4	180
0.4	0.7120	100	0.04	467.5	11,688.1	20.0	27.32	13.33	284.82	7.4	163
0.5	1.2079	250	0.125	1,033.1	8,264.7	10.0	17.08	8.33	603.97	11.1	109
0.5	0.8543	250	0.125	1,461.0	11,688.1	20.0	34.15	16.66	427.16	11.1	101

capacities. In terms of epistemology, we might consider how we think about starship design. In terms of methodology this pertains to the process we use to design a machine. In terms of motivational philosophy, this is manifest of our dreams, and vision for the future we want to engineer.

This author often acts as a peer reviewer on others technical papers for academic journals. In particular, one will frequently read a new paper on a starship design that has been newly conceived, or it may be a perturbation of an existing idea. When reading such papers, one is in a similar position to a patent clerk, and several questions comes to mind. Does the propulsion rely on validated experimental physics? Does the design rely on extrapolations within known scaling laws or are we in the domain of complete uncertainty? Does the design rely on untested but physics theorems which are mathematically consistent? Does the design rely on unknown physics that has no proven basis in reality?

Looking into the future can be particularly difficult since it is filled with unknowns. In general, it is acceptable to extrapolate out to a few decades but beyond that the performance of technology is difficult to imagine. The further into the future one attempts to imagine, the larger the uncertainty. Having thought about this a little, it may be related to the two different types of uncertainties, known as systematic and random. For technologies in the near-term extrapolation, they will likely be dominated by systematic uncertainties in attempting to predict their likely performance. But as one goes further into the future, the random uncertainties become more dominant. This is also because other technologies are being invented and they may have an influence on each other. An example would be imaging technology in an optical lens system integrating with electrostatic printing technology to produce a photocopier or Xerox machine. Another example would be the integration of a mobile phone, computer, camera and GPS to produce a smartphone. Similarly, it is difficult to say how one technology will influence another, so that advanced propulsion systems for space mature earlier than envisioned. NASA created the NASA Horizon mission methodology as a way of looking into the future [12]. We should also keep in mind that famous quote from the writer Arthur C Clarke "*The only way to discover the limits of the possible is to go beyond them into the impossible*" [13].

When pondering the balance between the scientifically rigorous and the speculative imagination, one can think of a case in point where this situation has occurred before. That is in the life and work of the Renaissance man Leonardo da Vinci during the 15th Century. During his lifetime not only did he create amazing pieces of artwork, but he also designed fantastic machines that could traverse land, sea and air such as the Ornithopter shown in Fig. 14.3. Yet, due to the immaturity of the technology of his day those machines could not be built during his lifetime, but many centuries later some of them did appear. This included tanks, helicopters, parachutes, aircraft. This is how the starship designer must see themselves, as they work on machines and spacecraft concepts, they are unlikely to see in their lifetimes. They must believe that one day, generations hence, their ideas may be built by someone. In this regard, the starship designer is a child of the Renaissance man.

Fig. 14.3 Illustration of Leonardo da Vinci Ornithopter[KFL]

The design of conventional spacecraft design systems is a well-trodden field of study that has produced all the spacecraft we see in the modern age. This largely depends on a process of systems engineering and the application of our physics knowledge in the construction of innovative machines for the domain of space. The principles upon which such machines are made, is also dependent upon the known laws of physics and our current technological maturity. Although we must not forget the modern innovations being made by companies like SpaceX with an emphasis on reusability.

In this section, we discuss some of these issues, with a particular emphasis on bridging the gap between plausible reality and the imagined conjecture, the latter of which is the domain of science fiction. Since whatever machines we imagine today, we can be sure that in the future they will likely be far surpassed by principles we are yet to uncover within the laws of nature.

A subject that should be addressed is the fidelity of the calculations conducted in any assessment of a starship machine. At one extreme there is turning the 'handle' on the ideal rocket equation $R = e^{\Delta v/v_{ex}}$ to work out the mass ratio required for a given mission and then the total masses consistent with that value. This is the starting basis of any calculations, and we might refer to this as the pre-concept definition level, otherwise known as scoping the design space.

At the other extreme there is taking a spacecraft, breaking it up into its multiple systems and sub-systems, calculating the performance, margins and uncertainties of each part, and then computing how they all fit together as a part of a fully integrated machine; hopefully much of this supported by experimental validation. This would be a detailed definition and would include a specification for each subsystem and its environmental regimes of operation.

In between these two extremes is the preliminary definition phase of design. Where many calculations have been conducted on the vehicle to appropriately sketch out its configuration and mass budget and even system performances, but where there are still several sub-system areas that are not addressed or are incomplete. It could be argued that the Project Daedalus study in parts achieved a preliminary level of vehicle design and for this reason we might refer to it as the first ever systems integrated starship design study in history.

There have been many other concepts and ideas presented in published papers, books or even at conference presentations, but covering most of the engineering systems and calculating how they integrate is often a key element that is lacking. Yet, going through these different phases is a necessary pathway towards a starship blueprint: (1) *Mission requirement* (2) *Concept Design* (3) *Preliminary Design* (4) *Detailed Design*. This must also be accompanied by design traceability, failure mode analysis where possible, sensitivity analysis such as using numerical codes. Then there is building models which can be adjusted at ease to observe scaling behaviour as the design evolves.

In considering the overall motivation behind the design of innovative starships, it is useful to quote the British astronomer Arthur Eddington, who emphasised the value of pushing assumptions until they break:

> *Of Icarus; In ancient days two aviators procured to themselves wings. Daedalus flew safely through the middle air and was duly honoured on his landing. Icarus soared upwards to the sun till the wax melted which bound his wings and his flight ended in fiasco. The classical authorities tell us, of course, that he was only "doing a stunt"; but I prefer to think of him as the man who brought to light a serious constructional defect in the flying-machines of his day. So too in science. Cautious Daedalus will apply his theories where he feels confident they will safely go; but by his excess of caution their hidden weaknesses remain undiscovered. Icarus will strain his theories to the breaking-point till the weak joints gape. For the mere adventure? Perhaps partly, this is human nature. But if he is destined not yet to reach the sun and solve finally the riddle of its construction, we may at least hope to learn from his journey some hints to build a better machine.* [14].

This underscores the fundamental purpose of any spacecraft design initiatives, which is to design a better machine, learning the lessons from other similar projects. Although it also has an additional function of helping to train up the next generation of designers interested in calculations relating to interstellar studies. A beautiful synergy between physics and engineering, which has to arc over the guides of scientific rigour but allowing for optimistic speculation; a challenge much harder than one realises.

Hard science demands quantification of uncertainties, and a certain level of conservative thinking where a healthy level of scepticism on new insights is required until the burden of proof is laid bare, often supported by levels of standard deviation. Yet, any scientist that demands too high a standard will forever find themselves confined to a comfort zone and any possible worlds of the imagination will forever be hidden from their gaze. It is not necessarily the case that a very good scientist is also a great scientist. Whilst good science requires rigorous adherence to the scientific method, great science often involves leaps of the imagination and guesses that may not be fully justified, instead informed by instinct.

Starship engineering can also be referred to as *extreme aerospace engineering*, due to the extremes of physical properties often encountered hence the sub-title of this book. This includes temperature, velocity, power, energy, pressure to name a few. Anyone attempting to conduct this most unusual exercise of free speculation about the possible, feels a certain level of joy, unbounded by the constraints of the physical world, and only guided by the laws of physics. Indeed, they are not really seen as laws, but variables upon which in one type of universe or another they can be manipulated—and as God creates reality, starship engineers create the infinite worlds of the imagination. Yet, becoming lost in this game one can find oneself straying too far from the practical, or even what is reasonable speculation and into pure fantasy.

A case in point is the original conception of warp drive engines in the *Star Trek* world. The instructions of its creator Gene Roddenberry to the design team were highly interesting and worth highlighting, and potentially warn us of the risks of a purely aesthetic constraint that is divorced from physics:

> *We're a hundred and fifty or maybe two hundred years from now. Out in deep space, on the equivalent of a cruiser-size spaceship. We don't know what the motive power is, but I don't want to see any trails of fire. No streaks of smoke, no jet intakes, rocket exhaust, or anything like that. We're not going to Mars, or any of that sort of limited thing. It will be like a deep-space exploration vessel, operating throughout our galaxy. We'll be going to stars and planets that nobody has named yet. I don't care how you do it, but make it look like it's got power* [15].

This is great for science fiction, and such a concept is illustrated in Fig. 14.4. However, it is a matter of debate on whether such a concept could one day be made a practical machine that could fly. Although we must also acknowledge that this sort of creative approach eventually led to an entirely new field of mathematical physics research known as Warp Drive using the General Theory of Relativity [16] which continues into the modern era [17, 18]. Time will show if this research leads to a dark alley of nothingness, or to the invention of machines that can change the world and literally bring our imagination into existence.

Another interesting story is the development of the Discovery-I for the movie *2001: A Space Odyssey* , where the original design concept, developed by Arthur C Clarke and others, had large wings which served the purpose of waste heat radiators. However, the Director Stanley Kubrick apparently had these removed on aesthetic grounds which is why you don't see any in the movie. Yet as we have seen in the earlier chapter 8 for the NASA Discovery-II and this author's own Discovery-III, in the real world engineering really can't avoid the physics.

It should also be noted that some hard science fiction stories can in fact be based on reasonably credible physics and engineering. This includes Torch ships in *Time for the Stars* (Heinlein [19]), relativistic interstellar ramjets in *Tau Zero* (Anderson [20]) and laser beamers in *The Mote in God's Eye* (Niven and Pournelle [21]). This also includes quantum vacuum drives in *The Songs of Distant Earth* (Clarke [22])

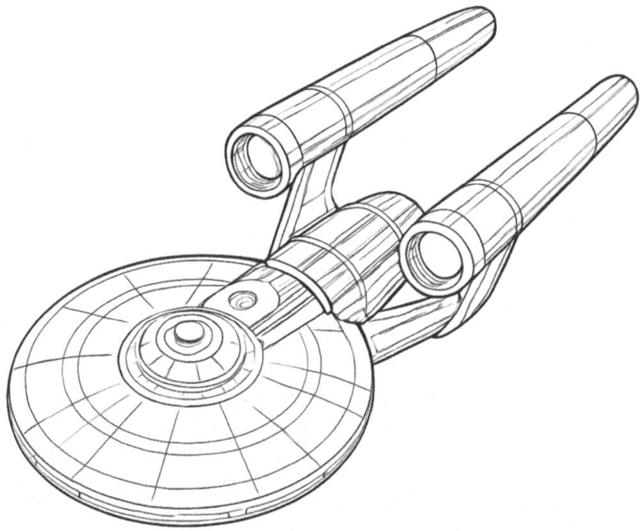

Fig. 14.4 Illustration of faster-than-light warp drive starship[KFL]

and microwave beamers in *Flight of the dragonfly* (Forward [23]), and antimatter annihilation reactions in *Flying to Valhalla* (Pellegrino [24]).

Looking around at designers of the interstellar community, one observes that indeed two categories of design animals exist. For want of a better word, let us call one the *Demons* and the other the *Angels*. The demon always criticises and aligns themselves with pessimism, with its favourite phrases uttered loudly to the chorus "*that is impossible*" and "*where is the evidence?*", killing all the hope of those that might have otherwise tried. Whereas the angel sees no limits in the mind of the imagination and what humanity can achieve, always aligning itself with positive optimism. Its favourite phrase "*It might just work*" and "*What will it take?*", whispered into the ear of those who would listen, and the advice of any nearby demons are simply treated as irrelevant background noise to a play that is destined to end with ambitious attainment.

The effective starship designer must then wear these angels and demons on each shoulder. To listen to both but not be overly persuaded by either. To search out the middle road of optimality and compromise. To boldly go where no-one has gone before, but to do so with careful navigation and leave a map so that others may follow the reasoning which led to the new solutions, inventions or innovations that may literally open up the stars and the far cosmos beyond.

For any design teams, the ideas and thinking straddle between the credible near-term and the speculative far-future, and all the way to the fantastic. At its heart, any starhip design projects became a creativity engine for ideas. In such a project there may be different approaches taken by designers. In Project Icarus for example [25–27], one team stayed as close to Daedalus as possible as a design philosophy and then perturbed systems from there through calculations, using a unique numerical programming tool developed just for this purpose. Another team involved the use and development of research tool within a university, and the progress of the design were integral to that. One team constructed a spreadsheet which included all the spacecraft systems. Another team focussed on a 'form follows function', as is standard practice, and considered each system in turn until the vehicle configuration began to emerge no matter what it looked like; although since some of their systems had already been defined elsewhere this drove the design layout. Another team allowed the configuration to be driven very heavily by the physics calculations.

The point about all the above, is that very few designers sat down and generated a brand-new configuration from scratch. Instead, vehicle configurations emerged out of a combination of the original Project Daedalus design, what had been achieved by others, what was out there in the literature, what areas of science got their most interest and therefore fine resolution of scrutiny, and imagination. It was very much an organic and somewhat random process, perhaps at times lacking in methodology, and in what some may interpret to have more in common with art and poetry than with the standard systems approach to engineering design. We might refer to this as an emergent configuration synthesis, which is arguably a legitimate design mode for engineering problems that are extremely unconstrained.

The starship designer is fundamentally a human being, and it has two control systems. The first is the mind and the second is the heart. The mind tells it what laws of science it is to be guided by and what is credible and is not, and how to calculate the numbers necessary to generate a machine from the imagination. It is mainly driven by the physics understanding of what is possible and the engineering appreciation for what is practical. The heart hears a different song and instead is inspired by personal experience, although can still be guided by heuristic exploration. One may be watching a television documentary for example and then you see a technology in a science fiction film and shout "*I gotta have that in my design*". This is not logical but is a source of creative capacity.

This then speaks again to the starship designer, that they must not be driven solely by the mind or solely by the heart, but to listen to both. For it is in the synergy between the ideas of the mind and the beating of the heart by what moves us into action that serendipity is to be found and from which the machines are to materialise into reality.

What does it mean to construct a machine? Even if that is just an idea from the imagination that is never actually built. It means a careful consideration of engineering systems, their moving parts and how they interact, using the principles of physics that we understand. That machine has the function to do work which involves energy in and energy out and this requires a high degree of numerical accounting to ensure consistency with conservation laws.

In fact, there are many paths to building a starship based on different types of propulsion methods. The paths taken are typically a function of the level of interest shown by individual designers, but it is likely that given time and sufficient maturation of technology, they would all work to some degree of performance. We can then conclude that interstellar flight by some form of propulsion is likely to be

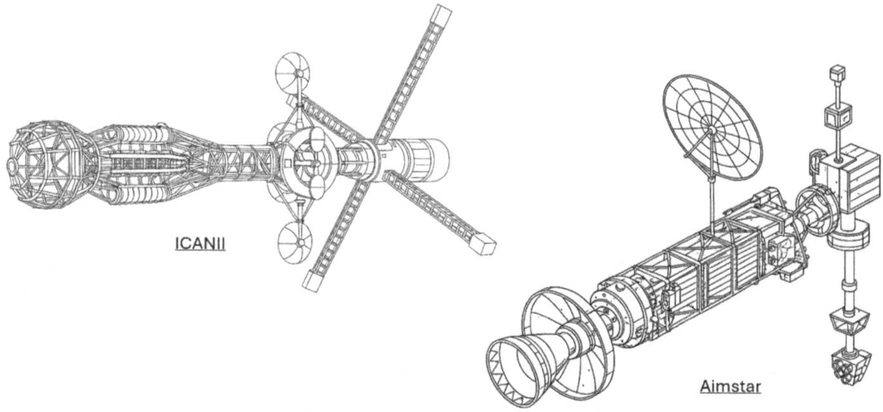

Fig. 14.5 Illustration of Antimatter-Catalysed-Fusion spacecraft[KFL]

a key feature of the human future.... one day... but possibly in the twenty-first and twenty-second Century. Although this author may not live to see such a launch, it excites me that my children may do so one day—the future is made from hope.

The author feels that he can say this with some confidence since there are so many ideas for the constructions of machines that can fly between the planets and the stars. So many in fact, that in this book we could not cover them all. Two promising concepts which deserve much more of an analysis is the antimatter catalysed fusion propulsion concepts called ICAN-II and Aimstar [28], which are illustrated in Fig. 14.5.

Another innovative concept is the Medusa sail, which marries solar sails with external nuclear pulse [29, 30]. Then there are smart ideas for deceleration like the Magnetic sail [31]. Many of these ideas were integrated together into a multi-modal propulsion system called Leviathan [32]. These sorts of concepts marry the visionary aspects of breakthrough physics with hard core engineering. Yet perhaps this will be the subject of another book by this author. Alternatively, you the reader may feel sufficiently inspired to go research them and pick up the baton of the designer. Can we really make such engines work? For if we could, this would most certainly be a game changer for our technological capability in space.

Producing designs in projects is great, but that is not the ultimate achievement of any project. Instead, it is motivating a human being to pick up a pencil and calculator and to consider an incredible future where the starship exists in some reality we may yet create. For this is the sure way towards inventing a self-fulfilling prophecy and is also the birthplace of all inventions. It is also how we manufacture a manifest destiny for ourselves out there among the stars.

Will humanity ever reach the stars? Will we build a spacecraft that can cross the vast expanse over light years distance and obtain images of exoplanets yet to be discovered? The future is not yet written and so nobody can say it will happen or it will not happen. Time must be allowed to unfold as it will, as the dance of both fate and destiny plays out to the background chorus of human geopolitical relationships. Yet, we perhaps can speak in terms of probability, and the existence of design projects over many years with all its associated noise and focus on the starship problem arguably makes such a future more probable.

Perhaps playing around with probabilities in an uncertain world is all we can do, but if starships are ever built, our hope is that the designers of that age will look back on what was done in the past and say we made a difference. This is ultimately the legacy of starship engineering, one for which we may not see ourselves, but will roll down through the centuries until the time is ready where the galaxy can finally be won. Someday, someone in the future may read some of the starship papers from the literature and then try to build such a machine.

This does of course raise the issue of if we can one day build such machines then others will too. This is a reference to the possibility of extraterrestrial intelligence that may exist out there. We may in fact find out that the galaxy was won by others a long time ago, and they are just waiting for us to reach 'Childhood's End' and join them among the stars. Perhaps the fact that the starship is both an engineering/physics challenge, but also a human challenge is itself a test. For to join any brothers and sisters among the stars we must first learn to conquer ourselves and attain the appropriate wisdom. This will also lead to the appropriate mindset that is required to design machines that can cross the stars to unseen worlds.

14.3 Implications of Advanced Propulsion

In terms of the peaceful application of nuclear fusion like technology, it clearly has the potential to generate clean and sustainable energy on our planet to help power our vast and expanding cities. Although there are other ways of achieving fusion ignition which are underway, such as the MCF method, the ICF method is also very clearly still in the running for how we might achieve 'stardom' on Earth.

On Earth the fusion facility NIF was developed to achieve ignition, but also to support the US nuclear stockpile program. This is needed due to the existence of the CTBT treaty which prevents further atmospheric and underground testing of nuclear weapons. NIF supports the US national weapons laboratories such as LANL and LLNL in underwriting their stockpile systems. NIF is a $\sim$2 MJ facility, but originally it was proposed to be a $\sim$10 MJ facility called the Laboratory Microfusion Facility (LMF). This was then reduced by a 1990 recommendation of the National Academy of Sciences (NAS) [33]. The French have also developed the Laser Megajoule (LMJ) which is a $\sim$1 MJ facility.

A successor facility may be along the lines of the originally proposed LMF and would perhaps operate at around $\sim$5–10 MJ. If such a facility were developed, given that NIF has already achieved ignition in 2022/2023 [34–36], it is perhaps more likely that this new facility would advance past the ignition phase to high gain and usher in a new era of inertial fusion energy facilities to power our cities, but also then attempts will be made to apply this new knowledge to other applications such as for space propulsion.

Space is often referred to as the last frontier, and many nations are already turning their attention to it in terms of the acquisition of materials which may be rare on Earth. This may include precious metals which are locked up within asteroids. Another is Helium-3 for which there are large sources of it deposited within the regolith soil layer of the Moon sourced from the solar wind. Although this will be of interest to commercial companies seeking to build a new profit line, some of those parties may also represent the interest of a particular nation state.

As nations seek to gain ground within this new frontier, the potential for conflict arises, be it over territory or materials. Nations are governed by the 1967 Outer Space Treaty (OST) [37] which is the principles upon which activities in outer space are formally regulated and this is an international legal agreement with many countries either a party or a signatory to the treaty. Specifically, the treaty prohibits the introduction of weapons of mass destruction in outer space.

Yet, it is an interesting observation that in principle any space propulsion technology can also be turned into a weapon and so how we manage this going forward will be a test of our ability to work co-operatively together. Indeed, it would be reasonable to say that a fusion propulsion system may be categorised as a 'dual-use'technology, in that it can be used for both peaceful nuclear power and military aims. In developing high performance space propulsion technology with high energy applications, the dual-use nature of this technology and its use in weapons of mass destruction is a real concern, but this should not prevent its development for the peaceful exploration of outer space provided this concern is well managed.

This would particularly be the case for very large fusion capsules which would come under the 1970 Non-Proliferation Treaty (NPT) (signed in 1968) otherwise known as the Treaty on the Non-Proliferation of nuclear weapons [38]. As we head towards becoming a fusion capable society these sorts of issues are likely to be more relevant. Yet the unavailability of relevant fusion physics data is also a hindrance to the advancement of the field given the potential benefit such technology may offer to humanity, provided improved co-operation between nation states can be facilitated around the shared goal of peaceful expansion out to the cosmos.

Historically, the US Air Force has kept an eye on developments with fusion propulsion [39]. Recently a report was issued on the state of the space industrial base 2021 [40] in which Helium-3 fuel was cited as being one of the motivations for both China and India to explore the Moon. The report states that "*the US, which first discovered the Helium-3 resource, and explained its significance to the world, should not cede the field*". The report also observes that the private sector is exploring fusion propulsion for fast and safe interplanetary space transportation.

The report stated that the "*demonstration of advanced propulsion capability would be a key enabler for quick expansion and proliferation of Cislunar infrastructure*". In a report written by prominent members of the US defence industry, it is quite clear that fusion fuels and fusion propulsion will become a military technological aim, if it is not already as a part of strategic defence plans, be it explicitly or implicitly. The authors are supporters of the vision for a US Space Force which now exists and will be an important part of our future, particularly in relation to nation state competition in space [41].

This underscores the importance of this book, which is making steps towards developing a designer capability to model advanced spacecraft propulsion systems. The better we can model those systems and describe how those future spacecrafts look and what their performance ranges are likely to be, the more information we have to inform our strategic thinking about the interplanetary civilisation that is emerging off world; to include civilian and military. Any nation that is not preparing like this will be ill-prepared for the future as the technologies mature to a high readiness level and appear a part of the real world. Good and reliable predictions now by various methods of extrapolation (rather than just science fiction) will help to de-risk potential conflicts for the future.

There is another implication for the development of an advanced propulsion capability that should be considered. It has been known for some time that certain technologies when introduced to society at too rapid a rate can have a disruptive impact on existing infrastructures. In 1960 a report was issued by the Brookings Institution [42] which discussed this very topic. The report stated the position that "*Certain potential products or consequences of space activities imply such a degree of change in world conditions that it would be unprofitable within the purview of this report to propose research on them. Examples include a controlled thermonuclear fusion rocket power source*".

The report also stated that "*The social and economic costs of rocket propulsion for humans and for freight need careful study before the implications, if any, can be evaluated....the implications, if any, for society will not be clear until the plans for and need of such projects are further clarified.*"

Further the report states "*...the development of a compact thermonuclear fusion power source for spacecraft would undoubtedly open up space for many kinds of large-scale activities - but it would very likely also change the political, social, and economic features of the Earth radically, since unlimited power for all uses would soon be available*".

These conclusions don't just apply to fusion technology but also any other technology which has a similar performance or even goes beyond it. It should be clear then that the development of an advanced propulsion capability will have huge implications for society at large. The fossil fuel industry will become irrelevant. The cost of energy will become cheap. People, companies and nations will be able to achieve more things for less effort. This will result in an accelerated step change in our technological progress in other areas. Society may change so rapidly that we are unable to keep up with that pace of change and therefore predict the ramifications. Just consider the implications of an efficient fusion reactor to powering an artificial intelligence supercomputer for example.

It would seem prudent then that whilst we are developing our capability to design and model fusion-based systems, that we also consider those technologies within a wider context. For Earth this is mostly in reference to inertial power generation for commercial reactors. For the interest of this book, it is mostly in the application to space propulsion systems. So that once we have derived viable spacecraft configuration and performance models, these are then considered in a mission context and how they would contribute to a broad strategic and technological roadmap within an interplanetary society.

The ability to send probes to distant destinations, be it interplanetary or interstellar, has the potential to significantly expand our knowledge base about the nature and content of the Universe. Just consider what organisations like the US space agency NASA and JPL has achieved from the 1950s through to the end of the last century in terms of the exploration of orbit, the Moon and all planets of our Solar System. What would such organisations achieve over a similar timescale if it had a fusion propulsion capability at its disposal?

If we can send robotic probes to the outer planets in a matter of months or weeks, then this opens the possibility of using the same technology to send human crewed missions to the same planets. This is because the significantly reduced ship time will also result in a reduced risk for the crew due to the reduced exposure time to space radiation, as well as the reduced bone density loss which increases the longer a mission is in a microgravity environment.

In addition, the human species faces some common threats in space for which may need this technology or others that are equivalent in scale. One of these is the threat of asteroids or Near-Earth-Objects (NEOs) impacting the Earth and causing mass devastation of life and infrastructure. As argued by others [43] the ability to send a spacecraft to an approaching object in rapid time is vital to deploying any successful deflection strategy. It also means that because the trip time is much shorter, we have longer to plan for how we might want to manage the risk and then spend time designing and constructing the probe that we need to deploy the appropriate mitigation strategy. The probability that our mitigation strategies in dealing with the threats from incoming Near-Earth Objects will be successful is increased if our space propulsion technology can travel incredibly fast.

Although some will argue this is less likely, as we continue to develop our technology that we may someday encounter intelligent extraterrestrial life. We would hope that such interstellar cultures are friendly in their approach towards us, but in any case, having a rapid propulsion technology gives us much more scope in terms of response options in the event of such contact.

For example, if we are to prevent the situation of being dominated by a superior species, then our technology must at least be comparable if we are to address them on equal terms. If we made contact it is likely to be with a species far older than ours and so our technology may also be inferior. This is based on astrophysical arguments, the age of the Universe and the ages of stars in our galaxy, as well as the plurality of exoplanets that have now been discovered which might suggest that life and intelligent life may be potentially ubiquitous in the Cosmos.

Closing that gap as much as possible may be critical to our survival or even just preventing the decay of our human culture when exposed to another that is more advanced technologically than our own. This opinion is informed by the many lessons of European cultures interacting with indigenous cultures and how those cultures developed subsequently. Some examples might include the British colonial interactions with the Aboriginal Australians, The North American colonist settler's interaction with the Native Americans, and the Spanish interaction with the Aztecs. The survival of our culture(s) and species at large when interacting with extraterrestrial civilisations will be increased if our technological prowess in space propulsion at the very least begins to approach that of a technologically advanced culture.

That said, we should also be aware that the human development of an interstellar capability has implications for how we might be perceived by observing extraterrestrial cultures. Let us take the example of the Project Daedalus probe [44] to highlight this point which is in relation to the existence of a spacecraft of large mass hurtling at high velocity towards another star system, which in flyby mode will have a significant kinetic energy. The classical (non-relativistic) kinetic energy is given as a function of the objects mass and velocity

$$E_{nr} = \frac{1}{2}mv^2. \tag{14.3}$$

In comparison, for the case of relativistic flight, the kinetic energy is given by

$$E_r = (\gamma - 1)mc^2. \tag{14.4}$$

Where the Lorentz factor $\gamma = 1/\sqrt{1-(v/c)^2}$, c is the vacuum speed of light. The energy can also be converted to a TNT equivalent (where Trinitrotoluene is a measure of explosive power) where 1 ton TNT $= 4.184 \times 10^9$ J of energy. Taking into account relativistic energy corrections, it will be equal to around 1.62×10^{18} J by the end of the first stage burn and 6.54×10^{17} J by the end of the second stage burn, the lesser value on the second stage because the first stage mass has been ejected and its propellant burned. This means that the probe when it arrives at Barnard's Star will have an approximate kinetic energy of $\sim$654 PJ. This is equivalent to around $\sim$156 Mtons TNT equivalent or $\sim$0.15 Gtons TNT.

For comparison Table 14.5 shows the quantity of events historically that an impacting Daedalus-like probe would have. This includes some of the first and the largest atomic and thermonuclear weapons ever tested but also with the Tunguska cometary impact event in Siberia which devastated vast areas of wilderness. The number given in the table is the quantity of these events that an incoming Project Daedalus probe would be equivalent to in energy if its kinetic energy were released suddenly such as in an impact event with a planetary body.

If such a probe were detected heading towards our planet, it is hard to believe we would not be extremely concerned. Although statistically the probability of a flyby probe of this nature hitting a planet or other solid astronomical object is incredibly low some on-board form of trajectory manoeuvring would be required in addition to long-range optical sensing of the flight path (both were included in the Project Daedalus spacecraft design) to minimise the risk.

Indeed, we should be searching for evidence of such artificial objects. Even if a spacecraft is restricted to fusion propulsion, we can in principle still detect

Table 14.5 Approximate calculations for kinetic energy equivalence (TNT) and quantity for a *Daedalus* probe approaching Barnard's star

Event	Energy	Number
Little Boy, Hiroshima, 06-Aug-1945	~15 ktons	10,420
Trinity, New Mexico, 16-Jul-1945	~25 ktons	6,252
Tunguska, Siberia, 30-Jun-1908	~15–30 Mtons	5–10
Tsar (Ivan), Novaya Zemlya, 30-Oct-1961	~50 Mtons	3

Table 14.6 Characteristics of ICF propulsion emission signatures

Reaction	Total energy (MeV)	T (keV)	$<\sigma_c v>$ (m^3/s)	$<\sigma_c v>/T^2$ (m^3/s keV^2)	ν (Hz)	λ (pm)
D(T,^{4}He)n	17.58	14	2.24×10^{-22}	1.14×10^{-24}	3.39×10^{18}	88.24
D(^{3}He, ^{4}He)p	14.67	60	7.45×10^{-23}	2.07×10^{-26}	1.45×10^{19}	20.59
D(D, ^{3}He)n	3.27	20	1.54×10^{-24}	6.85×10^{-27}	4.85×10^{18}	61.77
D(D, T)p	4.03	20	1.40×10^{-24}	6.21×10^{-27}	4.85×10^{18}	61.77
p(p, D)e$^+$	1.442	10	4.76×10^{-49}	4.76×10^{-49}	2.42×10^{18}	123.5

such vehicles as they approach our solar system. Such possibilities are discussed by other authors who suggested the detection of x-ray emissions associated with Bremsstrahlung and Cyclotron emission over ~1–100 keV plasma temperatures [45, 46]. For a given thermonuclear plasma temperature T this can be used to calculate the average energy of the radiation by $E = 3k_BT/2$ and the energy is converted via 1 eV $= 1.602\times10^{-19}$ J. The frequency ν of the radiation is then found from $E = h\nu$, where h is the Planck constant 6.626×10^{-34} Js. The wavelength is then given from $c = \lambda\nu$. Table 14.6 shows calculations by this author for when this occurs using some of the reaction cross sections discussed in Chap. 4 to even include the proton-proton chain reaction for application to an interstellar ramjet [8].

The frequencies and wavelengths shown in Table 14.6 for the different fusion reactions correspond to x-rays, which are characterised by a frequency range of $3 \times 10^{16} - 3 \times 10^{19}$ Hz and a wavelength range of 10–10,000 pm, where pico = 10^{-12}. This would be for a potential energy range of 145 eV up to 124 keV.

All the same, if any planets at the target star system are occupied by other life or even intelligent life, then an impact by a spacecraft of this mass and velocity would be devastating and this underscores the argument for a reconnaissance probe that conducts a full deceleration into the orbital system. One might speculate further that the maturation of human technology to the point of launching such missions would be of interest to any intelligent life existing on nearby star systems and may necessitate continued monitoring of human activity as a precaution. The development of the power, energy and propulsion technologies on Earth and in space would be of particular interest since any extrapolation requires a reliable point of origin. It also suggests that a high degree of computing intelligence (e.g. AI) and autonomous decision-making is required if a flyby probe is to be selected.

It should be clear that there are many benefits [47] to the development of an advanced propulsion technology, which includes fusion propulsion technology and alternatives methods discussed in this book. It should be a priority of the human civilisation at large to focus our research efforts on its maturation. That said, it is also clear that it should be approached with proper caution and wisdom so that we are mindful to the implications that may arise from its emergence in society. Perhaps all we can say for sure, is that a future where the starship emergence will be a spectacular one.

Problems of Chap. 14

14.1 Look up the 1967 Outer Space Treaty (OST) developed by the United Nations in the peaceful development of outer space. Identify in the text any words that might refer to advanced propulsion technology, even if indirectly or implied.

14.2 Calculate the kinetic energy of a 10 tons robotic probe entering the Solar System with a velocity equal to 0.2c. Assuming the probe impacted the Earth what would be its Gtons TNT equivalent in energy? Calculate it non-relativistic and relativistic and compare the difference in energy.

14.3 (i) In this book we have mostly discussed reaction engine type propulsion systems, although we have also mentioned some others such as solar sail and beamed propulsion systems. Do some research and list other methods that have been realised for travelling to the stars. (ii) Can you conceive of a method for getting to the stars using an advanced propulsion engine? This is a challenge for you the reader. It may be a completely new idea or a perturbation of an existing idea. Use the chapters and equations of this book to help you scope out the overall systems design in terms of energy, power, heat management and other engineering systems that may need consideration. You should start with a structured mission requirement such as a defined payload mass to a given target and design from there.

References

1. Forward RL (1984) Roundtrip interstellar travel using laser-pushed lightsails. J Spacecraft 21(2):187–195
2. Swinney RW, Long KF, Hein A, Galea P, Mann A, Crowl A, Obousy R (2012) Project Icarus: exploring the interstellar roadmap using the icarus pathfinder and starfinder probe concepts. JBIS 65(7/8):244–254
3. Long KF (2023) Development of the HeliosX mission analysis code for advanced ICF space propulsion. Acta Astronautica 202:157–173
4. Dewald E (2016) Status of indirect drive ICF experiments on the national ignition facility, LLNL, TR-686482
5. Meezan NB, Atherton LJ, Callahan DA (2010) National ignition campaign hohlraum energetics. Phys Plasmas 17:056304
6. Long KF (2022) On the application of pulse propulsion frequency in inertial fusion space mission design. JBIS 75(4):127–141
7. Williams CH, Dudzinski LA, Borowski SK, Juhasz AJ (2005) Realizing 2001: a space odyssey: piloted spherical torus nuclear fusion propulsion, NASA/TM-2005-213559, AIAA-2001-3805
8. Bussard RW (1960) Galactic matter and interstellar flight. Astronautica Acta 6:170–194, Fasc.4

9. Dyson FJ (1968) Interstellar transport. Phys Today 21(10):41–45
10. Martin AR (1984) World ships - concept, cause, cost, construction and colonisation. JBIS 37(6):243–253
11. Bond A, Martin AR (1984) World ships - an assessment of the engineering feasibility. JBIS 37(6):254–266
12. Anderson JL (1996) Leaps of the imagination: interstellar flight and the horizon mission methology. JBIS 49:15–20
13. Clarke AC (1962) Profiles of the future: an inquiry into the limits of the possible. Harper & Row, New York
14. Eddington A (1927) Stars and atoms. Oxford University Press, Oxford, p 41
15. Whitfield SE, Roddenberry G (1968) The making of star trek. Ballantine Books, New York City, p 79
16. Alcubierre M (1994) The warp drive: hyper-fast travel within general relativity. Class Quantum Grav 11:L73–L77
17. White H (2013) Warp field mechanics 101. JBIS 66(7/8):242–247
18. Taylor TS (2024) Modifications to the Alcubierre warp field metric in anisotropic matter and implications to detection of warp fields. JBIS 77(1):2–15
19. Heinlein R (1956) Time for the stars. Scribners, New York
20. Anderson P (1970) Tau zero. Doubleday, New York
21. Niven L, Pournelle J (1974) The mote in god's eye. Simon & Schuster, New York
22. Clarke AC (1986) The songs of distant earth. Del Rey Books, New York
23. Forward RL (1984) The flight of the dragonfly. Simon & Schuster, New York
24. Pellegrino C (1993) Flying to Valhalla. Avon Books, London
25. Long KF, Fogg M, Obousy R, Tziolas A, Mann A, Osborne R, Presby A (2009) Project Icarus: son of daedalus - flying closer to another star. JBIS 62(11/12):403–414
26. Swinney RW, Freeland RM, Lamontagne M (2020) Project Icarus: designing a fusion powered interstellar probe. Acta Futura 12:47–49
27. Swinney R, Freeland R, Lamontagne M, Long KF (2025) Project Icarus: a review of the project, spacecraft concepts and design process. JBIS 78(5):143–154
28. Gaidos G, Lewis RA, Schmidt MK, Smith GA (1999) AIMSTAR: antimatter initiated microfusion for precursor interstellar missions. Acta Astronautica 44(2–4):183–186
29. Solem JC (1993) Medusa: nuclear explosive propulsion for interplanetary travel. JBIS 46:21–26
30. Solem JC (1994) Nuclear explosive propulsion for interplanetary travel: extension of the medusa concept for higher specific impulse. JBIS 47:229–238
31. Zubrin RM, Andrews DG (1990) Magnetic sails and interstellar travel. JBIS 43:265–272
32. Long KF, Crowl A, Tziolas A, Freeland R (2012) Project Icarus: nuclear fusion space propulsion & the Icarus Leviathan concept. Space Chronicles 65(1):43
33. Welch RA (Chairman) (1990) National research council, review of the department of energy's inertial confinement fusion program: the national ignition facility, final report, National Academy of Sciences
34. Zylstra AB, Hurricane OA, Callahan DA et al (2022) Burning plasma achieved in inertial fusion. Nature 601(7894):542–548
35. Zylstra AB, Kritcher AL, Hurricane OA et al (2022) Experimental achievement and signatures of ignition at the national ignition facility. Phys Rev E 106:025202
36. Christopherson AR, Hurricane OA, Weber C et al (2023) Alpha-heating analysis of burning plasma and ignition experiments on the national ignition facility. Phys Plasmas 30(6):062705
37. Treaty OS (1967) Treaty on principles governing the activities of states in the exploration and use of outer space, including the moon and other celestial bodies, United Nations
38. United Nations. Office of Public Information (1970) Treaty on the non-proliferation of nuclear weapons, United Nations
39. Haloulakos VE, Borque RF (1989) Fusion propulsion study, final report for the period September 1987 to January 1989, McDonnell Douglas Space Systems Co and General Atomics, AD-A212 932, AL-TR-89-003

40. Olson J, Butow S, Felt E, Cooley T, Mozer J, Garretson P (eds) (2021) State of the space industrial base 2021, Infrastructure & Services for Economic Growth & National Security
41. Long KF (2025) The emergence of a fusion propulsion capability this century and implications for exploration and settlement strategies. In: Space education and strategic applications journal (SESA), conference proceedings, Spring
42. Michael DN (1960) Proposed studies on the implications of peaceful space activities for human affairs, A report prepared for the committee on long-range studies of the NASA, by the Brookings Institution
43. Wurden GA, Weber TE, Turchi PJ, Parks PB, Evans TE, Cohen SA, Cassibry JT, Campbell EM (2016) A new vision for fusion energy research: fusion rocket engines for planetary defense. J Fusion Energy 35:123–133
44. Bond A, Martin AR (1978) Project Daedalus: the mission profile, final study report. JBIS Special Supplement:S37–S42
45. Zubrin R (1995) Detection of extraterrestrial civilizations via the spectral signature of advanced interstellar spacecraft. In: Seth Shostak G (ed) Progress in the search for extraterrestrial life, ASP conference series, vol 74
46. Freitas RA (1983) The search for extraterrestrial artifacts (SETA). JBIS, 36(11):501–506
47. Long KF (2019) Interstellar probes: the benefits to astronomy & astrophysics. JBIS 72(11):370–375

Chapter 15
Imagining the Vision

> *To those who know the immensity of the void between the stars, the thought of interstellar flight must seem like visionary nonsense, utterly beyond the power of men to achieve ... To me, star flight appears as one of the great challenges of Nature, for I see the universe of stars as an arena that has been set for countless eons, patiently awaiting all comers. At any moment in time, any race - human or alien - that feels moved to pick up the gauntlet may do so. To whoever wins, the reward is survival*
>
> — *James Strong*, 1965 [1]

15.1 Images of Starships

In this book we have presented many advanced propulsion concepts, for spacecraft architectures that can help us explore the Solar System and beyond towards the nearest stars. Here we show some of these concepts visualised in colourful illustrations. These range from the near-term practical to the fantastic and speculative. Yet it is certainly interesting and thought provoking to consider these possibilities. In the next few pages, we will show colour images of some of these concepts so as to illustrate the ideas in more detail. These are shown in Figs. 15.1–15.20.

For some of you these images may represent actual machines that may indeed one day take flight to the stars, based on a reasonable extrapolation of existing principles in physics and engineering. For others, these will be viewed as mere fanciful creations of an overly imaginative mind and better left to the pages of science fiction. Yet, in either case, we must be reminded of the visions of the Renaissance man Leonardo da Vinci, who's machines were centuries ahead of his time. When taken in this context, the visions of the starship may in fact represent a form of prophecy on a future that is yet to be made. The only thing we can say for sure, is such machines will represent *extreme aerospace engineering* design challenges for those bold enough to attempt them.

K. F. Long, *Stellar Engines*, Springer Aerospace Technology,
https://doi.org/10.1007/978-3-032-18062-9_15

Fig. 15.1 NASA/JPL Voyager 2 spacecraft flypast of planet neptune[KFL]

Fig. 15.2 NASA/JPL proposed TAU spacecraft flypast of planet Uranus[KFL]

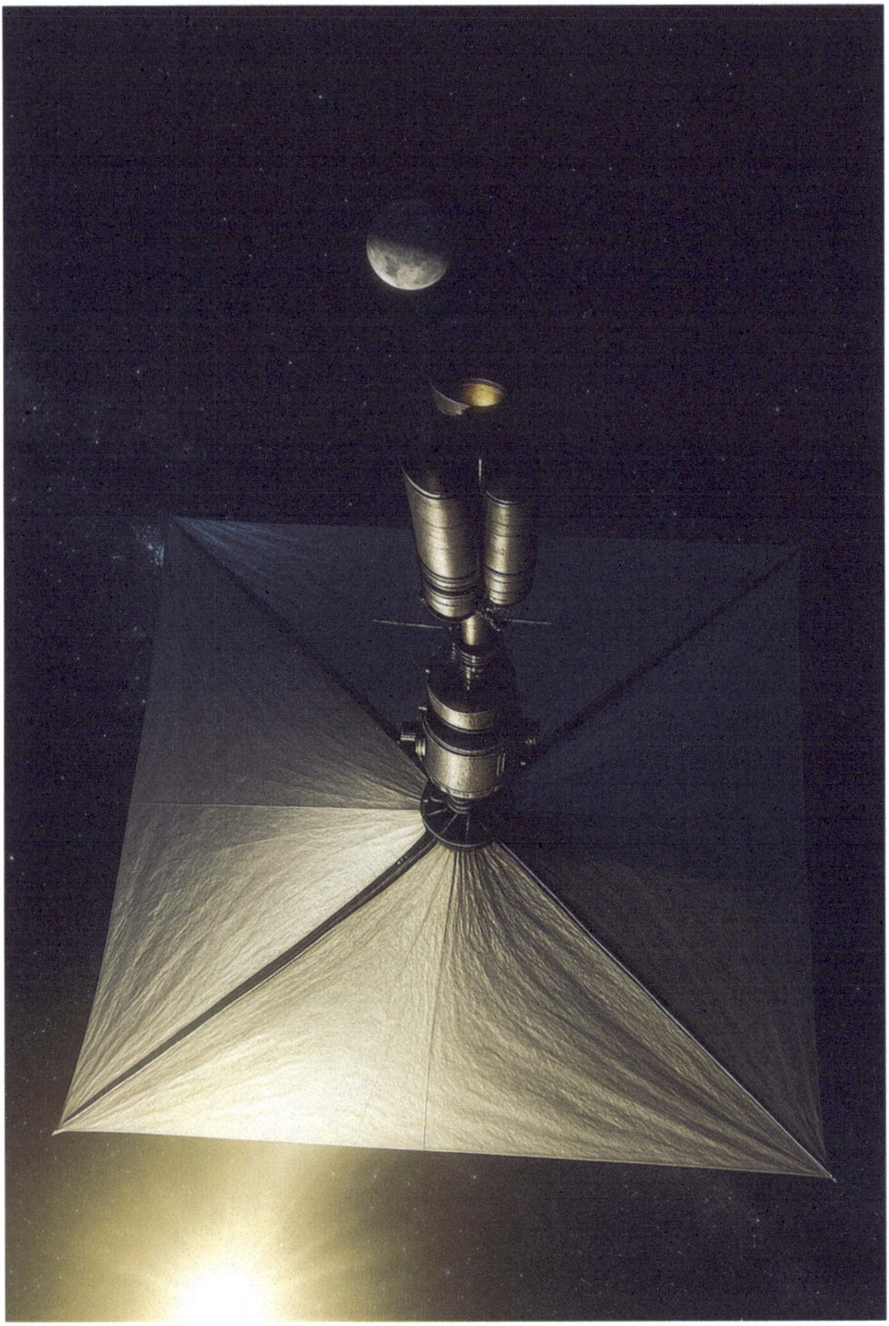

Fig. 15.3 Hypothetical hybrid design VAST spacecraft concept[KFL]

Fig. 15.4 LLNL theoretical VISTA spacecraft at planet Mars[KFL]

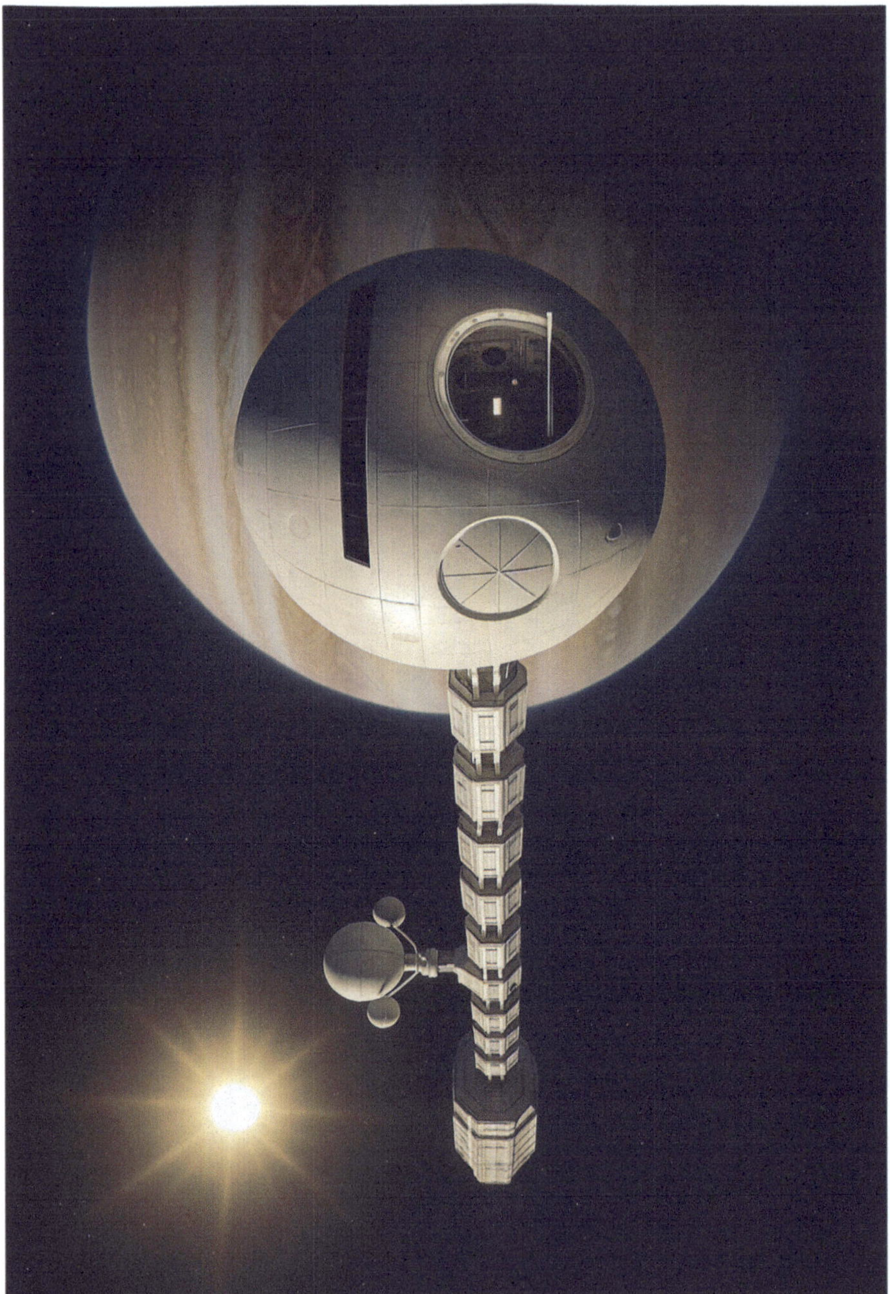

Fig. 15.5 Science fiction Discovery-I spacecraft concept at planet Jupiter[KFL]

Fig. 15.6 Speculative peacekeeper spacecraft concept USS Heinlein at planet Saturn[KFL]

Fig. 15.7 Theoretical SunVoyager spacecraft concept departs earth orbit[KFL]

Fig. 15.8 Theoretical project daedalus interstellar flyby spacecraft probe departing Titan orbit[KFL]

Fig. 15.9 Theoretical resolution interstellar flyby spacecraft probe in departure orbit[ML]

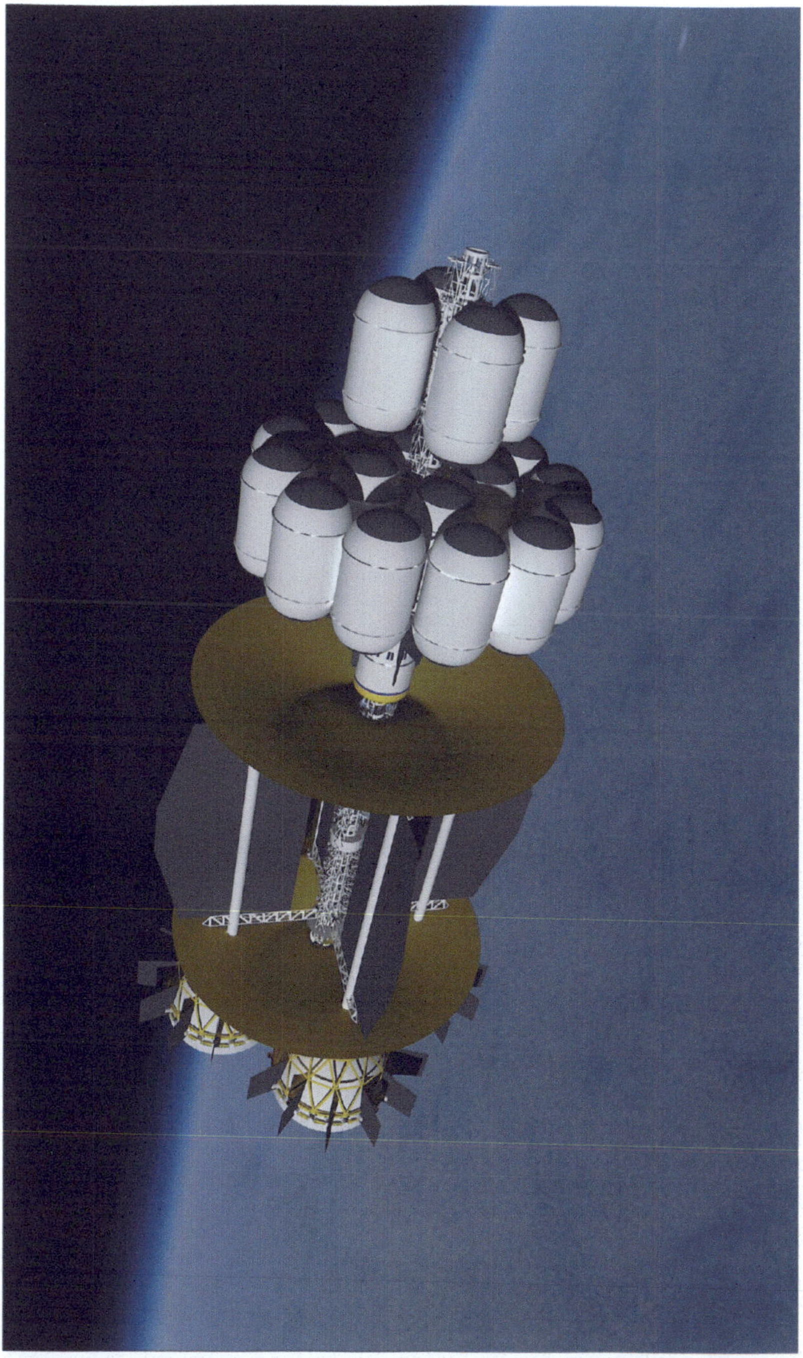

Fig. 15.10 Theoretical Endeavour interstellar Rendzesvous spacecraft probe departing earth orbit[ML]

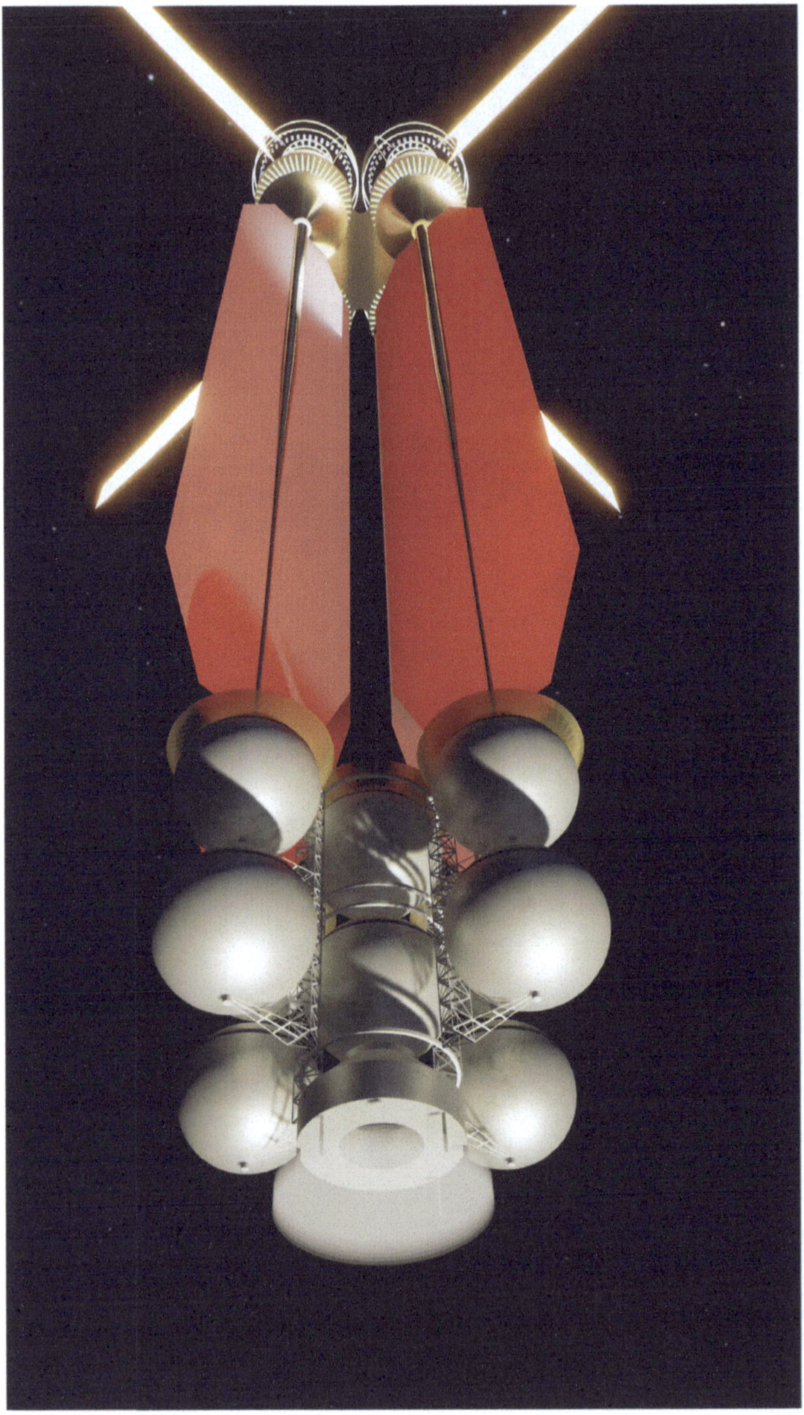

Fig. 15.11 Theoretical Pegasus interstellar rendezvous spacecraft probe under engine ignition[ML]

Fig. 15.12 Theoretical Pegasus spacecraft rendezvous probe at Jupiter frontal view[ML]

Fig. 15.13 Theoretical Pegasus spacecraft Rendezous probe at Jupiter rearward view[ML]

Fig. 15.14 Interstellar spacecraft under construction in low earth orbit[KFL]

Fig. 15.15 Theoretical Enzmann interstellar crewed spacecraft concept departs for another star[KFL]

Fig. 15.16 Speculative asteroid interstellar spacecraft worldship on a long journey[KFL]

Fig. 15.17 Speculative massive interstellar worldship arrives at a new star[KFL]

Fig. 15.18 Fantastic interstellar Ramjet spacecraft concept through the interstellar medium[KFL]

Fig. 15.19 Fantastic black hole evaporator engine interstellar spacecraft concept[KFL]

Fig. 15.20 Hypothetical advanced spacecraft approaches a passing near earth object[KFL]

Reference

1. Strong J (1965) Flight to the stars. Temple Press Books, Philadelphia

Solutions

Chapter 1: Introduction

Problem 1.1 Kinematics

(i) Time to 4.3 ly t = 75,879.95 years (using $1ly = 9.4607 \times 10^{15}$ m).
(ii) For a 100-year mission v = 12,899.6 km/s or 2,719.3 AU/Year.

Problem 1.2 Constant Acceleration

(i) S = 10,000 AU = 1.496×10^{15} m, t = 30 years = 9.467×10^{8} s, acceleration time = $t/2$, so using $S = at^2/2$, therefore

$$\frac{S}{2} = \frac{1}{2}a\left(\frac{t}{2}\right)^2,$$

and

$$a = \frac{2(S/2)}{(t/2)^2} = \frac{4S}{t^2} = 6.68 \times 10^{-3} \text{ m/s}^2.$$

(ii) $v_{peak} = at/2 = 3.16 \times 10^6$ m/s = 3,160 km/s = 1.05% c.

Problem 1.3 Non-constant Acceleration

(i) We start from $a = dv/dt$ with $a(t) = kt$ so that $dv/dt = kt$ and therefore

$$v(t) = \int_0^t kt dt = \frac{1}{2}kt^2.$$

K. F. Long, *Stellar Engines*, Springer Aerospace Technology,
https://doi.org/10.1007/978-3-032-18062-9

Then using $v = ds/dt$ we get $ds/dt = kt^2/2$ so that

$$s(t) = \int_0^t \frac{1}{2}kt^2 dt = \frac{1}{6}kt^3.$$

Then applying the condition for half the trip we get $s_{1/2} = kt_{1/2}^3/6$, therefore

$$k = \frac{6s_{1/2}}{t_{1/2}^3} = \frac{6 \times 7.48 \times 10^{14}}{(4.733 \times 10^8)^3} = 4.23 \times 10^{-11} \text{ m/s}^3.$$

(ii) The peak velocity would be given by $v_{peak} = kt_{1/2}^2/2 = 4.74 \times 10^6$ m/s = 4,740 km/s = 1.58%c. For the maximum acceleration this will be given by $kt_{1/2}$ = 0.02 m/s^2 or when expressed in units relative to Earth gravity 0.002 g.

Chapter 2: Interstellar Trajectory

Problem 2.1 Lagrangian

(i) Using the chain rule

$$\frac{d}{dt}\left(\frac{1}{2}\dot{\theta}^2\right) = \frac{d}{dt}\left(\frac{1}{2}u^2\right) = \frac{d}{du}\left(\frac{1}{2}u^2\right)\frac{du}{dt} = u\dot{u} = \dot{\theta}\ddot{\theta}.$$

Using the Product rule

$$\frac{d}{dt}\frac{1}{2}(\dot{\theta}\dot{\theta}) = \frac{1}{2}(\ddot{\theta}\dot{\theta} + \dot{\theta}\ddot{\theta}) = \dot{\theta}\ddot{\theta}.$$

Problem 2.2 Daedalus

(i) The value of $v_\infty \approx$ D/T = 5.9/50 = 35,398.877 km/s = 0.118c.
(ii) Assuming r_o = 5 AU, and where $v_o = (v_\infty^2 + v_{esc}^2)^{1/2}$, this leads to $v_o \approx v_\infty$+5 m/s, or $v_\infty \approx v_o$. Note the assumed cruise velocity for the Daedalus mission was 36,600 km/s and the mission was slightly less than 50 years.

Problem 2.3 Relativity

First we equate the line element to the proper time $c^2 d\tau^2 = c^2 dt^2 - d\theta^2$ then re-arrange to get $d\tau^2 = dt^2 - d\theta^2/c^2$. Re-arranging further we get

$$\frac{d\tau^2}{dt^2} = 1 - \frac{d\theta^2}{c^2 dt^2}.$$

And since $d\theta/dt = \dot{\theta}$ we get

$$\frac{d\tau}{dt} = \sqrt{1 - \frac{\dot{\theta}^2}{c^2}}.$$

Then since $\gamma = dt/d\tau$ from special relativity we invert the equation to find

$$\gamma = \frac{1}{\sqrt{1 - \dot{\theta}^2/c^2}}.$$

Chapter 3: History of Research

Problem 3.1 Parameters

(i) For the ICF driven systems the range of masses is $10^{-3} < m_{cap}(g) < 10s$ and the range of pulse frequencies is $1 < f_{hz}(Hz) < 10^3$. However, note that Orion/Helios goes as low as 0.3 Hz and up to 10 tons in unit mass. Note also for the interstellar ramjet technically $m_{cap} \to 0$ (individual protons) and $f_{hz} \to \infty$ (continuous propulsion).

Problem 3.2 TRLs

(i) From the NASA TRL definitions, most of these concepts come under 2-3. TRL2: Technology concept and/or application formulated TRL3: Analytical and experimental critical function and/or characteristic proof of concept. Some such as the interstellar ramjet will be TRL1: basic principles observed and reported. It could be argued that Orion is TRL4: Component and/or breadboard validation in laboratory environment.

Problem 3.3 Mission

(i) At 6 months constant acceleration at 0.0981 m/s^2 the velocity achieved will be 1,546.84 km/s. It would have reached a distance of $at^2/2 = 1.22 \times 10^{13}$ m or 81.52 AU.
(ii) It would then take a further 1.283 years to travel the remaining 418.48 AU distance. The total mission time to the target from launch would be 1.78 years.
(iii) At the defined cruise speed it would take 26.86 hours to traverse 1AU.

Chapter 4: Laser Fusion Physics

Problem 4.1 Reactions

(i) The reaction is $p + {}^{11}B \to 3\,{}^4He$, where for the masses ${}^1H$ = 1.007825, ${}^{11}B$ = 11.009305, ${}^4He$ = 4.002603 and 1uc^2 = 931.494 MeV. Then combined reactant mass m_{react} = 12.0171304, product mass m_{prod} = 12.0078097, therefore Δm = 0.00932062. Then $Q = \Delta mc^2 = 8.682$ MeV, $\alpha = \Delta m/m_{react} = Q/m_{react}c^2$ = 0.000776 (0.0776%). This compares to DT at 0.376%, D^3He at 0.392% and DD at 0.09–0.1%. Therefore $p^{11}B$ converts $\sim$5 times less rest mass energy compared to DT and D^3He and is equivalent in levels to DD.

Problem 4.2 QED Vacuum

(i) To generate pair production of e^+e^- we assume λ = 1nm. $\omega = 2\pi c/\lambda = 1.884 \times 10^{18}$ rad/s, $f = c/\lambda = 2.998 \times 10^{17}$Hz.

(ii) $I_L = c\epsilon_o E^2/2 = 2.243 \times 10^{33}$ W/m^2. For a diffraction limited spot size $P = I_L \times A = I_L \pi r^2 = I_L(\lambda/2)^2 = I_L \times (\pi/4)\lambda^2 = 1.762 \times 10^{15}$ W.

Problem 4.3 Plasma Frequency

(i) $n_e = 10^{13}$ m^{-3}, $\omega_{pe} = 1.78 \times 10^8$ rad/s, then $f_{pe} = \omega_{pe}/2\pi = 2.84 \times 10^7$ s^{-1} or 28.4 MHz.

(ii) On the basis $\omega = \omega_{pe}$ and $\lambda = 2\pi c/\omega_{pe} = c/f_{pe}$, we find that λ = 10.6 m, which is the cutoff wavelength. This is very long wavelength radio waves. As an exercise you might want to look up different plasma densities and compute the reflected wavelengths. Since $\omega_{pe} \propto \sqrt{n_e}$ a denser plasma will have a cut off at a shorter wavelength and block longer wavelengths.

Chapter 5: Inertial Confinement Fusion

Problem 5.1 Fuel Heating

(i) 1% of the fuell mass is m_{hs} = 2.88 mg, T = 50 keV. Then E_{heat}= 16.67 MJ.

(ii) For the entire fuel of 288 mg, E_{heat} = 1.67 GJ.

Problem 5.2 Engine Flow

(i) For a DT fuel α = 0.0038, $\eta_j = v_{ex}^2/2\alpha\phi c^2$ = 43.96%.

(ii) The capsule mass is defined by $m_{cap} = \dot{m}/f_{hz} = \dot{m}v_{ex}^2/(2n_e\eta_j\alpha\phi f_{hz}c^2) = 0.01$ grams.

Problem 5.3 Driver Energy

(i) $E_{dr} = (\alpha\phi m_{cap}c^2/4.385 \times 10^6)^{3/4} = 64.62$ MJ.

(ii) $E_{source} = E_{dr}/\epsilon = 430.85$ MJ.

Chapter 6: Reactor Engineering

Problem 6.1 Power

(i) Assuming $\epsilon = 0.3$, $v_{ex} = 1{,}500$ km/s, $T = 0.1$ MN, $P_e = (1.5 \times 10^6 \times 10^5)/(2 \times 0.3) = 250$ GWe.

(ii) Assuming $\eta = 0.01$, $\tau = 1$ hour, with $P_e = 250$ GWe, $E = P_e \tau / \eta = 90$ PJ. For a reactor mass $m_r = 50{,}000$ tons, rating $= P_e / m_r = 5$ kWe/kg.

Problem 6.2 Injection

(i) Detonation time of pellets $t_{det} = 1/500$ Hz $= 0.002$ seconds. Plasma expansion wave time $t_{wave} = R_{ch}/v_{wave} = 12.5\text{m}/15 \times 10^6$ m/s $= 0.83$ µs.

(ii) For the injection velocity the minimum value would be $v_{inj} = 5\text{m}/0.002$ s $=$ 2.5 km/s. Note however, one must allow for some margin and so a value of 3 km/s may be preferred.

(iii) The acceleration can be approximated by $a = v_{inj}^2/(2 \times 10m) = 312{,}500$ m/s^2 $= 31{,}855 g_o$, where g_o is acceleration due to gravity.

Problem 6.3 Magnetic Fields

(i) For $r_L = 1$m, $m = 6.6447 \times 10^{-27}$ kg, $v = 1.5 \times 10^7$ m/s, $q = 2e = 3.204 \times 10^{-19}$ C, and using $B_\alpha = mv/qr_L = 0.311$ T.

(ii) For protons $m = 1.67 \times 10^{-27}$ kg, $q = 1e = 1.6022 \times 10^{-19}$ kg, therefore $B_p = 0.156$ T.

(iii) Note that $B_\alpha / B_p \approx 2$, since ^{4}He has 2 protons whereas hydrogen has 1 proton. The ^{4}He particle has 4 times the mass and 3 times the charge. The neutrons are not relevant since they do not have charge.

Chapter 7: Principles of Aerospace Design

Problem 7.1 Engine Performance

(i) $v_{ex} = 1{,}551.31$ km/s, $\dot{m} = 0.01$ kg/s, $T = 15.51$ kN, $P_j = 12.03$ GW, $P_{sp} = 24.07$ kW/kg. Note we treated the entire pellet as explicit thermonuclear fuel mass and neglected any external ablator shells.

Problem 7.2 Laser Change

(i) For solar sail phase no change. For laser sail phase $S_2 = 310.5$ AU and $\Delta v_2 =$ 252.48 km/s. For fusion phase no change except at end of phase spacecraft at $S_3 = 2{,}722.44$ AU.

(ii) Total velocity end of all acceleration $v_{tot} = 2{,}035.86$ km/s. Spacecraft must then cruise 269,209 AU and will arrive at the destination after 648.29 years.

Problem 7.3 Fusion Change

(i) m_{fuel} = 5,000 tons, Δv_3 = 3,718.22 km/s, S_3 = 4,485.67 AU, cruise distance 267,436.21 AU.

(ii) Time to destination from launch is 362.2 years.

Chapter 8: Interplanetary Spacecraft Designs

Problem 8.1 Shadow Shield

(i) S_{dp} = 3m/sin(10°) = 17.27 m.

(ii) $D = 2 \times S_{dp} \times tan(15°)$ = 9.25 m.

Problem 8.2 Expellant

(i) m_{cap} = 0.1 g, m_{exp} = 100 g, f_{hz} = 100 Hz, $\dot{m}$ = 10.01 kg/s, α = 0.0039, ϕ = 0.333, ϵ_{mix} = 1.0, therefore η_j = 0.6 and v_{ex} = 185 km/s.

(ii) Just use the modified equation for v_{ex} that includes expellant propellant and you should get the same number.

Problem 8.3 Pellet Rate

(i) m_{prop} = 100,000 kg, $m_{pell} = 1 \times 10^{-5}$ kg, N_{pell} = 10 billion.

(ii) Rate = 10 billion / 5 years = 6.34 pellets/s or f_{hz} = 6.34 Hz.

(iii) If used in-situ the pellets would have to be manufactured at a rate exceeding 6.34 Hz, especially allowing for margins and detonation failures. But if the pellets are manufactured at base and then preloaded then can be formed over a period exceeding 5 years, although depending on the fuel type one must allow for radioactive gas decay and any associated heating effects.

Chapter 9: Interstellar Spacecraft Designs

Problem 9.1 Stage Engines

(i) m_{prop} = 35,000 tons, m_{cap} = 0.5 g, f_{hz} = 100 Hz, n_{eng} = 1, $t_b = 7 \times 10^8$ s = 22.19 years.

(ii) For n_{eng} = 7, $t_b = 1 \times 10^8$ s = 3.17 years.

(iii) With increased number engines there is increased mass and increased radiation dose and a higher thermal load for each pulse period which also suggests a higher shielding requirement, which also increases the mass. If the vehicle is much heavier, this will also influence the total mission time since it takes more energy to accelerate a higher mass.

Problem 9.2 Mean Free Path

(i) DT fuel, $A = 2.0141$, $\rho_o = 0.25$ g/cm^3, $\Lambda = 500$, $\rho_c = 125$ g/cm^3, $r_c = 0.15$ mm, $N_A = 6.02 \times 10^{23}$ mol^{-1}, $n = 3.74 \times 10^{25}$ cm^2, 14 MeV neutron, $\sigma_c = 1.2 \times 10^{-24}$ cm^2, $\lambda = 0.22$ mm, $\lambda / r_c = 1.48$, therefore the neutrons will escape the fuel.

(ii) 2.45 MeV neutron, $\sigma_c = 2.5 \times 10^{-24}$ cm^2, $\lambda / r_c = 0.71$, therefore the neutrons will be contained within the fuel.

(iii) 14 MeV: comes from the reaction D(T,^{4}He)n. 2.45 MeV: comes from the side reactions D(D,^{3}He)n or the Deuterium reacting with the DT fuel.

Problem 9.3 Lasers

(i) In Chap. 4 we already mentioned other types of lasers, such as Krypton Fluoride, Carbon Dioxide and Heavy Ion Beams. But perhaps Diode Pumped Solid State Lasers (DPSSL) deserve special attention since these have much potential for very high laser power without being accompanied by laser-plasma instabilities at high intensity. Within the different laboratory experiments around the world achievements include a pulse frequency of 100 Hz at 10 J, 1 kW highest average power at 10–100 Hz, and a peak pulse intensity up to 1.1×10^{23} W/cm^2. In terms of efficiency a high powered 940–980 nm pump diode will have a wall-plug efficiency 40–55% and some experiments report even higher values. This compares to optical flash pumps like NIF at <1%.

Chapter 10: Propulsion Power Management

Problem 10.1 Gordinier

(i) $\dot{m} = 1.5$ kg/s, $v_{ex} = 100$ km/s, $P = 150$ GW, $T = \dot{m} v_{ex} = 150$ kN, $P_j = 0.5 \dot{m} v_{ex}^2 = 7.5$ GW $= f' P$.

(ii) $f' = P_j / P = 0.05$, $\theta = f'/(1 - f') = 0.0526$.

Problem 10.2 PCLM

(i) Laser efficiency $\epsilon_1 = 10\%$, $f_{hz} = 100$ Hz, Fusion output $P_f = 300$ GW, Gain $G = 100$, input power $P_{in} = 3$ GW, Jet Power $P_j = 25.5$ GW, induction power $P_{ind} = 5$ GW, thrust power $P_{th} = 29$ GW, jet efficiency $\eta_j = 75\%$. From the fusion power 5% is laser coupling loss, 8% is neutron loss, 15% is x-ray loss, and 20% is the power to the plasma debris.

(ii) Efficiency of waste heat at inductor 40%, therefore $P_{in} = P_{ind} \times 0.4 = 2$ GW.

(iii) $E_{out} = P_f / G = 3{,}000$ MJ, therefore $E_{in} = E_{out} / G = 30$ MJ, and $E_o = E_{in} / \epsilon_1 = 300$ MJ. These calculations assume $P_f = 300$ GW. The quoted energies are per pulse equivalenced to 1-second output.

Problem 10.3 Radiators

(i) $\rho \Delta x = 2.5$ kg/m^2, $P_r = 25$ GW, $f_v = 7\%$, $F_{space} = 1 - f_v = 0.93$ $\epsilon = 80\%$, $T = 1{,}000$ K, therefore from $P/A = F_{space}\epsilon\sigma T^4$ and $A_{2-side} = P_r/(P/A)$ = 592,630.5 m^2 and A_{1-side} = 296,315.3 m^2.

(ii) m_{rad} = (2.5 kg/m^2 $\times$ A_{2-side} = 1481.57 tons, and $m_{coolant} = 0.6 \times m_{rad}$ = 888.94 tons. Total radiator mass then is 2,370.51 tons.

(iii) For the radiator length, assuming square panels, $L = \sqrt{A_{1-side}}$ = 544.35 m.

Chapter 11: Large Crewed Vessels

Problem 11.1 Phobos

(i) $m_o = 1.07 \times 10^{16}$ kg, ϵ_p =0.1% = 0.001, $m_{strip} = 1.07 \times 10^{13}$ kg, $n = 10^9$ drones, $R_m = 10^{-4}$ kg/s, ϵ_r= 0.9, U = 5/6, the strip mining time τ_m = 1,651.23 days or 4.52 years.

(ii) The throughput rate is $\dot{m}_{use}$ = 75 tons/s.

Problem 11.2 Enzmann Slow Boat

(i) m_{cap} = 0.05 g, $m_{exp} = 50m_{cap}$ = 2.5 g, f_{hz} = 100 Hz, n_e = 8 engines, therefore $\dot{m}_8 = m_{cap}n_e f_{hz}(1 + \theta)$ = 2.04 kg/s, and $\dot{m}_1 = \dot{m}_8/n_e$ = 0.255 kg/s.

(ii) Mass of fuel $M_{fuel} = m_{cap} f_{hz} t_b = \dot{m}_{cap} t_b$ = 157.68 tons per engine or 1,261.44 tons total. Mass of expellant $M_{exp} = m_{exp} f_{hz} t_b = \dot{m}_{exp} t_b$ = 7,884 tons per engine or 63,072 tons total.

Problem 11.3 Asteroid Ship

(i) For R = 5 km, this would give values for the membrane stress of 78.49 MPa (CRFP), 385.14 MPa (Steel), 132.47 MPa (Aluminium alloy) and 217.35 MPa (Titanium alloy).

(ii) For R = 10 km, this would give values for the membrane stress of 313.99 MPa (CRFP), 1,540.55 MPa (Steel), 529.87 MPa and 869.39 MPa (Titanium alloy).

(iii) For R = 10 km, assuming ω = 0.0886 rad/s, σ_{rot} = 1,255.99 MPa.

Chapter 12: Massive World Ships

Problem 12.1 Moon

(i) Radius $r = vT/2\pi$ = 571,816 km.

(ii) $M_m = 7.35 \times 10^{22}$ kg, Δv = 0.5 km/s, change in momentum $\Delta p = M_m \Delta v$ = 3.675×10^{25} kgm/s, Force $F = \Delta p/\Delta t = M_m \Delta v/\Delta t = 1.165 \times 10^{15}$ N.

(iii) Number of engines $N = F/T_n$ = 1,165 engines. Note for interest that holding the same period whilst increasing the orbital speed would imply a non-Keplerian forced orbit change.

Problem 12.2 Earth

(i) For the new orbital radius we use Kepler's third law of planetary motion where $T^2 \propto r^3$ or $(T_2/T_1)^2 = (r_2/r_1)^3$. Therefore $T_2/T_1 = 0.9856$ and $r_2 = 1.481 \times 10^{11}$ m.

(ii) It will be approximately 1% closer to the Sun than the present distance.

(iii) We use the equation for orbital velocity $v = (GM/r)^{1/2}$, where $G = 6.674 \times 10^{-11}$ m^3/kgs^2 and $M_s = 1.989 \times 10^{30}$ kg, therefore the new orbital speed is $v_2 = 29.93$ km/s, which compares to $v_1 = 29.78$ km/s by a difference of $\Delta v =$ 0.15 km/s.

Problem 12.3 Earth 2

(i) Change in momentum $\Delta p = M_e \Delta v = 8.958 \times 10^{26}$ kgm/s, Average force $F = \Delta p/\Delta t = 2.84 \times 10^{15}$ N.

(ii) $F_{satv} = 3.4 \times 10^7$ N, therefore number of rockets $= F/F_{satv} = 84$ million rockets.

(iii) Total number of rockets $N_{tot} = 84{,}000{,}000$, cost \$1.5 billion/unit, total nations $N_n = 200$, therefore per nation $N = N_{tot}/N_n = 420{,}000$ Saturn V rockets. Cost per nation = \$630 trillion, Total cost = \$126,000 trillion.

Chapter 13: Fantastic Propulsion Concepts

Problem 13.1 Ramjet

(i) $\beta = 0.66$c, $\gamma = 1.332$, $n = 10^7$ m^{-3}, $dm/dt = 1.2536 \times 10^{26}$ p/s = 0.20968 kg/s, $A_o = 6.272 \times 10^{10}$ m^2, $R = 141$ km.

(ii) For the same speed a ^{4}He particle would have a rest mass of 3,727 MeV, $\gamma =$ 1.332, $E_{kin} = (\gamma - 1)m_o c^2 = 1.237$ GeV.

Problem 13.2 Classical

(i) $R_s = 2.48 \times 10^{-54}$ m.

(ii) $T_H = 7.34 \times 10^{49}$ K, $P = 1.27 \times 10^{86}$ W, $\tau = 3.94 \times 10^{-97}$ s.

Problem 13.3 BAIR

(i) $\dot{m} = 50$ mg/s, $v_{ex} = 0.9$c, $T = 13.49$ kN.

(ii) $m = 10{,}000$ tons, $a = 0.00135$ m/s^2, $S = 4.067 \times 10^{16}$ m, $S/2 = 2.0336 \times 10^{16}$ m, $t_{tot} = 347.91$ years, $v_c = 7{,}410.88$ km/s = 2.47%c.

(iii) With 100 black holes per second, $v = 0.99c$, $A = 10{,}000$ km^2, $L = 10$ m, σ_{BH} = 100 pb = 10^{-38} m^2, $n = 18{,}355.75$ protons/cm^3.

Chapter 14: The Starship Engineer

Problem 14.1 Space Treaty

(i) "Refrain from placing in orbit around the Earth any objects carrying nuclear weapons or any other kind of weapons of mass destruction or from installing such weapons on celestial bodies" (Annex, but also repeated in Article IV). "The establishment of military bases, installations and fortifications, the testing of any type of weapons and the conduct of military manoeuvres on celestial bodies shall be forbidden" (Article IV).

Problem 14.2 Yield

(i) non-relativistic energy $E_{nr} = mv^2/2 = 1.7975 \times 10^{19}$ J.
(ii) Assuming 1 tons TNT = 4.184×10^9 J = 4.2961 Gtons TNT.
(iii) Relativistic energy $E_r = (\gamma - 1)mc^2 = 1.8533 \times 10^{19}$ J = 4.4295 Gtons TNT,where $\gamma = 1.02062$. This is a difference of 0.133 Gtons TNT or 133.33 Mtons TNT or around 3% higher when you include relativistic effects at this velocity.

Problem 14.3 Methods

(i) Some methods might include different types of space drives, warp drives, wormholes or krasnikov tubes. There are also other more fantastic methods such as teleportation through quantum tunnelling. Each is a potential book and indeed many have been written.
(ii) Hopefully you made some progress. If you did, congratulations you are now a starship designer. Consider writing it up and submitting a paper to the *Journal of the British Interplanetary Society* or elsewhere.

Index

K. F. Long, *Stellar Engines*, Springer Aerospace Technology,
https://doi.org/10.1007/978-3-032-18062-9

The manufacturer's authorised representative in the EU is Springer Nature Customer Service Centre GmbH, Europaplatz 3, 69115 Heidelberg, Germany. If you have any concerns regarding our products, please contact ProductSafety@springernature.com

Printed and bound by CPI Group (UK) Ltd, Croydon, CR0 4YY
21/09/2026
02220012-0014